T5-AQQ-848

GUANIDINES 2

Further Explorations of the Biological
and Clinical Significance of
Guanidino Compounds

GUANIDINES 2
Further Explorations of the Biological
and Clinical Significance of
Guanidino Compounds

Edited by
Akitane Mori
Okayama University Medical School
Okayama, Japan

Burton D. Cohen
The Bronx Lebanon Hospital Center
Bronx, New York

and

Hikaru Koide
Juntendo University School of Medicine
Tokyo, Japan

PLENUM PRESS • NEW YORK AND LONDON

Library of Congress Cataloging in Publication Data

International Symposium on Guanidino Compounds in Biology and Medicine (2nd: 1987: Schizuoka-shi, Japan)
 Guanidines 2: further explorations of the biological and clinical significance of guanidino compounds / edited by Akitane Mori, Burton D. Cohen, and Hikaru Koide.
     p.   cm.
 "Proceedings of the Second International Symposium on Guanidino Compounds in Biology and Medicine, held May 31–June 2, 1987, in Susono City, Schizuoka, Japan" – T.p. verso.
 Includes bibliographies and index.
 ISBN 0-306-43223-4
 1. Guanidino compounds – Physiological effect – Congresses. 2. Guanidino compounds – Pathophysiology – Congresses. 3. Guanidino compounds – Metabolism – Congresses. I. Mori, Akitane, 1930–   . II. Cohen, B. D. (Burton D.) III. Koide, Hikaru. IV. Title. V. Title: Guanidines two.
     [DNLM: 1. Guanidines – congresses. QU 60 I6075g 1987]
 QP801.G83I59 1987
 599'.01924 – dc20
 DNLM/DLC   89-8437
 for Library of Congress   CIP

QP
801
.G83
I59
1987

Proceedings of the Second International Symposium on Guanidino Compounds in Biology and Medicine, held May 31–June 2, 1987, in Susono City, Shizuoka, Japan

© 1989 Plenum Press, New York
A Division of Plenum Publishing Corporation
233 Spring Street, New York, N.Y. 10013

All rights reserved

No part of this book may be reproduced, stored in a retrieval system, or transmitted in any form or by any means, electronic, mechanical, photocopying, microfilming, recording, or otherwise, without written permission from the Publisher

Printed in the United States of America

FOREWORD

Guanidine is named for its similarity to the purine guanine which, in turn, is named for its principal source guano, which comes from the Inca word, huano, for dung. Guanidine, therefore, translates into dung-like, which is hardly a genteel way to introduce a subject.

On the other hand, texts are seldom inspirational, frequently crude and rarely literary and should be judged on how successfully they assemble, organize and present current data. I am impressed that the material which follows goes a long way toward successfully achieving those goals.

The International Guanidine Society is a synthesis of three groups of investigators: biologists studying guanidines as phosphagens, neurologists interested in guanidines as convulsants and nephrologists involved with guanidines as toxins. As a member of the latter group, I am gratified by the considerable progress this book represents.

To begin with, there now appears to be a common theme which unifies current speculation concerning the metabolic origin of the guanidines in uremia. At the First International Congress in 1983, evidence was presented which supported the theory that certain guanidines were products of the mixed function oxidation of urea. This year's meeting brings together overwhelming data showing that methylguanidine is an effect of active oxygen reacting with creatinine. An idolatry, worshipped throughout biochemistry, that urea and creatinine are inert byproducts of protein metabolism, is shown to have feet of clay.

More important and practical, however, is the evidence that urinary levels of guanidinoacetic acid (GAA) can be of clinical value as a highly sensitive warning of impending renal failure under a variety of conditions. Among the conditions reported are protein ingestion, renal transplantation, malignant hypertension and a number of cytotoxic drugs including gentamicin, tetracycline, cyclosporine and cisplatin. Additional support is obviously necessary before clinical chemistry laboratories rush out to purchase guanidine analyzers, but the current data are especially impressive since they are arrived at by a series of investigators working independently to solve different aspects of the same problem. It is an area to watch.

Once again we are deeply indebted to Professor Akitane Mori for pulling all this together. Without his dedication to guanidines and the Guanidine Society, this volume would not be possible. Professor Mori's unflagging zeal in support of this arcane subject is solely responsible for this text and the progress it represents.

B.D. Cohen

PREFACE

Guanidino compounds are known to be ubiquitous in both animal and plant Kingdoms. Ever since the discovery of guanidine by Strecker in 1861, more than 100 naturally occurring guanidino compounds have been identified. Some of these substances play important biological roles in animal life, such as the participation of arginine in ureagenesis, and of creatine in muscular contraction. On the other hand, toxic effects of guanidino compounds have also been observed. As early as 1886, Brieger identified methylguanidine in stale horse meat and reported that creatine was decomposed by Vibrio cholerae (Kommabacillen) to produce methylguanidine which could induce dyspnea, muscle fibrillation, and generalized convulsions. Although some other reports concerning the toxicity of guanidino compounds such as methylguanidine or guanidine were also made in the early 1900s, during the following more than half a century, there appears to have been little interest in the subject.

The finding of guanidinosuccinic acid in uremic patients by Cohen's group in 1963 has served as a major stimulus for the current wave of research on guanidino compounds as uremic toxins. In recent years, considerable information about guanidino compounds has accumlated not only in the field of nephrology, but also in other fields such as, analytical chemistry, biochemistry, physiology and neurotoxicology. These subjects were discussed at the first Symposium on Guanidino Compounds, held in Tokyo in 1983, (Guanidines: Historical, Biological, Biochemical and Clinical Aspects of the Naturally Occurring Guanidino Compounds, Plenum Press, New York and London, 1985).

The present monograph comprises the Proceedings of the Second Symposium on Guanidino Compounds held in Susono City, at the foot of Mt. Fuji, June 1 - 2, 1987, and covers the following main topics.

1. Analytical methods for guanidino compounds and new guanidino compounds:
   New developments in analytical methods for guanidino compounds utilizing special enzymes or high performance liquid chromatography have been introduced. Isolation of new guanidino compounds, such as γ-guanidinooxy-propylamine, N^G-methylagmatine, phascolamine and phascolasomine, is reported in this section.

2. Metabolism of guanidino compounds:
   A new idea for the existence of multiple forms of transamidinase has been presented by Dr. J. Van Pilsum. The relationship of superoxide anions or hydroxyl radicals to guanidino compounds has for the first time been highlighted in this symposium. Effect of arginine-free diet on ureagenesis, guanidino compounds in "sparse-fur" mutant mice with ornithine transcarbamylase deficiency, and biosynthesis of guanidinoethanol, are also discussed in this section.

3. Physiological, pharmacological and toxicological aspects of guanidino compounds:
   Dr. B.D. Cohen presented a special lecture entitled "Guanidines as drugs", and reviewed pharmacological and clinical effects of drugs, such as cimetidine, amiloride, guanethidine, guanbenz and guanfacine, which contain guanidino groups(s), on kidney, stomach and central nervous system. Alterations of guanidino compounds under unusual situations, such as anesthesia, diabetes, hypertension, hemorrhagic shock and immersion stress are discussed in this section. In addition, a digitalis like activity of dimethylarginine, inhibitory effect of adenosine and its potentiations on methylguanidine synthesis, and effect of guanidino compounds on GABA-stimulated chloride channel, are also presented.

4. Involvement of guanidino compounds in seizure mechanisms:
   Considerable information has been brought together concerning seizure mechanisms and guanidino compounds. Studies involving compounds such as, 2-guanidinoethanol, and guanidinoethanesulfonate in epileptic patients as well as in experimental animal models are reported. The involvement of catecholamines in the mechanism of seizures induced by α-guanidinglutaric acid is also discussed is this section.

5. Hyperargininemia:
   Dr. A. Lowenthal, one of the first investigators to report hyperargininemia, overviews "Hyperargininemia". Recent findings in hyperargininemia are reported by Dr. Lowenthal's colleagues and the Nagoya University group.

6. Involvement of guanidino compound in renal dysfunction:
   Involvement of guanidino compounds, especially of guanidinoacetic acid, methylguanidine and guanidinosuccinic acid, in renal dysfunction had been demonstrated not only in patients with acute and chronic renal failure and with kidney transplantation, but also in experimental animal models with renal failure and in an in vitro model system for nephropathy. The role of guanidine as an indicator of peroxidation in uremics, and of nonprotein nitrogen in progressive renal failure, is also discussed in this section.

   The pace of exploration of guanidino compounds has accelerated greatly in the past decade. The proceedings of the 1987 International Symposium on Guanidino Compounds compile a major part of current topics in biological and medical research as related to this group of substances. I hope that this volume will become a significant milestone in the study of the 'guanidine kingdom', and will serve as an important basis for futher research and collaboratian in this area.

                                                                    A. Mori

Okayama, June 1988

ACKNOWLEDGEMENTS

The editorial work was performed by Dr. I. Yokoi, the Department of Neurochemistry, Institute for Neurobiology, Okayama University Medical School. We wish to express our gratitude to Dr. I. Yokoi.

CONTENTS

IV. INVOLVEMENT OF SEIZURE MECHANISM

V. HYPERARGININEMIA

VI. INVOLVEMENT OF GIUANIDINO COMPOUNDS IN RENAL DYSFUNCTION

I. ANALYTICAL METHOD FOR GUANIDINO COMPOUNDS

AND NEW GUANIDINO COMPOUNDS

ENZYMIC DETERMINATION OF METHYLGUANIDINE IN SERUM AND PLASMA OF HEMODIALYSIS

PATIENTS AS A MARKER FOR HYDROXYL RADICALS

Moto-o Nakajima, Kazuo Nakamura, Yoshio Shirokane and
*Yoshihei Hirasawa

Bioscience Research Laboratory, Kikkoman Corporation
Noda; *Kidney Center, Shinrakuen Hospital, Niigata, Japan

INTRODUCTION

Methylguanidine (MG) is known to accumulate in body fluids of uremic and hemodialysis patients[1,2] and has been proved to be a strong uremic toxin[3]. It was reported that erythrocyte deformability and Na^+,K^+- ATPase activity of erythrocyte membranes decreased in hemodialysis patients and there was a significant negative correlation between erythrocyte deformability and MG level[4]. Recent reports reveal that MG is converted from creatinine (CRN) by the action of various species of active oxygen, especially hydroxyl radicals as produced in the Fenton reaction[5,6] and that free hemoglobin acts as a biological Fenton reagent to generate hydroxyl radicals[7], and as an iron promoter in the Fenton reaction[8]. Since free radicals were shown to play an unfavorable role renal failure[9-11], determination of MG in body fluids as a marker for hydroxyl radicals could prove useful in clinical practice.

MG has been determined by automated high-performance liquid chromatography (HPLC) with a fluorometric detection device[12-14]. However, HPLC is inappropriate for analyses of many samples. In clinical practice, enzymic determination of metabolites has widely been employed because of its simplicity and substrate specificity. To establish the enzymic determination of MG we found two new enzymes, methylguanidine amidinohydrolase (MGAH EC 3. 5. 3. 16)[15] and methylamine oxidase (MAOD), and developed an assay system. We report a preliminary enzymic method for the determination of MG in plasma, serum and urine of patients with chronic renal filure[16].

MATERIALS AND METHODS

Plasma and Serum Specimens

Plasma and serum in hemodialysis patients were collected just before hemodialysis. These samples were stored at -20°C until analysis.

3

Chemicals

The partly purified preparations of MGAH (30 kU/g) and MAOD (0.92 kU/g) were prepared by us from <u>Alcaligenes</u> sp. N-42[12] and <u>Bacillus</u> sp. N-104, respectively. Methyl-3-aminocrotonate (MAC) was obtained from Aldrich Chemical Co., (Milwaukee, WI). MG sulfate was purchased from Eastman Kodak Co., (Rochester, NY). The other chemicals were also commercially available.

Reagents

The deproteinization reagent was prepared by dissolving 75 g of trichloroacetic acid per liter in distilled water. Enzymic reagent (I) for MG was a Tris HCl buffer solution (10 mM, pH 7.5) containing 14 kU of MGAH and 4 kU of MAOD per liter. Enzymic reagent (II) for sample blank was a Tris HCl buffer solution (10 mM, pH 7.5) containing 4 kU of MAOD per liter. MAC solution was prepared by dissolving 23.2 g of MAC per liter in dimethyl sulfoxide. The MG standard solution was prepared by dissolving 3.34 mg of MG sulfate per liter in distilled water.

Procedure

To 1.0 ml samples in test tubes, add 2.0 ml of deproteinization reagent, mix, and keep in ice cold water for 5 min. Centrifuge (1,000 x g, 15 min) the mixture to yield supernatant free from protein. Into each of duplicate test tubes, place 0.5 ml of supernatant, add 0.5 ml, of a carbonate buffer solution (0.5 M, pH 10.2), mix, then add 0.1 ml of enzymic reagent (I) for MG to one tube and enzymic reagent (II) to other for sample blank, mix, and incubate at 37°C for 15 min. Into both test tubes, add successively 0.5 ml of phosphate buffer solution (0.5 M, pH 5.5) and 0.1 ml of MAC reagent, mix, and incubate at 37°C for 15 min. Cool the tubes in running water for 1 min and measure the fluorescence at an excitation wavelength of 375 nm and an emission wavelength 465 nm vs the sample blank. Determine the MG concentration in plasma and serum by the use of a standard curve (Fig.1).

RESULTS

Improved Enzymic Method

We replaced centrifugal ultrafiltration with a chemical reagent for deproteinization. The increase in the quantity of each enzyme and the alteration of the reaction pH to approximately 9 allowed us to reduce the two

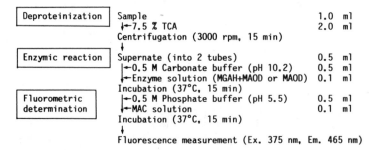

Figure 1.   Enzymic analysis of methylguanidine. TCA; trichloroacetic acid, MGAH; methylguanidine amidinohydrolase, MAO; methylamine oxidase and MAC; methyl-3-aminocrotonate.

steps of the enzyme reaction procedure to one. These modifications consider-
ably simplified the enzymic determination of MG.

Good linearity between fluorescent intensity and the amount of MG was
observed over a wide range from 5 μg to 1 mgMG/dl.

Analytical recovery was measured with serum and plasma samples at three
different MG values (0 to 47.8 μg/dl) by adding 50 or 100 μg of MG per
deciliter. The mean analytical recovery of MG was 94.3% (range 85.3-99.1%).
Within-run precision (CV) for MG and between-run (CV) for daily MG analyses
of serum and plasma stored at -20°C averaged 6.0% and 4.4%, respectively.

The enzymic method was compared with the HPLC method using 40 plasma
samples. Linear regression analysis revealed a good correlation (r = 0.974)
between the two methods (Fig. 2).

Enzymic Determination of MG and its Relation to some Clinical Data in Hemodialysis Patients

Clinical date from hemodialysis patients are shown in Table 1. The mean
MG level and SD in sera of 286 hemodialysis patients determined by the present enzymic method were 36.9 and 20.1 μg/dl, respectively. Table 2 demon-
strates the relationship between MG levels and other clinical data. A significant correlation was found between MG (y) and CRN (x) values in serum
with the linear regression equation of y = 4.435 x - 22.481 and a correlation
coefficient of 0.771 (Fig. 3). However, since the serum MG found in about one
third of the cases was only slightly correlated with serum CRN, ΔMG (the difference between the measured MG and the calculated MG from the above equation
with measured CRN) was introduced. We find that patients are divided into 3
groups based on ΔMG:the first group with 10 μgΔMG/dl or more, the second with
less than 10 μg ΔMG/dl and more than -10 μgΔMG/dl, and the third -10 μgΔMG/dl
or less. In the second group, a significant correlation was found between MG
and CRN with a linear regression equation of y = 4.542 x - 23.687 and correlation coefficient 0.940. In two groups with ΔMG of 10μg/dl or more and -10
μg/dl or less, significant correlations were also found between MG and CRN
with a linear regression equation of y = 4.998 x - 11.416, y = 2.676 x -
15.820 and correlation coefficient 0.854, 0.722, respectively (Fig. 4).

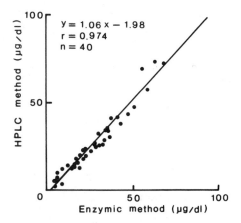

Figure 2.   Methylguanidine concentrations in plasma of hemodialysis patients
            measured by enzymic and HPLC methods.

Table 1. Clinical characteristics in serum of hemodialysis patients before
          dialysis

	n	Mean ± SD
Sex (M/F)	286	(180/106)
Age (years)	283	50.1 ± 15.2
Duration of hemodialysis (months)	271	81.1 ± 59.3
Methylguanidine (μg/dl)	286	36.9 ± 20.1
Creatinine (mg/dl)	266	13.5 ± 3.2
Urea nitrogen (mg/dl)	266	75.6 ± 15.7
Uric acid (mg/dl)	265	8.7 ± 1.4
Hematocrit (%)	262	25.2 ± 5.1
K (mEq/1)	265	5.4 ± 0.8
Ca (mg/dl)	265	8.7 ± 1.0
Pi (mg/dl)	265	6.4 ± 1.6
Iron (μg/dl)	263	69.7 ± 36.1
c-PTH (ng/ml)	248	6.1 ± 7.0

Table 2. Correlation between serum methylguanidine levels and other clinical
          data of hemodialysis patients

	n	r	P <
Creatinine (mg/dl)	266	0.711	0.001
Urea nitrogen (mg/dl)	266	0.382	0.01
Uric acid (mg/dl)	265	0.256	0.01
Hematocrit (%)	262	0.207	0.01
K (mEq/1)	265	0.194	0.01
Pi (mg/dl)	265	0.268	0.01
c-PTH (ng/ml)	248	0.163	0.05
Triglyceride (mg/dl)	247	0.194	0.01

Comparison of ΔMG in various Patient Groups with Complication

    In patients complaining of pain in various joints, infection and
gastrointestinal bleeding, ΔMG was often 10 μg/dl or more (Fig. 5). The mean
ΔMG level of this group was significantly high compared with that of the
patient group without complications. We found overproduction of MG in those
hemodialysis patients with inflammatory complications (pain in various joints
and infection). The patients on hemodialysis of long duration frequently
complained of pain in various joints.

Duration of Hemodialysis

    As the patients with hemodialysis of long duration frequently complained
of pain in various joints, we examined the relationship between MG and duration of hemodialysis. The serum MG level of the patients was weakly correlated with the duration of hemodialysis, r = 0.324 (p < 0.01).

    As shown in Table 3, CRN levels were almost similar in the patients with

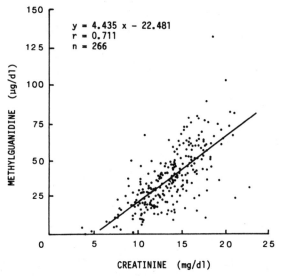

Figure 3. Correlation between methylguanidine and creatinine in sera of hemodialysis patients.

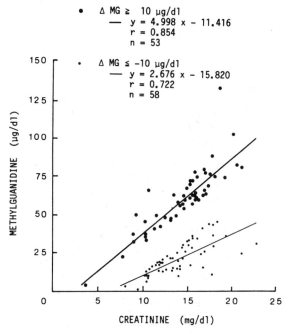

Figure 4. Correlation between methylguanidine (MG) and creatinine in the two patient groups with ΔMG of 10 μg/dl or more and -10 μg/dl or less.

hemodialysis of both 2 to 5 year and of 8 year duration or more. However, MG and ΔMG were significantly increased in patients with hemodialysis of 8 year duration or more.

Relationship between MG and Clinical Data Concerned with Iron or Copper

To ascertain whether hydroxyl radicals promote MG production from CRN in hemodialysis patients, we examined the relationship between MG levels and clinical data concerned with iron or copper, since iron and iron compounds may enhance hydroxyl radical generation from activated oxygen species.

Table 4 shows that the MG levels in patients on hemodialysis for 8 years or more was correlated with the parameters involved in iron or copper metabolism. Significant positive correlations were found between MG levels and the number of red blood cells, hematocrit values or hemoglobin levels. Less significant negative correlations were found between MG levels and serum iron, serum copper or ceruloplasmin.

As hemoglobin-binding iron is Fe (II) and transferrin-binding iron (= serum iron) is Fe (III), we assumed that the ratios of hemoglobin to serum iron responded to the quantity of iron concerned with the generation of hydroxyl radicals. Fig. 6 shows a significant correlation ($r = 0.594$, $p < 0.01$) between MG levels and the ratio of hemoglobin to serum iron in patients on hemodialysis for 8 years or more and showing a ΔMG of 10 µg/dl or more. However, no such significant correlation was found in patients with ΔMG of -10 µg/dl or less.

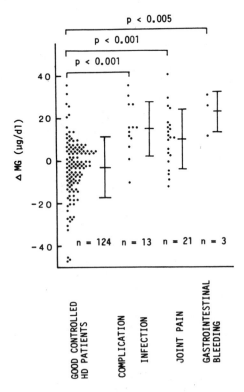

Figure 5.   Comparison of ΔMG in various patient groups with complication.

Table 3. Comparison of methylguanidine (MG) levels, ΔMG values and creatinine levels in patients with hemodialysis of 2 to 5 year and 8 year duration or more

	Duration of hemodialysis		P <
	2 - 5 years (n = 71)	8 years (n = 94)	
Methylguanidine (μg/dl)	33.8 ± 16.6	44.6 ± 18.7	0.001
ΔMG (μg/dl)	-3.6 ± 14.8	3.3 ± 13.0	0.005
Creatinine (mg/dl)	13.5 ± 2.9	14.4 ± 3.1	NS

Mean ± S.D. and NS; not significant

Table 4. Correlation between serum methylguanidine levels and creatinine or other clinical data relating to iron and copper of patients with hemodialysis of 8 year duration or more

	n	r	p <
Creatinine (mg/dl)	98	0.764	0.001
RBC ($x10^4/mm^3$)	95	0.480	0.01
Hemoglobin (g/dl)	97	0.439	0.01
Hematocrit (%)	98	0.414	0.01
Transferrin (μg/dl)	98	0.341	0.01
Unsaturated iron binding capacity (μg/dl)	97	0.383	0.01
Serum iron (μg/dl)	98	-0.335	0.01
Ferritin (ng/ml)	95	-0.292	0.01
Serum copper (μg/dl)	97	-0.300	0.01
Ceruloplasmin (mg/dl)	97	-0.245	0.05

DISCUSSION

Good correlations were found between MG and CRN in the three stratified patient groups based on ΔMG as described above. This finding suggests that MG is directly produced from CRN in hemodialysis patients. Nagase et al.[6] showed that CRN was exquisitely sensitive to oxidation to MG in the presence of hydroxyl radicals in vitro and that the formation of MG was stimulated by generators of hydroxyl radicals. Each slope of the linear regression equations between MG and CRN (Fig. 3,4) may reflect quantities of various species of active oxygen, especially hydroxyl radicals. Hydroxyl radical generation must be enhanced in the highest ΔMG groups.

In the Fenton reaction, Fe (II) and H_2O_2 generate hydroxyl radicals (Fig. 7). In plasma, Fe (II) is rapidly oxidized to Fe (III) by the copper-containing protein, ceruloplasmin, and Fe (III) is incorporated into the specific iron-binding protein, transferrin. Most serum iron is present as transferrin-binding iron, Fe (III). A decrease of ceruloplasmin may result in an

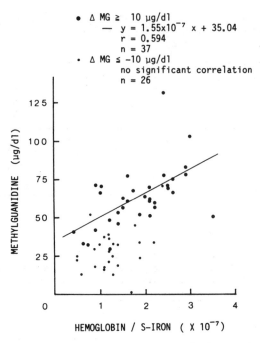

Figure 6. Correlation betwee methylguanidine (MG) and (hemoglobin/serum
           iron) in the two patient groups with ΔMG of 10 μg/dl or more and
           of -10 μg/dl or less.

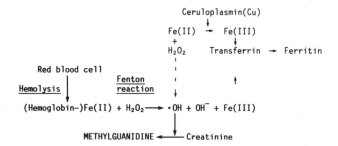

Figure 7. Hypothetical relationship between methylguanidine synthesis and
           hydroxyl radical generation.

increase in the ratio of Fe (II) and Fe (III). The degree of iron saturation
of transferrin decreases as serum iron decreases. In the case of rheumatoid
arthritis, it has been known that a decrease in serum iron closely correlates
with activity of the inflammatory process which depends on hydroxyl radical
formation. The true generator of hydroxyl radicals in the Fenton reaction is
considered to be free hemoglobin or iron released from hemoglobin[7,8]. In
addition, it is believed that an increase of free hemoglobin derived from
hemolysis occurs in patients with hemodialysis. We found a positive correlation between MG and hemoglobin levels. Moreover, a negative correlation was
found between MG levels and serum iron or ceruloplasmin. Furthermore, a
significant correlation was found between MG levels in patients with ΔMG of
10 μg/dl or more and the hemoglobin/serum-iron ratio (Fig. 6). These results

are also consistent with the hypothesis that hydroxyl radicals promote MG production from CRN in patients on hemodialysis. Our findings suggest that the production of MG from CRN in hemodialysis patients is caused by an increase in iron, especially hemoglobin-Fe (II), which is involved in the generation of hydroxyl radicals. Therefore, the measurement of MG in serum or plasma of hemodialysis patients and the calculation of ΔMG may give us useful information concerning the generation of the highly reactive and biologically damaging hydroxyl radical in vivo.

REFERENCES

1. Stein, I.M., Perz, Johnson G.R. and Cummings, N.B., Several levels and urinary excretion of methylguanidine in chronic renal failure, J. Lab. Clin. 77 (1971) 1020-1024.
2. Menichini, G.C. and Giovannetti, S., A new method for measuring guanidine in uremia, Experientia 29 (1973) 506-507.
3. Giovannetti, S. and Barsotti, G., Dialysis of methylguanidine, Kidney Int. 6 (1974) 177-183.
4. Mikami, H., Ando, A., Fujii, M., Okada, A., Imai, E., Kokuba, Y., Orita, Y. and Abe, H., Effect of methylguanidine on erythrocyte membranes, in: "Guanidines," A. Mori, B.D. Cohen and A. Lowenthal Eds., Plenum Press, New York, (1985) pp. 205-212.
5. Nagase, S., Aoyagi, K., Narita, M. and Tojo, S., Biosysnthesis of methyl-guanidine in isolated rat hepatocytes and in vivo, Nephron 40 (1985) 470-475.
6. Nagase, S., Aoyagi, K., Narita, M. and Tojo, S., Active oxygen in methyl-guanidine synthesis, Nephron 44 (1986) 299-303.
7. Sadrzadeh, S.M., Graf, E., Panter, S.S., Hallaway, P.E. and Eaton, J.W., Hemoglobin - a biologic Fenton reagent, J. Biol. Chem. 259 (1984) 14354-14356.
8. Gutterridge, J.M., Iron promoters of the Fenton reaction and lipid per-oxidation can be released from haemoglobin by peroxides, FEBS Lett. 201 (1986) 291-295.
9. Rehan, A., Johnson, K.J., Wiggings, R.C., Kunkel, R.G. and Ward, P.A., Evidence for the role of oxygen radicals in acute nephrotoxic nephritis, Lab. Invest. 51 (1984) 396-403.
10. Paller, M.S., Hoidal, J.R. and Ferris, T.F., Oxygen free radical in is-chemic acute renal failure in rat, J. Clin. Invest. 74 (1984) 1156-1164.
11. Diamond, J.R., Bonventre, J.V. and Karnovsky, M.J., A role for oxygen free radical in aminonucloside nephrosis, Kidney Int. 29 (1986) 478-483.
12. Yamamoto, Y., Manji, T., Saito, A., Maeda K. and Oka, K., Ion-exchange chromatographic separation and fluorometric detection of guanidino compounds in physiologic fluids, J. Chromatogr. 162 (1979) 327-340.
13. Hiraga, Y. and Kinoshita, T., Post-column derivatization of guanidino compounds in high-performance liquid chromatography using ninhydrin, J. Chromatogr. 226 (1981) 43-51.
14. Hung, Y., Kai, M., Nohta, H. and Ohkura, Y., High-performance liquid chromatography of guanidino compounds using benzoin as a fluorogenic reagent, J. Chromatogr. 305 (1984) 281-294.
15. Nakajima, M., Nakamura, K., Shirokane, Y. and Mizusawa, K., A new amidinohydrolase, methylguanidine amidinohydrolase from Alcaligenes sp. N-42, FEBS Lett. 110 (1981) 43-46.
16. Nakajima, M., Nakamura, K. and Shirokane, Y., Enzymic determination of methylguanidine in urine, in:"Guanidines," A. Mori, B.D. Cohen and A. Lowenthal Eds., Plenum Press, New York, (1985) pp. 39-46.

MEASUREMENT OF CREATINE AND PHOSPHOCREATINE IN MUSCLE TISSUE USING HPLC

Katsuo Haruki, Shuichi Hatakeyama, Masayoshi Hirata, Hiroaki Muramoto, Yohei Tofuku and Ryoyu Takeda

The 2nd Department of Internal Medicine, School of Medicine, Kanazawa University, 13-1 Takaramachi, Kanazawa 920, Japan

Most uremic subjects usually show muscle weakness and a decreased activity of daily living. Since the ability of the kidney to synthesize guanidinoacetic acid (GAA) decreases in uremia, the impaired metabolism of GAA may affect muscle energy metabolism, because creatine (CTN), an energy source of muscle, is a metabolite of GAA. CTN is produced from GAA in the liver, transfered to muscle tissue for energy storage, and stored as phospho-creatine (PC). In order to evaluate CTN metabolism in uremia, it is necessary to evaluate the muscle content of CTN and PC as well as that in blood. The usual method to measure CTN is an enzymatic spectrophotometric method and the sensitivity is not high enough to apply to muscular content. We investigated a more reliable method to measure CTN and PC in muscle tissue using HPLC.

We used the JASCO G-520 Guanidine Autoanalyzer. The column is Guanidinopak, and the eluent, Na citrate buffer, is run through the column with a flow rate of 1.0ml/min at 60°C. The final reaction with alkaline ninhydrin is measured fluorophotometrically. In this analytical system, the recovery rate is $93.2 \pm 5.5\%$, and the coeficient of variation is 5.7%. The minimal sensitivity is about $3\mu g/dl$.

PC is easily transformed into CTN by acidification. At high pH levels, PC transformation was little. As the pH levels were lowered from 10 to 2, the CTN peak increased. At about pH 2, the CTN peak became maximal, and the transformation of PC into CTN was 94% of equimolar CTN.

The procedure for measuring CTN and PC in muscle is as follows. Muscle samples were frozen immediately after withdrawal by immersing in liquid nitrogen at the melting point and then freeze-dried. A portion of the sample was stored at -80°C. After freeze-drying, connective tissue and blood were removed as far as possible, and the remainder powdered after measurement of dry weight. The muscle powder was dissolved in 50mM sodium bicarbonate solution and deproteinized through an Amicon membrane filter. Half the filtrate was applied to the JASCO G-520 Guanidine Autoanalyzer for the measurement of CTN. The other half was acidified at pH 2 by the addition of

0.1N HCl solution for conversion of PC to CTN and then also applied to the Autoanalyzer. This HPLC method was adequate for the measurement of CTN. The minimal sensitivity was as low as 3μg/dl, and the recovery rate and coefficient of variation was also satisfactory. We measured serum CTN, muscle CTN and muscle PC in 5 untreated rabbits. Muscle tissues were sampled from the quadriceps femoris. Serum CTN levels were 1603 ± 337 μg/dl, and there was only a narrow variation. Muscle CTN and PC levels were 53.5 ± 24.5mg/100g dry weight and 253 ± 122 mg/100g dry weight, respectively. In spite of a relatively unvariable concentration of serum CTN, there was a wider variety in the levels of CTN and PC in muscle tissue, and we could not find any significant correlations between serum CTN and muscle CTN or PC. These results indicate that serum levels of CTN do not reflect muscular contents of CTN and PC, and in order to evaluate CTN metabolism of muscle tissue it is necessary to measure muscle CTN and PC as well as blood CTN.

γ-GUANIDINOOXYPROPYLAMINE AND N^G-METHYLAGMATINE:NOVEL GUANIDINOAMINES FOUND

IN SEEDS OF LEGUMINOUS PLANTS

Shigeru Matsuzaki, Koei Hamana* and Kimiyasu Isobe**

Department of Physiology, Institute of Endocrinology and
*College of Medical Care and Technology, Gunma University
Maebashi 371 and **Amano Pharmaceutical Co. Ltd., Aichi
181, Japan

INTRODUCTION

A number of guanidino compounds have been found in plants. Seeds of
Leguminosae are especially rich sources of these compounds[1]. Among them are
arginine, N^G-monomethylarginine, N^G, N^G-dimethylarginine, N^G,N'^{G}-dimethyl-
arginine,γ-hydroxyarginine, homoarginine, γ-hydroxyhomoarginine and canavanine. Arginine and homoarginine are decarboxylated to form agmatine and
homoagmatine respectively, both of which have already been found in some
plants[1,2]. Decarboxylation products of other guanidino amino acids, however,
have never been detected in plants.

In the course of study of polyamines in leguminous plants using high-
performance liquid chromatography (HPLC), we observed several unknown
compounds which reacted with o-phthalaldehyde but corresponded to none of the
polyamines and guanidino compounds which are known to be present in nature.
In the present study attempts were made to identify these unknown guanidinoamines in various seeds. Since leguminous plants contain various guanidino
amino acids including L-canavanine[1], we speculated that the unknown compounds
would be decarboxylation products of some of these amino acids.

MATERIALS AND METHODS

Chemicals

Agmatine, homoarginine N^G, N^G-dimethylarginine, N^G, N'^{G}-dimethylargi-
nine, canavanine·H$_2$SO$_4$, o-methylisourea and palladium-barium sulfate were
purchased from Sigma (St. Louis, MO) and N^G-monomethylarginine from Calbio-
chem-Behringer Corp. (La Jolla, CA). 3-Hydroxypropylamine was obtained from
Wako Pure Chemicals (Osaka). Ten aliphatic pentaamines which contain amino-
propyl and/or aminobutyl groups were kindly supplied by Dr. K. Samejima of
Josai University. N-Carbamylputrescine was synthesized in our laboratories.

Synthesis of Guanidinoamines

The synthesis of homoagmatine (monoamidinocadaverine) and audouine (diamidinocadaverine) was carried out by guanidation of cadaverine with o-methylisourea[2]. Briefly, cadaverine·2HCl was treated with o-methyl-isourea-HCl in 2M NaOH at pH 10.5 for 5 days at 25°C. Homoagmatine and audouine were separated from unreacted cadaverine on a column (9 x 120 mm) of ion-exchange resion (Hitachi Custom 2612). The conversion of the sulfate forms of canavanine, agmatine and o-methylisourea to the hydrochloride forms was achieved by passing through a column of Dowex 1-X8 before use.

N^G-Methylagmatine was enzymatically synthesized from N^G-monomethylarginine. Briefly, N^G-monomethylarginine was incubated with the lysate of Escherichia coli 7020 (IFO 3544) at 30°C for 2h. After centrifugation, the supernatants were mixed with 1N HClO$_4$. Methylagmatine in HClO$_4$ extracts were separated on a column (9 x 100 mm) of Hitachi Custom 2612.

γ-Guanidinooxypropylamine was also enzymatically synthesized from canavanine. Canavanine·HCl was incubated with acetone-dried E. coli 7020 (IFO 3544) in 0.2 M acetate buffer (pH 5.25) at 30°C for 1 h[3]. After termination of the reaction with cold 0.5 N HClO$_4$, the decarboxylation product of canavanine was purified by column chromatography as described above.

Materials

Seeds of Canavalia gladiata and other leguminous plants were purchased from a local seed dealer. Wistaria floribunda seeds were collected in the botanical gardens in Maebashi. After water imbibition for 12 h, the seeds were germinated and allowed to grow in the dark on moistened filter paper discs.

Amine Analysis

Seeds and seedlings were homogenized in 3 volumes of 0.5 N HClO$_4$. After centrifugation of the homogenates, the supernatants were passed through a column (2.5 x 2 cm) of Dowex-50 W to concentrate basic compounds including polyamines and to eliminate amino acids[4]. Polyamines and guanidinoamines were analyzed by high-performance ion-exchange column chromatography as described previously[5] with slight modifications. The identity of amines was also confirmed by thin-layer chromatography (TLC) on cellulose and silica gel[6]. The presence of amines on the chromatograms was monitored by visualizing them with either 0.1% ninhydrin, Sakaguchi or alkaline nitroprusside-ferricyanide (pentacyanoferrate) reagent[7].

Degradation of Methylated Guanidinoamines by Alkaline Hydrolysis

Samples were heated in 2M Ba(OH)$_2$ at 120°C for 20 h in sealed tubes[8]. Guanidinoamines and their products were analyzed before and after alkaline hydrolysis.

Degradation of Guanidinoamines by Agmatine Oxidase

Agmatine oxidase purified from Pencillium chrysogenum[9] cleaves not only agmatine but also homoagmatine, N^G-methylagmatine and γ-guanidinooxypropyl amine. This enzyme was used to characterize these compounds.

Large Scale Isolation of Guanidinooxypropylamine

Approximately 1 kg of <u>Wistaria</u> seeds and <u>Canavalia</u> seedlings were homogenized in 10 ℓ of 0.5 N HClO₄. Polyamines and unknown compounds in HClO₄ extracts were concentrated by the use of a Dowex-50 W column, and then separated by a Hitachi Custom 2612 column[6].

RESULTS

When analyzed with the use of our highly sensitive HPLC, putrescine,
spermidine, homospermidine, spermine and agmatine were detected in the seed
of <u>Wistaria</u> <u>floribunda</u> (Fig. 1A). Besides these polyamines a prominent peak
X1 was found between spermine and agmatine. This peak eluted a little prior
to canavalmine. The same peak as XI appeared in <u>C</u>. <u>gladiata</u> seedlings (Fig.
1D), though not in the seeds. Since seeds of both <u>Wistaria</u> and <u>Canavalia</u> are

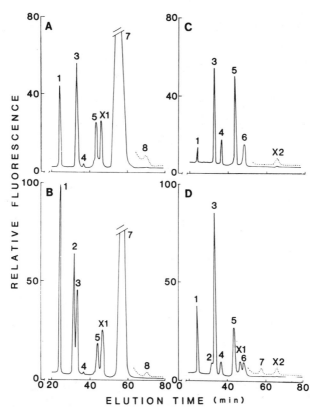

Figure 1.   Elution profiles of polyamines found in <u>Wistaria</u> <u>floribunda</u> and
            <u>Canavalia</u> <u>gladiata</u>. Polyamines were separated by high-performance
            ion-exchange chromatography from the seeds (A and C) and seedling
            (B and D) of <u>W. floribunda</u> and <u>C. gladiata</u>, respectively. Elution
            patterns of the polyamines were followed by o-phthalaldehyde. The
            numbers above the peaks correspond to:(1) putrescine; (2) cadave-
            rine; (3) spermidine; (4) homospermidine; (5) spermine; (6) cana-
            valmine; (7) agmatine and (8) homoagmatine. X1 and X2, unknown
            compounds. Dotted lines indicate the curves at 25 times standard
            sensitivity.

rich in canavanine, we speculated that XI might be a decarboxylation product
of this amino acid. Indeed XI from both <u>Wistaria</u> seeds and <u>Canavalia</u> seedlings behaved chromatographically like authentic guanidinooxypropylamine
whenever different columns and solvent systems were used. A minor peak which
corresponded to homoagmatine was observed in both seeds and seedlings of
<u>Wistaria</u>, but not in <u>Canavalia</u>. Several other minor peaks including X2 were
found in <u>Canavalia</u> seeds and seedlings.

When pooled fractions A-I from <u>Wistaria</u> seeds and B-I from <u>Canavalia</u>
seedlings, both of which contained X1 (Fig. 2), were analyzed by TLC, X1
behaved just like guanidinooxypropylamine (Fig. 3). Unlike agmatine and
homoagmatine, X1 was not Sakaguchi-positive. It gave a positive nitroprusside-ferricyanide reaction, which is specific for guanidinooxy compounds.
When reduced in the presense of palladium-barium sulfate[7], X1 disappeared
and new compounds which corresponded to guanidine and 3-hydroxypropylamine
were produced. These two products were detected by TLC and HPLC. All these
findings show that X1 is indeed identical to guanidinooxypropylamine.

Another unknown peak was found in seeds of <u>Glycine max</u> and <u>Psophocarpus</u>
<u>terragonolobus</u> (Fig. 4). This peak was much higher in mature seeds than in
immature seeds. This compound was resistant to acid hydrolysis and periodate
oxidation, excluding conjugated and hydroxylated polyamines, respectively.
It was not cleaved by reduction in the presence of palladium-barium sulfate,
excluding γ-guanidinooxyplopylamine. Compound X is not only o-phthalaldehyde-
and ninhydrin-positive but also Sakaguchi positive, suggesting that it is a
guanidino compound. The retention times of the compound on our chromatograms
were indeed identical to those of N^G-methylagmatine. On thin-layer chromatograms, compound X behaved exactly like N^G-methylagmatine. In order to confirm
the identity, the compound X enriched fraction was incubated with agmatine
oxidase. The X peak disappeared after incubation and the oxidation products
were identical to those from the authentic N^G-methylagmatine. Furthermore,
peak X was degraded to form N-carbamylputrescine and then putrescine by
alkaline hydrolysis (Fig. 5). All these findings show that compound X is

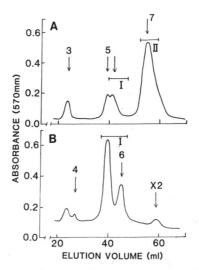

Figure 2.   Large-scale separation of polyamines and X1 from <u>W. floribunda</u>
(A) and <u>C. gladiata</u> (B). Polyamines were detected by ninhydrin
reagent at 570 nm. Pooled fractions A-I, A-II and B-I were used
for further analysis.

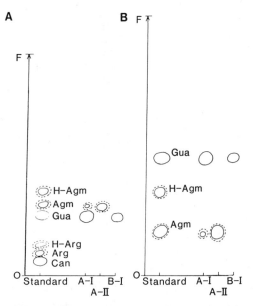

Figure 3.   Thin-láyer chromatographic separation of A-I, A-II and B-I on
            cellulose (A) and silica gel G (B). Arginine (Arg), homoargi-
            nine (H-Arg), canavanine (Can), agmatine (Agm), homoagmatine
            (H-Agm) and guanidinooxypropylamine (Gua) were run simulataneous-
            ly. Spots on chromatograms were visualized by spraying alkaline
            nitroprusside-ferricyanide (open circle) or Sakaguchi (dotted
            circle) reagent. Solvent systems used were isopropanol-ammonia
            (7:3, V/V) (A) and chloroform-methanol-ammonia (2:2:1, V/V) (B).
            O;origin, F; solvent front.

actually identical to N^G-methylagmatine. The peak which corresponded to
N^G-methylagmatine was found not only in pea seeds but also several other
beans tested. These beans included Phaseolus vulgaris L., Vicia sativa L.,
Medicago sativa L.

DISCUSSION

     The present study has shown that seeds of leguminous plants contain many
different polyamines and basic guanidinoamines. L-Canavanine (α-amino-γ-
guanidinooxybutyric acid) is a nonprotein amino acid first found in the jack
bean Canavalia ensiformis[10]. It can be converted to guanidinooxypropylamine
by decarboxylation in these plants. In fact, this amine has been shown to be
formed from canavanine by E. coli 7020[11], though this bacterium normally
contains no guanidinooxypropylamine. It is of interest to note that the content of this amine increases in Canavalia seedlings. Probably a decarboxylase
for canavanine is induced during germination as seen with decarboxylases for
other amino acids[12]. Since W. floribunda seeds contain enormous amounts of
agmatine and large amounts of guanidinooxypropylamine, it is possible that
a common enzyme is responsible for the formation of both these amines. Although canavanine is a substrate for the arginine decarboxylase of E. coli[13],
it does not serve as a substrate but rather as an inhibitor for the arginine
decarboxylase purified from oat seedling[14]. Thus, it remains unknown whether
or not guanidinooxypropylamine is formed by arginine decarboxylase or by a
specific decarboxylase in leguminous plants. Probably the decarboxylation of

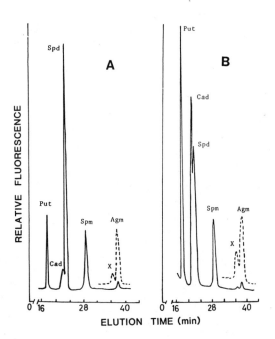

Figure 4. Separation of polyamines and guanidinoamines in soybean <u>Glycine</u>
<u>max</u> (A) and seedlings of winged bean <u>Psophocarpus</u> <u>tetragonolobus</u>
(B). Abbreviations: Put; putrescine, Cad;cadaverine, Spd; spermidine, Spm; spermine, Agm; agmatine and X; unknown compound.

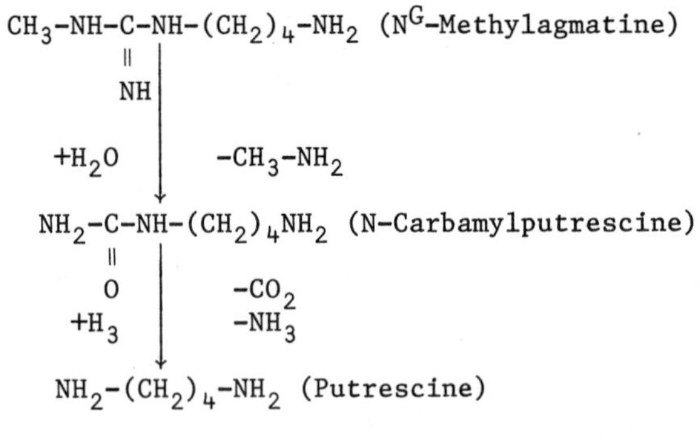

$$CH_3-NH-C-NH-(CH_2)_4-NH_2 \quad (N^G-Methylagmatine)$$
$$\overset{\|}{NH}$$

$$+H_2O \qquad -CH_3-NH_2$$

$$NH_2-C-NH-(CH_2)_4NH_2 \quad (N-Carbamylputrescine)$$
$$\overset{\|}{O} \qquad -CO_2$$
$$+H_3 \qquad -NH_3$$

$$NH_2-(CH_2)_4-NH_2 \quad (Putrescine)$$

Figure 5. Alkaline hydrolysis of N^G-methylagmatine.

methylarginine(s) can be catalyzed by arginine decarboxylase, for the levels
of both N^G-methylagmatine and agmatine increase in parallel during germination when this decarboxylase activity is high. It is not known, however,
whether this guanidino compound plays a specific functional role or is mere-
ly a by-product of arginine decarboxylase. Furthermore, it remains to be
elucidated in N^G-methylagmatine is further metabolized to form putrescine by
the action of agmatine imidohydrolase.

SUMMARY

γ-Guanidinooxypropylamine, a decarboxylation product of canavanine was found in seeds of Wistaria floribunda and seedlings of sword bean Canavalia gladiata. Another novel guanidinoamine N^G-methylagmatine was found in seedlings of soybean Glycine max, pea Pisum sativum, winged bean Phophocarpus tetragonolobus and others. Agmatine was detected ubiquitously in seedlings examined, while homoagmatine only in seeds and seelings of W. floribunda. These findings suggest that certain guanidino amino acids are metabolized by decarboxylases particularly in germinating seeds of leguminous plants.

REFERENCES

1. Robin, Y. and Marescau, B., Natural Guanidino Compounds, in: "Guanidines," A. Mori, B.D. Cohen and A. Lowenthal Eds, Plemum Press, New York and London (1985) pp 383-438.
2. Ramakrishna, S. and Adiga, P.R., Homoagmatine from Lathyrus sativus seedlings, Phytochemistry 12 (1973) 2691-2695.
3. Hamana, K. and Matsuzaki, S., Natural occurrence of guanidinooxypropylamine in Wistaria floribunda and the sword bean Canavalia gladiata Biochem. Biophys. Res. Commun. 129 (1985) 46-51.
4. Inoue, H. and Mizutani, A., A new method for isolation of polyamines from animal tissues, Anal. Biochem. 56 (1973) 408-416.
5. Matsuzaki, S., Hamana, K, Imai, K. and Matsuura, K., Occurrence in high concentrations of N^1-acetylspermidine and sym-homospermidine in the hamster epididymis, Biochem. Biophys. Res. Commun. 107 (1982) 307-313.
6. Hamana, H. and Matsuzaki, S., Unusual polyamines in slime molds Physarum polycephalum and Dictyostelium discoideum, J. Biochem. (Tokyo) 95 (1984) 1105-1110.
7. Natelson, S., Canavanine in alfalfa (Medicago sativa). Experientia 41 (1985) 257-259.
8. Kakimoto, Y., and Akazawa, S., Isolation and identificantion of N^G, N^G- and N^G, N'^G-dimethylarginine, N^ε-mono-, di- and trimethyllysine, and glucosylgalactosyl-and galactosyl-δ- hydroxylysine from human urine, J. Biol. Chem. 245 (1970) 5751-5758.
9. Isobe, K. Tani, Y. and Yamada, H., Crystallization and characterization of agmatine oxidase from Penicillium chrysogenum, Agr. Biol. Chem. (Tokyo) 46 (1982) 1353-1359.
10. Kitagawa, M. and Tomiyama, T., A new amino-compound in the jackbean and a corresponding new enzyme, J. Biochem. (Tokyo) 11 (1929) 265-271.
11. Makisumi, S., Decarboxylation of hydroxyarginine and canavanine by Escherichia coli, J. Biochem. (Tokyo) 49 (1961) 292-296.
12. Altman, A., Friedman, R. and Levin, N., Arginine and ornithine decarboxylases, the polyamine biosynthetic enzymes of mung bean seedlings, Plant Physiol. 69 (1982) 876-879.
13. Blethen, S.L., Boeker E.A. and Snell, E.E., Arginine decarboxylase from Escherichia coli I. Purification and specificity for substrates and coenzyme, J. Biol. Chem. 243 (1968) 1671-1677.
14. Smith, T.A., Arginine decarboxylase of oat seedlings, Phytochemistry 18 (1979) 1447-1452.

PHASCOLINE AND PHASCOLOSOMINE:NEW CATABOLIC PRODUCTS OF PYRIMIDINES IN

SIPUNCULID WORMS

Yvonne Robin and Yvonne Guillou*

Laboratoire du Métabolisme Minéral des Mammiféres (E.P.H.E.)
Faculté de Pharmacie, 92290 Chatenay-Malabry, France, and
*Département de Biochimie et Génétique Moléculaire, Institut
Pasteur, 75015 Paris, France

INTRODUCTION

The diversity of guanidino compounds found in invertebrates and the pecularities of their metabolism have been reviewed recently[1,2]. A good many of these compounds were isolated from various worm phyla and consisted, besides several muscular phosphagens, of a number of highly basic compounds of unknown biological function mainly localized in the viscera. Among the latter group, two monosubstituted guanidino amides, phascoline (N-(3-guanidinopropionyl)-2-hydroxy-n-heptylamine) (GPHHA) and phascolosomine (N-(3-guanidinoisobutyryl)-2-methoxy-n-heptylamine) (GIBMHA), isolated from the sipunculid worms Phascolion strombi and Phascolosoma vulgare respectively[3], have retained our attention owing to their abnormally high concentration in the host viscera and to the possibility that their guanidino moieties are derived from pyrimidine precursors. Since ß-aminopropionic acid (ß-APA) and ß-aminoisobutyric acid (ß-AIBA) are known products of the degradation of the pyrimidines in living organisms, it could be assumed that they were further metabolized and incorporated into the GPHHA and GIBMHA molecules. Previous assays in our laboratory supported this hypothesis. In the present work, we confirm the rôle of uracil and thymine in the biosynthesis of the ß-aminoalkyl chains of GPHHA and GIBMHA, respectively, and that of arginine in providing their amidine group. Free ß-guanidinopropionic acid (ß-GPA) and ß-guanidinoisobutyric acid (ß-GIBA), also identified in the tissues of the worms, were found to be labeled from the corresponding pyrimidine precursors. The preferential distribution of ß-GPA and GPHHA in the genus Phascolion and that of ß-GIBA and GIBMHA in the genus Phascolosoma are specified. The high concentration of the guanidino amides in the viscera of the worms is confirmed and the changes in concentration of the major guanidino compounds in viscera and muscle of Phascolosoma vulgare upon fasting are investigated. The biological significance of the new compounds is discussed.

23

MATERIALS AND METHODS

Animals

The sipunculid worms Phascolion strombi and Phascolosoma vulgare were collected in Britanny (France) as previously reported[3]. They were kept for about 18 h in running sea water before starting the experiments. The in vivo assays were performed on similar groups of animals consisting of 20 specimens (total average weight about 2.8g) for Phascolion strombi and of 5 specimens (total average weight about 12.5g) for Phascolosoma vulgare. During the assays, each group of worms was kept in 750 ml of filtered sea water in a cold room at + 6°C, the water being renewed every 48 h. For the biogenetic studies, 3 groups of animals were utilized for each radioactive precursor assayed. The 3 groups received the same amount of radioactivity by intra-coelomic injection and were sacrified after 1, 2 and 5 days, respectively. The viscera were collected, pooled for each group of worms and stored at -25°C. For the study of the changes in concentration of the guanidino compounds in Phascolosoma vulgare tissues upon fasting, 7 groups of animals were utilized. A standard group was sacrified at 0 time, the other groups being sacrified after 1, 2, 4, 8, 14 and 28 days, respectively. For each group of worms, body-wall muscle and viscera were collected separately, pooled and stored at -25°C.

Tissue Extracts

For the isolation of the crystallized compounds, the extracts were prepared according to Guillou and Robin[3]. In all the other assays, the tissues (viscera or muscle) were homogeneized with 3 vol of 2% acetic acid. The homogenate was heated for 5 min at 100°C in a boiling water bath, cooled and centrifuged. The pellet was reextracted one time with 1 vol of 1% acetic acid. The supernatents were pooled and stored at -25°C.

Purification Procedures

GPHHA, GIBMHA, ß-GPA, ß-AIBA and ß-GIBA were purified by ion exchange column chromatography as previously described[3]. The extracts were filtered on columns of Amberlite IRC 50 resin, 100-200 mesh, pyridinium form, equili-brated in a 1M pyridine-1M acetic acid buffer (pH 6.0). GPHHA and GIBMHA were retained on the resin and eluted with 2M acetic acid in 50% ethanol. ß-GPA, ß-AIBA and ß-GIBA, not retained on the Amberlite IRC 50 colums, were further purified by binding to columns of Dowex 50 X2, 100-200 mesh, H^+ form, and fractional elution with 2N HCl. HCl was removed in vacuo. All the steps of the purification were monitored by thin layer chromatography. GPHHA and GIBMHA were isolated as the crystallized picrates and ß-GPA and ß-GIBA as the crystallized hydrochlorides.

Taurocyamine (TC) and hypotaurocyamine (HTC) were purified from muscle extracts as described previously[3].

Degradation of Radioactive Phascolosomine and Separation of the Fragments

Hydrochloric acid hydrolysis of radioactive phascolosomine and the sepa-ration of the products, ß-GIBA and 2-methoxy-n-heptylamine, were effected according to Guillou and Robin[3]. Subsequent deamidination of ß-GIBA yielding ß-AIBA was obtained by heating the compound for 3 h at 100°C in the presence of barium hydroxide. The barium was eliminated as its insoluble sulfate.

Table 1. In vivo labeling of phascoline from radioactive precursors in Phascolion strombi

Labeled precursor [a]	Time (days)	Radioactivity recovered[b] (cpm x 10^{-3})		GPHHA[b]	
		Viscera	GPHHA	Viscera (mg)	Sp. activity (cpm/μmole)
[^{3}H-5 Uracil]	1	2230	345	1.34	58950
	2	2158	413	0.96	105120
	5	442	65	0.06	264710
[^{14}C-amidino] L-Arginine	1	1750	96	1.09	21520
	2	1201	108	0.74	35660
	5	839	52	0.24	52940

[a]The total radioactivity administered per group of animals was 8 μCi for [^{3}H-5] uracil (specific activity 28 kCi/mole) and 6 μCi for [^{14}C-amidino] L-arginine (specific activity 45 Ci/mole).[b]The values are expressed per group of animals treated in each assay. GPHHA; phascoline

Analytical Methods

Thin-layer chromatography was carried out on cellulose-coated plastic or aluminium sheets. The solvents were pyridine, isoamyl alcohol, acetic acid, water (8:4:1:4) for the resolution of GPHHA, GIBMHA and TC, and n-butanol, diethylamine, acetone, water (10:2:10:5) for that of ß-GPA, ß-AIBA and ß-GIBA. Electrophoresis on cellulose/plastic sheets in a mixture of 1M pyridine-1M ammonia (1:1) for 1 h under 500 V was used for the resolution of 2-methoxy-n-heptylamine. Guanidino compounds were visualized with the Sakaguchi reagent and amino compounds with ninhydrine. Standard guanidine derivatives were obtained as reported previously[3]; ß-AIBA was the commercial product.

Guanidino compounds were estimated in the chromatographically pure fractions, using the diacetyle-α-naphthol method[4]. According to the lability of HTC, the total [TC + HTC] concentration of Phascolosoma vulgare muscle was estimated as TC after oxidative treatment of the extract[3].

Radioactivity was measured using Bray[5] scintillation mixture and an Intertechnique SL 3000 liquid scintillation spectrophotometer. Aliquots of the purified fractions were chromatographied on cellulose/aluminium sheets, the compounds were visualized with the appropriate reagent and the radio-activity of the spots was estimated.

RESULTS

Biogenesis of the Guanidine Moieties of GPHHA and GIBMHA

The biosynthesis of GPHHA and GIBMHA after in vivo administration of radioactive precursors to the worms was studied in the viscera, which contain the major part of these compounds[3]. The results reported in Table 1 for

Table 2. In vivo labeling of ß–AIBA, ß–GIBA and phascolosomine from radio-active precursors in Phascolosoma vulgare

| Labeled precursor[a] | Time (days) | Radioactivity recovered[b] (cpm x 10^{-3}) | | | | GIBMHA[b] | |
		Viscera	ß–AIBA	ß–GIBA	GIBMHA	Viscera (mg)	Sp. activity (cpm/µmole)
[^{14}C–methyl] Thymine	1	22253	16983	92	3383	13.79	66825
	2	8446	3591	862	3782	10.56	97558
	5	6238	1887	1629	1620	3.94	112451
[^{14}C–amidino] L–Arginine	1	7381			1753	14.83	32199
	2	5279			1554	6.58	64332
	5	3048			790	2.50	86078

[a]The total radioactivity administered per group of animals was 15 µCi in all the assays. The specific activity of ^{14}C–thymine was 47 Ci/mole and that of ^{14}C–L–arginine was 43.5 Ci/Mole.
[b]The values are expressed per group of animals treated in each assay. GIBMHA; phascolosomine, ß–AIBA; ß–aminoisobutyric acid and ß–GIBA; ß–guanidinoisobutyric acid.

GPHHA and in Table 2 for GIBMHA show that the two compounds incorporate radioactivity from the corresponding pyrimidines labeled in their ß–amino-alkyl chains, and from arginine labeled in its amidine group. The incorporation profile is similar in all the series of assays: one day after the administration of the labeled precursor, a significant amount of radioactivity is found in the GPHHA or GIBMHA fractions; this value slightly increases within 2 days, then decreases; however, on account of the rapid disappearance of the guanidino amides from the viscera, their specific radioactivity reaches its higher value at the end of the experiment (5 days). Degradation studies performed on labeled GIBMHA isolated from Phascolosoma vulgare viscera show that the radioactivity incorporated from [^{14}C–methyl] thymine is exclusively distributed in the ß–isobutyryl chain of GIBMHA, while that incorporated from [^{14}C–amidino] arginine is localized in the amidine group.

Labeled ß–GPA was found in the viscera of Phascolion strombi after the administration of [^{3}H–5] uracile, but the results were not quantified. A mor detailed study was effected on Phascolosoma vulgare injected with[^{14}C–methyl] thymine. As shown in Table 2, radioactive ß–AIBA and ß–GIBA were found in the viscera of the worm. One day after the beginning of the assay, 76% of the total radioactivity of the viscera extracts were found in the ß–AIBA fraction, while the radioactivity of the ß–GIBA fraction was very low. From the second day, the radioactivity of the ß–AIBA fraction decreased rapidly and that of the ß–GIBA fraction increased significantly to reach a value similar to that of the ß–AIBA fraction on the 5th day of the experiment. This value is close to that of the GIBMHA fraction at the same time of the experiment. A comparison of the amounts of radioactivity recovered in the ß–GIBA and GIBMHA fractions throughout the assay (Fig. 1) shows that there is some proportionality between the increase of the former and the decrease of the latter.

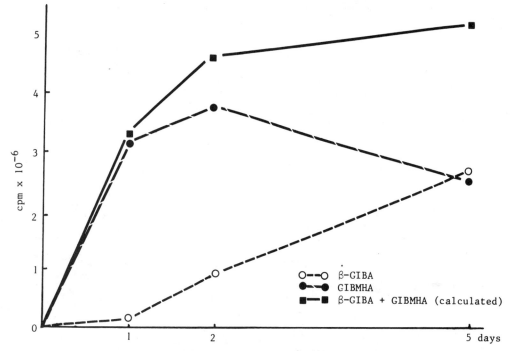

Figure 1.   Radioactivity incorporated from ^{14}C-thymine into guanidino
            compounds in <u>Phascolosoma vulgare</u> viscera. ß-GIBA;
            ß-guanidinoisobutyric acid and GIBMHA; phascolosomine.

        The specific radioactivity of ß-AIBA and ß-GIBA could not be determined
since the low concentration of the two compounds in the <u>Phascolosoma</u> extracts
obliged us to add unlabeled standards as tracers during the purification
steps. However, the high radioactivity which accumulates in the ß-AIBA fraction during the first day gives good evidence that the amino compound is an
intermediate in the biogenetic pathway from thymine to ß-GIBA and to GIBMHA.

Biogenesis of the Amine Moiety of GIBMHA

        Several possible precursors of the amine part of GIBMHA ([^{14}C-1]formate,
[^{14}C-U] acetate, [^{14}C-2] glycine, [^{14}C-U] serine and [^{14}C-methyl] methionine)
were administered to <u>Phascolosoma vulgare</u>. No radioactivity was found to be
incorporated in the GIBMHA molecule from any of these compounds.

Changes in the Levels of the Guanidino Compounds in <u>Phascolosoma vulgare</u> Tissue during Fasting

        In <u>Phascolosoma vulgare</u>, the guanidine content of viscera consists
essentially of GIBMHA[3] and that of muscle of a mixture of TC and HTC in
their free and phosphorylated forms[6]. Their changes in concentration were
studied on freshly collected worms kept for 4 weeks in filtered sea water.
The results are reported in Table 3. At the beginning of the experiment
(Time 0), the GIBMHA content of viscera was very high, representing about 3
times the total phosphagen guanidines content of muscle evaluated as TC.

Table 3.   Changes in concentration of phascolosomine and of taurocyamine[a] in
           Phascolosoma vulgare tissues upon fasting

Time	Weight of tissie[b]		Guanidino compound[c]	
(days)	Viscera (g)	Muscle (g)	Phascolosomine (mg/g wet viscera)	Taurocyamine (mg/g wet muscle)
0	3.3	2.2	7.3	2.5
1	3.0	2.4	5.8	2.3
2	3.2	2.0	5.6	2.5
4	3.0	2.2	4.4	2.4
8	3.2	2.2	3.3	2.3
14	2.0	1.9	0.7	1.8
28	2.2	2.4	0.0	1.2

[a]The amount of taurocyamine estimated represents the sum of the free and
phosphorylated [taurocyamine + hypotaurocyamine] content of muscle.
[b]Total wet weight per group of animals.
[c]Results expressed per group of animals.

Within the first week, about 50% of the GIBMHA content of viscera disappeared, but no significant change was observed in the other values. After 4
weeks of captivity, GIBMHA had disappeared totally while nearly half of the
muscular guanidines were still present. All the animals survived and appeared to be in good condition, the weight of the muscular tissues was
unchanged and that of the viscera was about 65% of the initial value.

Distribution of ß-GPA, ß-GIBA, GPHHA and GIBMHA in Phascolion strombi and in Phascolosoma vulgare

The distribution of GPHHA and GIBMHA according to the genus of the worms
has been specified. The former is preferentially distributed in Phascolion
strombi and the latter in Phascolosoma vulgare, and about 90% of these compounds are found in the viscera versus 10% in the muscle, as was previously
reported[3]. However, their distribution is not as specific for the zoological
genus as we first thought. Small amounts of GPHHA were found in Phascolosoma
vulgare and GIBMHA was identified as traces in Phascolion strombi.

The presence of free ß-GPA and ß-GIBA has been established in the two
worms. ß-GPA was isolated in small amounts from the viscera of Phascolion
strombi and characterized as traces in the muscle of the same animal and in
the tissues of Phascolosoma vulgare. Conversely, ß-GIBA was isolated from
Phascolosoma vulgare viscera and characterized in minute amounts in the
muscle of the worm and in the tissues of Phascolion strombi.

DISCUSSION

In invertebrates, the guanidine content of viscera is usually very low.
The only tissue containing substantial amounts of guanidino compounds in
their free and phosphorylated forms is the muscle, in which the phosphagen
constitutes a pool of energy freely available for muscular contraction. The
intriguing point with GPHHA and GIBMHA was therefore their accumulation in

the viscera of a number of sipunculid worms and their rapid disappearance upon fasting[3-7], which suggested that they could be storage products.

An approach to a better understanding of this problem was to obtain informations on the biogenetic origin of GPHHA and GIBMHA. The results obtained in the present study (Tables 1 and 2) give evidence that, in the biosynthesis of the guanidine moiety of each compound, the alkyl chain is derived from the pyrimidines and the amidine group is provided by arginine. On the other hand, we could not establish the biosynthesis of the amine moiety of GIBMHA in Phascolosoma vulgare from several possible labeled precursors assayed, questioning whether the worm was able to synthesize this part of the molecule or if it was taken from the diet. Thus, the limiting factor in the synthesis of GIBMHA when the phascolosomes are kept for several days in filtered sea water might be the lack of the amine moiety, accounting for the rapid disappearance of the guanidino amide from the worms viscera upon fasting (Table 3). On the contrary, the synthesis of the guanidine part from the pyrimidines does not seem to be involved in this process. After the administration of labeled thymine to Phascolosoma vulgare, the radioactivity of the ß-GIBA fraction increases throughout the assay as that of the GIBMHA fraction declines, and the over-all synthesis of the guanidines formed from the pyrimidine precursor is still positive after 5 days (Fig. 1).

According to these results, GIBMHA does not seem to be a storage product to be reutilized by the worm in defective nutritional conditions. Another argument is the fact that the total disappearance of the guanidino amide from the viscera after a 28 days fasting period (Table 3) does not significantly affect the phascolosomes, which suggests that GIBMHA is not indispensable to their survival.

In fact, GPHHA and GIBMHA possess interesting physiological properties which could account for their presence in the worms. They show a strong lowering effect on the beatings of the heart of the Daphnie[8] and a high negative chronotrope activity on cultured heart cells[9]. The activity seems to be located in the aminoalcohol or aminoether part of the molecule[9]. The two compounds exhibit also a moderate inhibitory effect on acetylcholine esterase of rabbit brain[10]. Administration of GIBMHA to Phascolosoma vulgare by intracoelomic injection caused death in a few days[7]. Since the guanidino amides are rather toxic substances, Baslow[11] and Chevolot[12] have suggested that there could be some connection between these compounds and the toxins detected by Chaet[13-14] in the coelomic fluid and coelomic cells of Golfingia (Phascolosoma) gouldii, a closely related sipunculid worm. This suggestion deserves to be further investigated. It should account for the high concentration of the compounds and for the pecularities of their distribution in a restricted number of sipunculid and annelid worms[3-15].

We have established in this study that ß-GPA, ß-GIBA, GPHHA and GIBMHA were formed from the pyrimidines and from arginine in a number of sipunculid worms, and that ß-AIBA was an intermediate in this process in Phascolosoma vulgare. These findings are in good agreement with the pathway proposed for synthesis of ß-GPA in the vertebrates[16-17], which consists in a transamidination reaction between ß-APA and L-arginine and is based on the observation that ß-APA is an acceptor for the amidine group of arginine in the presence of hog kidney amidinotransferase. They also support the comment by Chang et al.[18] that the synthesis of ß-GPA seems to be regulated in a different way from that of many other guanidino compounds, which are considered as normal products of protein catabolism. However, recent studies have shown that no radioactivity was found to be incorporated into ß-GPA, neither from ^{14}C-amidino arginine by perfused rat liver[19], nor from ^{15}N-amidino arginine in the liver of mice administered the labeled compound[20]. Thus, the synthetic origin of ß-GPA in mammals remains a controversial problem.

REFERENCES

1. Robin, Y., Metabolism of arginine in invertebrates:Relation to urea cycle and to other guanidine derivatives, in:"Urea Cycle Diseases," A. Lowenthal, A. Mori and B. Marescau Eds., Plenum Press, New York (1982) pp407-417.
2. Robin, Y. and Marescau, B., Natural guanidino compounds, in: "Guanidines," A. Mori, B.D. Cohen and A. Lowenthal Eds., Plenum Press, New York (1985) pp383-438.
3. Guillou, Y. and Robin, Y., Phascoline (N-(3-guanidinopropionyl)-2-hydroxy-n-heptylamine) and phascolosomine (N-(3-guanidinoisobutyryl)-2-methoxy-n-heptylamine), two new guanidino compounds from sipunculid worms, J. Biol. Chem. 248 (1973) 5668-5672.
4. Rosenberg, H., Ennor, A.H. and Morrison, J.F., Estimation of arginine, Biochem. J. 63 (1956) 153-159
5. Bray, G.A., A simple efficient liquid scintillator for counting aqueous solutions in a liquid scintillation counter, Anal. Biochem. 1 (1960) 279-285.
6. Robin,Y. and Thoai, N.V., New monosubstituted biological guanidine derivative, hypotaurocyamine (2-guanidinoethanesulfinic acid) and the related phosphagen, Biochim. Biophys. Acta 63 (1962) 481-488.
7. Robin,Y. and Guillou, Y., Unpublished data.
8. Robin,Y. and Michel, C., Unpublished data.
9. Auclair, M.C. Adolphe M., Guillou, Y. and Robin, Y., Effet de la phascoline et de la phascolosomine, nouveaux dérivés guanidiques naturels, sur les cellules cardiaques de rat en culture, Compt. Rend. Soc. 170 (1976) 65-70.
10. Matsumoto, M., Fujiwara, M., Mori, A. and Robin, Y., Effet des dérivés guanidiques sur la cholinacétylase et sur l'acétylcholinestérase du cerveau de lapin, Compt. Rend. Soc. Biol. 171 (1977) 1126-1129.
11. Baslow, M.H., "Marine Pharmacology. A Study of Toxins and Other Biologically Active Substances of Marine Origin," Krieger Publ., Huntington, New York (1977).
12. Chevolot, L., Guanidine Derivatives, in "Marine Natural Products," P.J. Scheuer Ed., Acad. Press, London (1981) pp53-91.
13. Chaet, A.B., Further studies on the toxic factor in Phascolosoma, Biol. Bull. 109 (1955) 356.
14. Chaet, A.B., Demonstration of a toxic factor in thermal death, Proc. Soc. Exptl. Biol. Med. 91 (1956) 599-602.
15. Robin, Y., Presence de l'acide ß-guanidinoisobutyrique libre et combiné chez des vers marins, Biochim. Biophys. Acta 93 (1964) 206-208.
16. Pisano, J.J., Mitoma, C. and Udenfriend, S., Biosynthesis of γ-guanidinobutyric acid from γ-aminobutyric acid and arginine, Nature 180 (1957) 1125-1126.
17. Shimoyama, M., Arginine and other guanidine derivatives. XVI. The influence of the addition of several kinds of acceptors on enzymic transamidination from glycocyamine by a hog kidney preparation, Bull. Osaka Med. School 7 (1961) 105-110.
18. Chang, J.S., Watanabe, Y. and Mori, A., The effect of sodium valproate on blood ammonia levels and the concentrations of guanidino compounds in mouse tissue, in:"Guanidines," A. Mori, B.D. Cohen and A. Lowenthal Eds. Plenum Press, New York (1985) pp125-133.

19. Perez, G.O., Rietberg, B., Owens, B. and Schiff, E.R., Effect of acute
    uremia on arginine metabolism and urea and guanidino acid production by
    perfused rat liver, Pfluegers Arch. 372 (1977) 275-278.
20. Shindo, S. and Mori, A., Metabolism of L-[amidino-^{15}N]-arginine to
    guanidino compounds, in:"Guanidines," A. Mori, B.D. Cohen and A.
    Lowenthal Eds., Plenum Press, New York (1985). 71-81.

II. METABOLISM OF GUANIDINO COMPOUNDS

EFFECTS OF ARGININE-FREE DIET ON UREAGENESIS IN YOUNG AND ADULT FERRETS

Devendra R. Deshmukh

Department of Pediatrics and Biological Chemistry, University
of Michigan, Ann Arbor, MI 48109, USA

INTRODUCTION

The major function of arginine is to protect mammals against ammonia
intoxication via urea synthesis. In addition, arginine has several other
important functions. Arginine is used by mammals for the synthesis of tissue
proteins. It is the only amino acid that provides the amidino group for
synthesis of creatine, a major source of high energy phosphate in muscle.
Arginine is also a precursor for the synthesis of polyamines, which play an
important role in cell division, tissue growth and differentiation.

Because rats are capable of storing some nitrogen without dietary
arginine, L-arginine is often classified as an intermediate between the
indispensible and dispensible amino acids. Most young mammals require
dietary arginine for optimum growth whereas adult mammals are assumed to
synthesize enough arginine for maintaining optimum growth and nitrogen
balance. However, according to Visek[1] adult mammals including humans may
not, under certain conditions, meet all of their arginine needs by
endogeneous synthesis. Visek also suggested that growth and nitrogen balance
may not be the only criteria which determine arginine need and other factors
such as orotic acid excretion, circulating hormone levels and immune
responsiveness should be evaluated as possible determinants of arginine
requirement.

Recent studies indicate that the requirement of arginine varies among
species. For example, feeding an arginine-free diet to rats for 13 days
caused growth retardation without signs of hyperammonemia[2,3] whereas a
single meal of an arginine-free diet to juvenile ferrets and near-adult cats
produced severe hyperammonemia and encephalopathy[4-8]. Similar results were
obtained when young dogs were fed arginine-free diets[9]. These studies
suggest that a dietary source of arginine is essential for young cats, dogs
and ferrets and probably for other carnivorous animals[10,11].

Our recent study[7] supported the view that the effects of arginine-free
diet differ not only between species but also between different age-groups
of the same species. For example, although a single meal of an arginine-free
diet caused hyperammonemia in young ferrets, it did not alter serum ammonia
levels in adult ferrets.

For the past few years, we have been investigating the arginine and ammonia metabolism in ferrets. In the present paper, the results are summarized and our recent study of the effects of a single feeding of an arginine-free diet on the concentration of plasma amino acids and the activities of urea cycle enzymes in liver and kidney of young and adult ferrets are reported.

MATERIALS AND METHODS

Ferrets

Two-month-old (350-400g) or eighteen-month-old, male sable-coated ferrets, vaccinated for canine distemper were purchased from Marshall Research Farm, North Rose, New York. They were housed in groups of two or three in cages with grid flooring, in an isolation room with controlled light and temperature (22-23°C, 12 hr light/dark cycle).

Diets

Water and stock diet (Cat chow, Ralston Purina Co., St. Louis, MO) were provided ad libitum. A synthetic diet containing free amino acids, vitamins, corn starch, sucrose and salt mixture was prepared as discribed previously[4-8] except that corn oil was used instead of turkey fat. An arginine-free diet was prepared by substituting arginine with an isonitrogenous amount of alanine. The slight increase in weight caused by the excess alanine was compensated for by decreasing the carbohydrate content in the arginine-free diet.

Experimental Design

Ferrets were fasted overnight and fed ad libitum, at 8.00 A.M. either the arginine containing or arginine-free diet. At 11.00 A.M., 3 ml of blood was collected by cardiac puncture from the comatose or lightly anesthetized animals into chilled heparinized tubes (Vacutainer) and centrifuged at 8000g for 15 min at 0-4°C. A portion of the plasma was stored at -20°C. Between 11.00-11.30 A.M. animals were killed. Livers and kidneys were removed, frozen in liquid nitrogen and stored at -20°C.

Amino Acid Analysis

Plasma was deproteinized with 3% sulfosalicylic acid for amino acid determination. A portion of liver was pulverized in liquid nitrogen and a 10%(W/V) homogenate was prepared in 3% sulfosalicylic acid. Amino acids in plasma and tissues were determined using a high performance liquid chromatography (Waters Associates) with a gradient programmer and integrater. Chromatographic separations were performed with a reverse phase C18 column as described previously[12].

Ammonia levels in plasma and tissues were determined on fresh specimens by an enzymatic method[13]. A 10% homogenate in either distilled water or distilled water containing 1 % triton X-100 was prepared and the activities of following enzymes in liver and kidney were determined at 37°C: Arginase[14] (EC 3.5.3.1.), carbamyl phosphate synthetase I[15](EC 6.3.4.16), ornithine-oxo-acid aminotransferase[16](EC 2.6.1.13) argininosuccinate synthetase[17](EC 6.3.4.5.), argininosuccinase[17](EC 4.3.2.1.) and ornithine carbamyl transferase[18] (EC 2.1.3.3.). The activity of L—arginine:glycine transamidinase (EC 2.1.4.1., commonly called transamidinase) was determined by the method of Van Pilsum et al[19]. Protein was measured by Lowry's method[20]. Student's t-test was used to calculate the statistical significance. P values < 0.05 were considered statistically significant.

RESULTS

The effects of diets containing varying amounts of arginine on the serum ammonia levels in two months-old ferrets is shown in Figure 1. Ferrets fed diets containing 0.3% or less arginine developed hyperammonemia within 2h of ingesting the diet. Symptoms such as lethargy, seizures and coma were observed only in ferrets with serum ammonia levels exceeding 1200 µM[6]. The degree of hyperammonemia and the severity of sickness decreased with increased arginine in the diet.

Young ferrets fed the arginine-free diet became hyperactive 2h after ingesting the diet. Within 30 min of the onset of hyperactivity, they became prostrate and subseqently became comatose. Seizures and convulsions were observed in several ferrets. Ferrets remained sick for about 4h and then recovered. Serum ammonia returned to normal levels in the recovered animals. Young ferrets fed the arginine-free diet had higher serum levels of glucose and orotic acid as well as higher urinary orotic acid than those fed the arginine-containing diet[7] (Table 1).

Hiperammonemia and encephalopathy were preveted in young ferrets by supplying dietary arginine and abbreviated by ornithine or arginine injection given during encephalopathy[6]. Therefore, young ferrets were assumed to be unable to synthesize ornithine from precursors other than arginine.

Adult ferrets did not develop hyperammonemia and encephalopathy even after eating an arginine-free diet for 7 days. Therefore adult ferrets appear capable of synthesizing a sufficient amount of arginine/ornithine from the sources other than dietary arginine. However serum and urinary levels of orotic acid were significantly elevated in adult ferrets fed an arginine-free

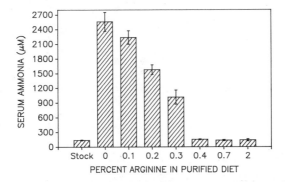

Figure 1. Effects of dietary arginine on serum ammonia levels in young ferrets. Two months-old ferrets were starved overnight and fed diets containing varying amounts of arginine. Blood was collected 3 hours after feeding the diet. Compiled from Deshmukh and Shope[6].

diet as compared to those fed arginine-containing meals. The greater orotic
aciduria and orotic acidemia in young ferrets suggested a greater diversion
of ammonia into pyrimidine synthesis than in the adult ferrets[7] (Table 1).

Table 2 shows the effects of the arginine-free diet on the urea cycle
enzyme activities in the livers of young and adult ferrets. In general, the
activities of enzymes in the livers of young and adult ferrets were
comparable to those in rats and humans[17-21]. The activities of ornithine
carbamyl transferase, carbamyl phosphate synthetase, argininosuccinate
synthetase and argininosuccinase were slightly higher in adult ferrets fed
the control diet (containing arginine) than similarly treated young ferrets

Table 1. Effects of arginine-free diet on clinical parameters in serum and
urine of young and adult ferrets

Parameter	2-mo-old ferrets fed		18-mo-old ferrets fed	
	Control	Arginine-free	Control	Arginine-free
		mg/L		
Orotate (Serum)	1030 ± 140	3730 ± 540[a]	340 ± 120[b]	1440 ± 370[ab]
Orotate (Urine)	20 ± 3	318 ± 34[a]	11 ± 3[b]	73 ± 14[ab]
Citrate (Urine)	51 ± 21	79 ± 36	29 ± 11	20 ± 8[b]
Creatinine (Urine)	239 ± 43	197 ± 28	854 ± 262[b]	711 ± 204[b]

Results are means ± SEM. n > 6. Statistically significant difference a as
compared with control ferrets in the same age-group and b as compared with 2
months old (2-mo-old) ferrets in the same diet group. Compiled from Thomas
and Deshmukh[7].

Table 2. Urea cycle enzyme activities in livers of young and adult ferrets
fed arginine-containing or arginine-free diet

Enzyme	2-mo-old ferrets		18-mo-old ferrets	
	Control	Arginine-free	Control	Arginine-free
		nmol/min/mg protein		
Ornithine carbamyl transferase	911 ± 43	890 ± 10	1163 ± 29[b]	899 ± 43[a]
Arginase	981 ± 60	977 ± 39	1055 ± 57	1060 ± 79
Ornithine amino-transferase	10.0 ± 0.8	9.5 ± 0.5	13.3 ± 1.1[b]	11.5 ± 0.5[b]
Argininosuccinate synthetase	6.6 ± 0.4	8.6 ± 1.0	4.5 ± 0.6[b]	4.6 ± 0.6[b]
Argininosuccinase	70.5 ± 1.5	95.5 ± 7.2[a]	73.2 ± 6.0	50.4 ± 6.8[ab]
Carbamyl phosphate synthetase	13.5 ± 1.0	10.3 ± 1.6	21.8 ± 1.3[b]	19.5 ± 2.4[b]

For experimental details, see caption of Table 1. Values are mean ± SEM of
n = 5. The letter a indicates a statistically significant difference (P <
0.05) when compared with arginine-containing diet in the same age-group and
b indicates statistically significant difference as compared with 2 months
old ferrets in the same diet group.

Table 3. Urea cycle enzyme activities in kidney of young and adult ferrets

Enzyme	2-mo-old ferrets	18-mo-old ferrets
	nmol/min/mg protein	
Arginase	6.4 ± 1.0	7.4 ± 1.1
Ornithine amino-transferase	35.0 ± 3.4	32.2 ± 3.4
Argininosuccinate synthetase	1.7 ± 0.2	2.9 ± 0.2[a]
Argininosuccinase	4.7 ± 0.6	5.6 ± 0.2

Experimental conditions were as described in the caption of the Table 1. Values are mean ± SEM of n = 5. The letter a indicates a statistically significant difference (P < 0.05) compared with 2 months old ferrets.

Table 4. Amino acid concentrations in plasma of young ferrets arginine-containing or arginine-free meals

Amino acid	Fasting	90 min		180 min	
		Control	Arg-free	Control	Arg-free
		μM			
Arginine	72 ± 5.0	125 ± 15[a]	50 ± 10[b]	102 ± 7.0[a]	24 ± 3[abc]
Ornithine	5.1 ± 0.2	25 ± 4.0[a]	7 ± 0.9[b]	16 ± 2.2[a]	3.6 ± 0.4[abc]
Citrulline	11 ± 0.4	17 ± 0.9[a]	16 ± 0.8[a]	22 ± 2.2[a]	20 ± 1.8[a]
Glutamate	53 ± 2.4	60 ± 4.0	75 ± 7.2[a]	46 ± 4.5	158 ± 23[abc]
Aspartate	2.1 ± 0.2	4.1 ± 0.4[a]	6.1 ± 1.1[a]	3.2 ± 0.4[a]	68 ± 11[abc]

Experimental conditions are outlined in the caption of Table 1. Values are means ± SEM, N > 5. Statistically significant differences (p < 0.05) are indicated by a when compared to fasting group, b when compared to control group at same time interval and c when compared to 90 min level in the same diet group.

(Table 2). Argininosuccinase activity was increased in the young ferrets fed an arginine-free diet whereas the activity was decreased in the similarly treated adult ferrets (Table 2).

The activities of urea cycle enzymes in the kidney of young and adult ferrets are shown in Table 3. The activity of argininosuccinate synthetase was higher in adult ferrets than that in young ferrets. In general, the activities of most enzymes were comparable to those reported for rat kidney[22]. However, the activity of arginine:glycine transamidinase in the kidney of young and adult ferrets (Table 3) was significantly lower than that in rat kidney[19].

The effects of an arginine-free diet on the plasma amino acid levels in young and adult ferrets are shown in Table 4. Fasting values for plasma amino acids in young ferrets were not significantly different from those in adult ferrets. However, fasting values for plasma ornithine in young and adult ferrets were significantly lower than those in rat, man or cat[10,23]. Young ferrets fed the arginine-containing (control) diet had higher levels of plasma arginine, ornithine and citrulline than similarly treated adult ferrets. Plasma arginine concentration was much lower in young ferrets fed an arginine-free diet than similarly treated adult ferrets. Ornithine levels were similar in these two groups whereas the concentration of citrulline was much higler in young ferrets. In addition plasma glutamine and aspartic acid levels were increased in young animals fed an arginine-free diet.

DISCUSSION

The increase in orotic acid excretion in the young and adult ferrets could be due to low levels of ornithine in the liver. The observation that young and adult ferrets fed the arginine-containing diet excreted higher amounts of orotic acid than other animals suggest that ferrets have limited or impaired ability to detoxify ammonia into urea via urea cycle.

Although, the general response to an arginine-free diet in young ferrets was similar to that in cats, there were some differences. Serum ammonia levels following a single meal of an arginine-free diet were much higher in young ferrets than that in cats[4~6‚10]. Most cats died after eating an arginine-free diet whereas > 80% of the ferrets survived. Young ferrets not only tolerated extremely high ammonia levels but also detoxified the ammonia within 4h. These results indicate that young ferrets may be more resistant to ammonia toxicity than cats and may have greater capacity to detoxify ammonia.

Several possibilities have been suggested to explain the metabolic basis for the rapid development of hyperammonemia in cats following a single meal of an arginine-free diet[10]. These include decreased or nonfunctional enzymes, enzyme inhibition, impaired activation of enzymes, decreased substrate concentration or an impaired transport of ornithine across the mitochondrial membrane. According to Stewart et al.[10], low levels of hepatic ornithine are probably responsible for making cats susceptible to hyperammonemia following an arginine-free diet. Similar to cats, both young and adult ferrets had low levels of plasma ornithine. However, only young ferrets became hyperammonemic after ingesting an arginine-free diet. Studies on the amino acid levels in the livers of young and adult ferrets fed arginine-containing or arginine-free diets are currently underway in our laboratory.

Rogers and Phang reported that the activity of pyrroline-5-carboxylate synthase in the intestinal mucosa of cat was very low as compared to that of rat[24]. They concluded that the limitation or lack of de novo synthesis of ornithine from glutamate in cats may be the metabolic basis for the rapid onset of hyperammonemia following an arginine-free meal. The same explanation may hold true for the development of hyperammonemia in young ferrets.

Although arginine is synthesized in large amounts by the liver, the high activity of hepatic arginase normally prevents release of any arginine from the blood[25]. Among extrahepatic tissues, the kidney has the greatest capacity to convert citrulline into arginine, and the level of activity varies among species. The citrulline required for this process is synthesized in the small intestine, released in the blood and taken up by the kidney. It has been shown that the daily production of arginine by the kidney and the daily requirement of additional dietary arginine to achieve high growth rates in young rats are the same[26]. An impaired uptake of citrulline by the kidney, a decreased synthesis of arginine or an increased catabolism of arginine in young ferrets may explain the development of hyperammonemia resulting from ingestion of an arginine-free diet.

Our results show that the activities of the enzymes required for arginine synthesis in ferret kidney are similar to those in rat kidney. To study the degradation of arginine by kidney, we measured the activity of arginine:glycine transamidinase activity in the kidney of young and adult ferrets and compared the levels with those in young rats. The transamidinase activity in young ferrets was not different from that in adult ferrets. However, the activity of this enzyme in young and adult ferrets was significantly lower than that in rats. This could be due to low levels of arginine or due to the high levels of creatine, (an inhibitor of transamidinase) in ferret kidney. Further studies are necessary to define the physiological significance of low transamidinase activity in ferret kidney.

Differences in response to the ingestion of an arginine-free diet among and within species are evident. For example, near-adult cats consistently developed hyperammonemia after eating an arginine-free diet[5]. Similar consistency in the response to the arginine-free diet was not seen in adult cats[10], although few adult cats did develop hyperammonemia. In contrast, none of the adult ferrets developed ammonia toxicity even after eating an arginine-free diet for 7 days. Since adult ferrets developed orotic aciduria and orotic acidemia, it appears that they also require dietary arginine, but the need is not as acute as in young ferrets. Additional studies on the effect of arginine-free diets in various age-groups of ferrets, cats and dogs may help to elucidate the species specificity of arginine requirements in carnivores.

Ferrets appear to be a good model for studying the metabolic basis of the hyperammonemia in carnivores because, unlike cats and dogs, only young ferrets develop hyperammonemia after ingesting the arginine-free diet. Therefore, metabolic differences between young and adult ferrets may help define the susceptibility of young ferrets to the arginine-free diet.

The disorder of an arginine deficiency also occurs naturally in the animals that resemble ferrets. Minks fed a purified diet containing adequate arginine developed a severe fatal disease only during the period of active fur growth[27]. Since the supplementation of arginine prevented the disease, production of fur, an arginine-rich cellular product, appears to have caused the arginine deficiency.

Dietary deficiencies generally require weeks or months to produce serious health defects other than growth depression. Arginine deficiency in ferrets and cats is unique in causing an immediate serious disorder. Because of the suddenness, severity and consistency of the response to the arginine-free diet, young ferrets are well suited as an animal model for the metabolic study of arginine and its regulation. Since arginine is a precursor for guanidines, ferrets could also be useful as an animal model for studying the metabolism of guanidino compounds.

ACKNOWLEDGMENTS

I thank Cindy D. Rusk for excellent technical assistance and Dr. Manuel Portoles for the analysis of amino acids in plasma. This work was supported in part by a grant from the National Institutes of Health, AI 20236 and by a research career development award.

REFERENCES

1. Visek, W.J., Arginine needs, physiological states and usual diets. A reevaluation, J. Nutr. 116 (1986) 36–46.
2. Milner, J.A., Mechanism for fatty liver induction in rats fed arginine deficient diet, J.Nutr. 109 (1979) 663–670.
3. Milner, J.A. and Visek, W.J., Orotate, citrate and urea excretion in rats fed various levels of arginine, Proc. Soc. Exp. Biol. Med. 147 (1974) 754–759.
4. Deshmukh, D.R., Maassab, H.F. and Mason, M., Interactions of aspirin and other potential etiologic factors in an animal model of Reye syndrome, Proc. Natl. Acad. Sci. (U.S.A.) 79 (1982) 7557–7560.
5. Morris, J.G. and Rogers, Q.R., Arginine:An essential amino acid for the cat, J. Nutr. 108 (1978) 1944–1953.
6. Deshmukh, D.R. and Shope, T.C., Arginine requirement and ammonia toxicity in ferrets, J. Nutr. 113 (1983) 1664–1667.
7. Thomas, P.E. and Deshmukh, D.R., Effect of arginine-free diet on ammonia metabolism in young and adult ferrets, J. Nutr. 116 (1986) 545–551.
8. Deshmukh, D.R., Kao, W., Mason, M. and Baublis, J.V., Serum enzyme alterations in arginine deficient, influenza infected ferrets. A potential animal model for Reye's syndrome, Enzyme 27 (1982) 52–57.
9. Czarnecki, G.L. and Baker, D.H., Urea cycle function in the dog with emphasis on the role of arginine, J. Nutr. 114 (1984) 581–590.
10. Stewart, P.M., Batshaw, M., Valle, D. and Walser, M., Effects of arginine-free meals on ureagenesis in cats, Am. J. Physiol. 241 (1981) E 310–E 315.
11. Burns, R.A., Milner, J.A. and Corbin, J.E., Arginine: An indispensible amino acid for mature dogs, J. Nutr. 111 (1981) 1020–1024
12. Portoles, M., Minana, M., Jorda, A. and Grisolia, S., Caffeine-induced changes in the composition of the free amino acid pool of the central cortex, Neurochem. Res. 10 (1985) 887–895.
13. Mondzack, A., Ehrlich, G.E. and Sigmiller, J.E., An enzymatic determination of ammonia in biological fluids, J. Lab. Clin. Med. 66 (1965) 526–531.
14. Tarrab, R., Rodriguez, J., Haitron, C., Palacois, R. and Soberon, G., Molecular forms of rat liver arginase. Isolation and characterization, Eur. J. Biochem. 49 (1974) 457–468.
15. Schimke, R.T., Adaptive characteristic of urea cycle enzymes in the rat, J. Biol. Chem. 237 (1962) 459–468.
16. Peraino, C. and Pitot, H.C., Ornithine δ transaminase, Biochem. Biophys. Acta 73 (1963) 222–231.
17. Nazum, C.T. and Snodgrass, P.J., Multiple assays of the five urea cycle enzymes in human liver homogenates, in: "The urea cycle," S. Grisolia, R. Baguena and F. Mayer Eds., John Wiley and Sons, New York (1975) pp 325–349.
18. Cerriotti, G., Ornithine carbamyl transferase, in: "Clinical Biochemistry," H.C. Curtius and M. Roth Eds., Walter de Gruyter, Berlin (1978) pp 1151–1156.
19. Van Pilsum, J.F., Taylor, D., Zaikis, B. and McCormick, P., Simplified assay for transamidinase activities of rat kidney homogenates, Anal. Biochem. 35 (1970) 277–286.
20. Lowry, O.J., Rosenbrough, N.J., Farr, A. and Randall, R.J., Protein measurement with folin-phenol reagent, J. Biol. Chem. 193 (1951) 265–275.
21. Schimke, R.T., Differential effects of fasting and protein-free diets on levels of urea cycle enzymes in rat liver, J. Biol. Chem. 237 (1962) 1921–1924.

22. Ratner, S., Enzymes of ornithine and urea synthesis, in: "Advances in enzymology," A. Meister Ed., John Wiley and Sons, New York, Vol. 39 (1973) pp 1-90.
23. Stewart, P.M. and Walser, M., Short-term regulation of ureagenesis, J. Boil. Chem. 255 (1989) 5270-5280.
24. Rogers, Q.R. and Phang, J.M., Deficiency of pyrroline-5-carboxylate synthase in the intestinal mucosa of the cat, J. Nutr. 115 (1985) 146-150.
25. Windmuller, H.G., Glutamine utilization by the small intestine, in: "Advances in enzymology," A. Meister Ed., John Wiley and Sons, New York, Vol. 53 (1982) pp 201-237.
26. Milner J.A. and Visek, W.J., Dietary protein intake and arginine requirement in the rat, J. Nutr. 108 (1978) 382-391.
27. Leoschke, W.L. and Elvehjem, C.A., The importance of arginine and methionine for the growth and fur development of minks fed purified diets, J. Nutr. 69 (1959) 147-150.

SERUM AND URINARY GUANIDINO COMPOUNDS IN "SPARSE-FUR" MUTANT MICE WITH

ORNITHINE TRANSCARBAMYLASE DEFICIENCY

Ijaz A. Qureshi, Bart Marescau*, Maurice Levy, Peter P.
De Deyn*, Jacques Letarte and Armand Lowenthal*

Pediatric Research Center, Ste-Justine Hospital and University
of Montreal, Montreal, Canada H3T 1C5, *Laboratory of
Neurochemistry, Born-Bunge Foundation and Laboratory of
Neuropharmacology, University of Antwerp, U.I.A. 2610
Antwerp, Belgium

SUMMARY

Various quanidino compounds were measured in serum and 24-hrs. urinary
samples from adult male hemizygous "sparse-fur" (spf/Y) mice, with a chronic
X-linked ornithine transcarbamylase (OTC) (E.C.:2.3.1.3.) deficiency. Normal
CD-1/Y males were used as controls. The results indicate no significant
differences in the serum guanidinosuccinic acid (GSA) values between spf/Y
$(0.34 \pm 0.06\mu M)$ (mean $\pm$ SEM) and CD-1/Y (0.28 ± 0.02), or its urinary excretion (spf/Y:34.4 $\pm$ 8.9 nmol/24 h vs CD-1/Y:33.2 $\pm$ 8.2). Serum and urinary
values of arginine, guanidinoacetic acid (GAA) and creatine were significantly lower ($p < 0.05 - < 0.01$) in spf/Y mice as compared to controls,
indicating an impairment of creatine synthesis caused by OTC deficiency.
Serum methylguanidine (MGua) was below our detection limits, but urinary
levels were significantly lower ($p < 0.05$) in spf/Y mice. Serum urea, citrulline and the amino acids related to the urea cycle were also measured and
the implications of their levels in relation to normal GSA production in
spf/Y mutant mice with chronic OTC deficiency are discussed.

INTRODUCTION

The X-linked "sparse-fur" (spf) mutation arose spontaneously in the
progeny of an irradiated mouse at Oakridge[1]. Demars et al.[2] discovered an
abnormal form of liver OTC in spf/Y hemizygotes, and reported that the spf
and abnormal OTC phenotypes were transmitted in the same manner as the OTC
deficiency hyperammonemia in humans. The suitability of spf mouse as animal
model of human disease in respect of orotic acid excretion, hyperammonemia,
hyperglutaminemia and mutant enzyme expression in the small intestine has
already been reported by us[3-5]. The present studies were planned to see if
the OTC deficiency in the mouse model had any effect on the serum and urinary
levels of guanidino compounds. Previously published work on some human
patients of OTC deficiency has shown an absence of GSA in urinary samples[6-7].

MATERIALS AND METHODS

Animals and their Breeding

The parent stock for the colony of spf mutant mice, maintained at the animal house of the Pediatric Research Center, Sainte-Justine Hospital, Montreal, was originally supplied by Dr. L.B. Russel of Oakridge National Laboratories, Oakridge TN[2]. The mutant gene was subsequently transferred to the CD-1 strain (Canadian Breeding Farms, ST-Constant, Que), to improve viability of the spf/Y hemizygous males[8]. Mutant spf/Y mice required for the present studies were obtained by crossing spf/spf homozygous females with normal +/Y males from the above commercial source. The progeny consistig of spf/Y males and spf/+ females were separated by simple sexing. Adult CD-1/Y males of the same age obtained from Canadian Breeding Farms were used as normal controls. All experimental mice were fed ad libitum on Purina mouse chow no. 5015 (Ralston Purina, St. Louis, MO) containing 17% protein. The animals were cared for in accordance with the principles outlined by the Canadian Council of Animal Care[9].

Animals used in this study were approximately 3 months of age with a weight range of 32-42 g. For collection of urine, individual animals were kept in plastic metabolic cages (Econo-metabolism unit, Maryland Plastics Inc., Federalsburg, MD) for 24 h with provision of food and water. Aliquots from urinary collections were analyzed for orotic acid by the colorimetric method of Adachi et al.[10], with modifications proposed by Rogers and Porter[11] A non-brominated blank was kept with each sample tube to eliminate interference from other urinary constituents[12].

At the end of the 24hrs. urinary collection, individual animals from the normal and mutant groups were decapitated to obtain serum collections. Pieces of liver tissue were kept on dry ice to verify the mutant status of animals by the measurement of OTC activity. The colorimetric method for the measurement of the liver OTC activity has already been described[13]. Liver protein was measured by the method of Lowry et al.[14], to express OTC activity in µmol of citrulline produced/mg liver protein/h.

Measurement of Guanidino Compounds and Amino Acids

The concentration of various guanidino compounds was determined using a Biotronic LC 6001 amino acid analyser adapted for guanidino compound determination. The guanidino compounds were separated over a cation exchange resin column using sodium citrate buffers, and were detected with the fluorescence-ninhydrin method as has been reported in detail earlier[15]. The procedure of sample preparation has also been described[16]. Serum urea nitrogen was measured by the technique of Ceriotti[17]. Amino acids related to the urea cycle were measured by ion exchange column chromatography using the same analyser.

Statistical Methods

The data were processed on an Apple IIc microcomputer with the help of the statistical program "Stats plus" (Human System Dynamics). Descriptive statistics, unpaired t-test and Mann Whitney's Rank Sum Test values were compiled for each group and comparison.

RESULTS

Comparison of Liver OTC Activity and Urinary Orotate in Experimental Mice

The mutant spf/Y group showed a mean liver OTC activity of 10.8 ± 1.2 (mean $\pm$ SD) $\mu mol/h/mg$ liver protein as compared to 112.8 ± 11.8 μmol for the CD-1/Y group. The mean excretion of urinary orotate for the spf/Y group was 6165 ± 4561 as compared to 266 ± 55 nmol/mg creatinine for the controls. Both these comparisons were significantly different at $p < 0.001$.

Comparison of Serum and Urinary Guanidino Compounds

Table 1 gives the serum concentration of 8 different guanidino compounds as measured in both the experimental groups, with statistical comparison on the basis of the t-test and the Mann Whitney's Rank Sum Test. The only statistically significant differences are seen in serum arginine, creatine and GAA, which were decreased in the mutant spf/Y group. GSA and the other guanidino compounds in spf mice were not significantly different from the CD-1/Y group.

Table 2 gives a similar picture of guanidino compounds excreted in the 24 hrs. urine in both experimental groups. Arginine, GAA, creatine and MGua values are significantly lower in spf/Y mice, whereas no significant differences are seen in respect of the excretion of GSA and other guanidine derivatives between the two groups.

Comparison of Serum Urea and Related Amino Acids

Table 3 represents the concentration of serum urea and selected amino-acids related to the urea cycle. Only citrulline values are significantly

Table 1. Serum concentration[a] of various guanidino compounds in adult spf/Y mutant mice as compared to normal CD-1/Y controls

Compound	spf/Y mice	Normal mice	P value t-test	P value Mann Whitney's test
GSA	0.34 ± 0.06[b]	0.28 ± 0.02[b]	0.40	0.53
Creatine	152 ± 22	208 ± 21	0.09	0.05*
GAA	1.34 ± 0.28	2.01 ± 0.25	0.09	0.03*
N-α-acetylarginine	2.47 ± 0.20	2.80 ± 0.16	0.23	0.48
Argininic acid	0.06 ± 0.02	0.04 ± 0.01	0.24	0.43
GPA	< 0.025	< 0.025	–	–
Creatinine	7.39 ± 0.98	9.95 ± 1.24	0.14	0.11
GBA	0.26 ± 0.09	0.13 ± 0.02	0.21	0.48
Arginine	78.4 ± 14.0	140.6 ± 13.3	0.01**	0.01**
Homoarginine	0.30 ± 0.08	0.33 ± 0.04	0.69	0.63
Guanidine	< 0.2	< 0.2	–	–
Methylguanidine	< 0.1	< 0.1	–	–

[a]μM; [b]Mean $\pm$ SEM (n = 6); *Significantly different at $P < 0.05$; **$p < 0.01$. GSA; guanidinosuccinic acid, GAA; guanidinoacetic acid, GPA; guanidino propionic acid and GBA; γ-guanidinobutyric acid.

Table 2.   Urinary excretion[a] of various guanidino compound in mutant spf/Y
           mutant mice as compared to normal CD-1/Y controls

Compound	Mutant spf/Y mice	Normal CD-1/Y mice	P value t-test	P value Mann Whitney's test
GSA	34.4 ± 8.9[b]	32.2 ± 8.2[b]	0.57	0.30
Creatine	483 ± 216	3030 ± 936	0.02*	0.01**
GAA	212 ± 53	607 ± 172	0.05*	0.05*
N-α-acetylarginine	27.1 ± 8.3	38.1 ± 12.5	0.49	0.63
Argininic acid	8.77 ± 2.32	17.2 ± 5.39	0.18	0.20
GPA	2.01 ± 0.50	3.15 ± 0.69	0.21	0.30
Creatinine	1762 ± 320	2161 ± 520	0.53	0.69
GBA	86.5 ± 25.6	118 ± 38.6	0.52	0.53
Arginine	11.4 ± 4.8	35.1 ± 10.4	0.06	0.05*
Homoarginine	< dl	< dl	-	-
Guanidine	28.6 ± 8.1	50.5 ± 14.7	0.22	0.03
Methylguanidine	22.8 ± 6.9	64.7 ± 20.5	0.08	0.05*

[a]nmol/24 hrs. urinary sample; [b]Mean ± SEM (n = 6); *Significantly different
at *$P < 0.05$; **$p < 0.01$. The abbreviations of guanidino compounds are as
same as shown in Table 1.

lower in the spf/Y group, whereas no significant differences are seen in the
serum urea values and other amino acids reported. Mean glutamine and homocitrulline values are higher and ornithine values lower in the spf mice, but
not statistically different than the control CD-1/Y group.

DISCUSSION

     The experimental group of the spf/Y mutant mice which showed less than
10% of liver OTC activity and a highly increased urinary orotate excretion,
corresponds to a representative animal model of the congenital hyperammonemia type II syndrome as seen in children[18]. These animals also showed
significantly lower values of serum citrulline and arginine (Tables 1 and 3),
which has frequently been reported in male hemizygotes of X-linked human
ornithine transcarbamylase deficiency[19].

     The observation that serum and urinary GSA values were normal in spf/Y
mice, hemizygous for OTC deficiency is interesting, in view of the fact that
the two earlier reports based on human patients had shown an absence of
urinary GSA[6-7]. The lack of difference in serum urea values of spf/Y mice
compared to CD-1/Y, seen in Table 3 is also quite intriguing. Two different
explanations may be offered as to why spf/Y mice with less than 10% of liver
OTC activity would show a normal production of urea. It may be due to an
increased synthesis of the mutant enzyme protein, as reported by Briand et
al.[20], which would provide a functional availability of a higher level of
total OTC activity. It may also be due to the ability of the mutant enzyme
to better utilize lysine as a substrate in the presence of a reduced
utilization of ornithine[21], or an increased activity of a putative entity,
lysine transcarbamylase as discussed by Carter et al[22]. The present studies

Table 3. Serum concentration of urea[a] and various amino acids[b] related to
          the urea cycle in spf/Y mutant mice as compared to normal CD-1/Y
          controls

Compound	Mutant spf/Y mice	Normal CD-1/Y mice	P value t-test	P value Mann Whitney's test
Urea	7.15 ± 0.65[c]	7.18 ± 0.27[c]	0.68	0.67
Glutamine	900 ± 102	773 ± 119	0.44	0.43
Citrulline	19.3 ± 5.7	69.2 ± 5.2	< 0.01**	< 0.01**
Homocitrulline	1.67 ± 0.42	1.25 ± 0.77	0.65	0.46
Ornithine	89.5 ± 11.1	113.3 ± 7.6	0.10	0.11
Lysine	223 ± 26	235 ± 19	0.68	0.30

[a]mM; [b]µM; [c]Mean ± SEM (n = 6); **Significantly different at $P < 0.01$

do not show any clear evidence for that, since an alternate "lysine-urea
cycle" scheme[23] should also show an increased turnover through the intermediate steps of homocitrulline and homoarginine. This may have to be tested
by loadig tests of these mice with lysine. A third explanation for normal
urea synthesis may simply be the role of hepatic arginase acting upon
dietary arginine received through the portal circulation.

Tables 1 and 2 also show that metabolites involved in the synthesis of
creatine, i.e. arginine, GAA and creatine itself were significantly lower in
spf/Y mice, in both serum and urinary samples. This indicated that arginine
deficiency seen in spf/Y mice may be the cause of an impairment of creatine
synthesis. Creatine serves as a metabolic reserve of ATP in the form of
phosphocreatine during muscle contraction, and as a reserve of ATP in nerve
tissue[24].

In the case of a serious deficiency of one of the enzymes involved in
the synthesis of arginine, this amino acid becomes essential. There will be
three general demands on whatever arginine is available: 1) hydrolysis to
ornithine and urea, 2) protein synthesis and 3) creatine synthesis. There
are, therefore, functional implications for both muscle and nerve tissue in
relation to the amount of arginine synthesized and available in the OTC
deficiency disease state. As the kidney is thought to be a more important
source of metabolic arginine than the liver[25] the decreased conversion of
citrulline to arginine in OTC deficiency is also of some significance to the
present discussion. It is now known that transamidination of glycine to form
GAA mostly takes place at the renal level, and the transmethylation of GAA
is principally seen in the hepatocyte[26]. Both these steps could be affected
adversely in ornithine transcarbamylase deficiency.

MGua was below our detection limits in sera from all experimental mice
(Table 1). The urinary levels were however, significantly reduced in spf/Y
mice (Table 2). The biosynthesis of this metabolite is not completely understood. The current hypothesis implicates creatine as its metabolic source[27].
The effect on the excretion of MGua in spf/Y mice may also therefore be
related to the metabolic implications of arginine deficiency, as explained
above.

The normal mouse serum guanidino compound levels reported in the present study differ slightly with an earlier work by Marescau et al.[16], in respect of certain specific compounds. This may be due to the differences in breed characteristics, developmental changes[28] and the level of nutrition[29] in the experimental animals used.

The present studies would be useful in understanding the mechanisms of adaptation of nitrogen metabolism in the chronic form of urea cycle disorders, particularly in partial ornithine transcarbamylase deficiency.

ACKNOWLEDGMENT

This work was supported by the Fondation Justine-Lacoste-Beaubien, the NATO Research grant #83-0913, the Quebec-Belgium Exchange program, the University of Antwerp and the Born-Bunge Foundation.

REFERENCES

1. Green, M.C., Mutant genes and linkages, in:"Biology of the Laboratory Mouse, 2nd., "E.L. Green Ed., McGraw-Hill, New york (1966).
2. Demars, R., Levan, S.L., Trend, B.L. and Russel, L.B., Abnormal ornithine carbamyl transferase in mice having the sparse-fur mutation, Proc. Natl. Acad. Sci. USA 23 (1976), 1693-1697.
3. Qureshi, I.A., Letarte, J. and Ouellet, R. Ornithine transcarbamylase deficiency in mutant mice. I. Studies on the characterization of enzyme defect and suitability as animal model of human disease, Pediatr. Res. 13 (1979). 807-811.
4. Qureshi, I.A., Letarte, J. and Ouellet, R., Spontaneous animal medels of ornithine transcarbamylase deficiency:Studies on serum and urinary nitrogenous metabolites, in: "Urea Cycle Diseases," A. Lowenthal, A. Mori, and B. Marescau Eds., Plenum Press, New York (1982) pp. 173-183.
5. Qureshi, I.A., Letarte, J. and Ouellet, R., Expression of ornithine transcarbamylase deficiency in the small intestine and colon of sparse-fur mutant mice, J. Pediatr. Gastroenterol. Nutr. 4 (1985) 118-124.
6. Stein, J.M., Cohen, B.D. and Kornhauser, R.S., Guanidinosuccinic acid in renal failure, experimental azotemia and inborn errors of the urea cycle, New Engl. J. Med. 280 (1969) 926-930.
7. Lowenthal, A. and Marescau, B., Urinary excretion of mono substituted guanidines in patients affected with urea cycle diseases, in: "Neurogenetics and Neuroophthalmology," A. Huber, and D. Klein Eds., Elsevier/North-Holland Biomedical Press, Amsterdam (1981) pp. 347-350.
8. Qureshi, I.A., Letarte, J. and Qureshi, S.R., Congenital hyperammonemia-model 235, in: "Handbook: Animal Models of Human Disease, Fasc. 11," C.C. Capen, D.B. Hackle, and G. Migaki Eds., Registry of Comparative Pathology, Armed Forces Institute of Pathology, Washington D.C. (1982) pp. 2.
9. Canadian Council on Animal Care, "Guide to the Care and Use of Experimental Animals, "vol. 1, Canadian Council on Animal Care, Ottawa (1981).
10. Adachi, T., Tanimure, A. and Asahina, M., A colorimetric determination of orotic acid, J. Vitaminol. 9 (1963) 217-226.
11. Rogers, L.E. and Porter, E.S., Hereditary orotoic aciduria. II. A urinary screening test, Pediatrics 42 (1968) 423-428.
12. Goldstein, A.S., Hoogenraad, N.J. and Johnson, J.D., Metabolic and genetic studies of a family with ornithine transcarbamylase deficiency, Pediatr. Res. (1974) 5-12.

13. Qureshi, I.A., Letarte, J. and Quellet, R., Study of enzyme defect in a case of ornithine transcarbamylase deficiency, Diabet. Metabol. 4 (1979) 239-241.
14. Lowry, O.H., Rosebrough, N.J. and Farr, A.L., Protein measurement with the folin-phenol reagent, J. Biol. Chem. 193 (1951) 265-275.
15. Marescau, B., Qureshi, I.A., De Deyn, P., Letarte, J., Ryba, R. and Lowenthal, A., Guanidino compounds in plasma, urine and cerebrospinal fluid of hyperargininemic patients during therapy, Clin, Chim. Acta 146 (1985) 21-27.
16. Marescau, B., De Deyn, P., Wiechert, P., Van Gorp, L. and Lowenthal, A., Comparative study of guanidino compounds in serum and brain of, mouse, rat rabbit and man, J. Neurochem. 46 (1986) 717-720.
17. Ceriotti, G. Ultramicrodetermination of plasma urea by reaction with diacetylmonoxime antipyrine without deproteinization, Clin. Chem. 17 (1971) 400-403.
18. Levin, B., Oberholzer, V.G. and Sinclair, R.L., Biochemical investigation of hyperammonemia, Lancet 2 (1969) 170-174.
19. Batshaw, M.L., Hyperammonemia, Curr. Probl. Pediatr. 14 (1984) 1-69.
20. Briand, P., Cathelineau, L., Kamoun, P., Gigot, D. and Penninckz, M., Increase of ornithine transcarbamylase protein in sparse-furmice with ornithine transcarbamylase deficiency, FEBS Lett. 130 (1981) 65-68.
21. Hommes, F.A., Eller, A.G., Scott, D.R. and Carter, A.L., Separation of ornithine and lysine activity of the ornithine transcarbamlase catalyzed reaction, Enzyme 29 (1983) 271-275.
22. Carter, A.L., Eller, A.G., Rufo, S., Metoki, K. and Hommes, F.A., Further evidence for a separate enzyme entity for the synthesis of homocitrulline distinct from regular ornithine transcarbamylase, Enzyme 32 (1984) 26-36.
23. Scott-Emuakapor, A., Higgins, J.V. and Kohrman, A.F., Citrullinemia: A new case with implications concerning adaptation to defective urea synthesis, Pediatr. Res. 6 (1962) 626-633.
24. Hird, F.J.R., Davuluri, S.P. and Mclean, R.M., Evolutionary relationship between arginine and creatine in muscle, in: "Urea Cycle Diseases," A. Lowenthal, A. Mori, and B. Marescau Eds., Plenum Press, New York (1982) pp 401-406.
25. Perez, G.O., Epstein, M., Reitberg, B. and Loutzenhiser, R., Metabolism of arginine by the isolated perfused rat kidney, Am. J. Physiol. 235 (1987) F 376-381.
26. Funahashi, M., Kato, H., Shimaka, S. and Nakagama, H., Formation of arginine and guanidinoacetic acid in the kidney in vivo. Their relation with liver and their regulation, J. Biochem. 89 (1981) 1347-1356.
27. Orita, Y., Tsuhakihara, Y., Ando, A., Nakata, K., Takamitsu, Y., Fukuhara, Y. and Abe, H., Effect of arginine or creatine administration on the urinary excretion of methylguanidine, Nephron 22 (1978) 328-336.
28. Watanabe, Y., Shindo, S. and Mori, A., Developmental changes in guanidino compound levels in mouse organs, in: "Guanidines," A. Mori, B.D. Cohen and A. Lowenthal Eds., Plenum Press, New York (1985) pp 49-58.
29. Shindo, S., Watanabe, Y. and Mori, A., Effects of starvation on guanidin compound metabolism in mice, Res. Commun. Chem. Pathol. Pharmacol. 54 (1986) 73-78.

BIOSYNTHESIS OF 2-GUANIDINOETHANOL

Yoko Watanabe, Isao Yokoi and Akitane Mori

Department of Neurochemistry, Institute for Neurobiology
Okayama University Medical School, 2-5-1 Shikatacho
Okayama 700, Japan

INTRODUCTION

2-Guanidinoethanol (GEt) was first identified in human urine by high performance liquid chromatography (HPLC), thin layer chromatography and gas chromatography-mass spectrometry[1]. GEt was also observed to be excreted in urine of vertebrates like rabbits, mice, rats and cats[2]. Previously, we reported that ethanolamine (EA) is a precursor of GEt biosynthesis in intact mice and that GEt is synthesized from Arg and EA by a transamidination reaction in isolated rabbit kidney[2]. In this study, we describe some characteristics of GEt biosynthesis from Arg and EA in rat kidney.

METHODS

Animals

Male Sprague Dawley rats were used. Rat kidneys were removed immediately after decapitation and homogenized in 10 volumes of ice cold distilled water.

Assay for Activities of GEt Synthesis

The activity of GEt synthesis was determined by measuring the formation of GEt in the assay mixture. The standard assay mixture (in a final volume of 2ml) contained : 50µl of kidney homogenate (about 5mg wet weight of kidney tissue), 15mM Arg and 300mM EA in the $H_3BO_3-KCl-Na_2CO_3$ buffer (pH 8.8) prepared by mixing 0.2 M H_3BO_3 containing 0.2 M KCl and 0.2 M Na_2CO_3. Incubation was carried out at 37°C. The assay mixture was preincubated for 10min before adding kidney homogenate. The reaction was started by addition of kidney homogenate and stopped by the addition of 1 ml of 30% trichloroacetic acid 5min after. After the centrifugation at 3,000 rpm for 10min, the supernatant was applied to a column of the HPLC system for the determination of GEt.

To examine the effect of pH on GEt biosynthesis, the activity was test-

53

ed at different pH values. The following buffer systems were used; citrate-
Na_2HPO_4 buffer (pH 5.0-7.5) prepared by mixing 0.1 M citrate and 0.2 M
Na_2HPO_4, 0.05 M Tris-HCl buffer (pH 7.5-8.7) and H_3BO_3-KCl-Na_2CO_3 buffer
(pH 8.5-10.0) prepared as described above.

For the kinetic studies of GEt formation, the concentrations of Arg and
EA in the assay mixture were varied as indicated in figures of the results
section. To examine the inhibitory effects of ornithine (Orn), Gly, guanidinoacetic acid (GAA), creatine and creatinine, each chemical was added to
the assay mixture at the indicated concentration. Protein concentrations
were determined by the method of Lowry et al.[3], with bovine serum albumin as
standard.

HPLC

GEt was fluorometrically analyzed by an HPLC system, Fully Automated
Guanidino Compounds Analyzer (Jasco, Tokyo, Japan)[4], on the basis of the
reaction of 9,10-phenanthrenequinone. However, a modification was employed
for elution buffers. The column was equilibrated with 0.4 M sodium citrate
buffer (pH 10.0) for 2min, and the elution was begun with the same buffer for
8min after sampling, followed with 1 M NaOH for 10min. Authentic GEt was
synthesized from EA and O-methylisourea sulfate[1].

Analytical Method of Kinetics

The results were analysed by nonlinear regression[5,6], and calculation
was carried out by a NEC 9801 Computor System. Nonlinear regression curves
were analysed by the following equations:

1. Without inhibitor

$$Vobs = \frac{[S] \cdot Vmax}{[S] + Km} + R \cdot [S]$$

Where Vobs is the observed velocity of formation; [S]
is the concentration of the substrate and R is the rate
of nonspecific formation.

2. Competitive inhibition

$$Vobs = \frac{[S] \cdot Vmax}{Km \cdot \left(1 + \frac{[I]}{Ki}\right) + [S]} + R \cdot [S]$$

Where [I] is the concentration of inhibitor.

3. Uncompetitive inhibition

$$Vobs = \frac{[S] \cdot Vmax}{Km + [S] \cdot \left(1 + \frac{[I]}{Ki}\right)} + R \cdot [S]$$

4. Non-competitive inhibition

$$Vobs = \frac{[S] \cdot Vmax}{\left[Km + [S]\right] \cdot \left[1 + \dfrac{[I]}{Ki}\right]} + R \cdot [S]$$

5. Mixed type inhibition

$$Vobs = \frac{[S] \cdot Vmax}{Km \cdot \left[1 + \dfrac{[S]}{Ki}\right] + [S] \cdot \left[1 + \dfrac{[I]}{\alpha \cdot Ki}\right]} + R \cdot [S]$$

Where α is the factor for the enzyme species to be at equilibrium.

In the inhibition study, each group of data using different concentrations of inhibitor was calculated using equations 2-5. Finally, the inhibition model that gave the lowest sum of squared residual was chosen as the type of inhibition, because the lower sum of squared residual fits better the model for inhibition kinetics.

RESULTS

Fig. 1 shows the chromatograms of authentic GEt and assay mixtures. GEt was produced after the kidney homogenate was incubated for 5min at 37°C in the presence of 15mM Arg and 300mM EA. Fig. 2 shows the effect of pH on GEt formation. The optimal activities of GEt synthesis was observed between

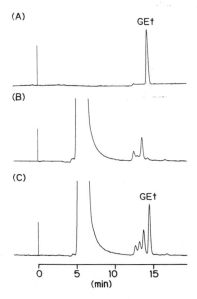

Figure 1.   Chromatograms of (A) authentic 2-guanidinoethanol (GEt) (1μM) (B) assay mixture before incubation (C) assay mixture after incubation for 5min at 37°C. Each assay mixture contains kidney homogenate (about 5mg wet weight of tissue), 15mM Arg and 300mM EA in the 2ml of the H_3BO_3-KCl-Na_2CO_3 buffer (pH 8.8).

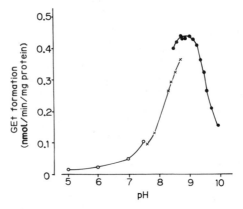

Figure 2.   Effect of pH on 2-guanidinoethanol (GEt) formation. Kidney homogenates (about 5mg wet weight of tissue) were incubated for 5min
at 37°C in the presence of 15mM Arg and 300mM ethanolamine (EA)
at different pH values. All pH values indicated represent the
final pH in the assay mixture. Citrate-Na_2HPO_4 buffer, ○; Tris-
HCl buffer, x; H_3BO_3-KCl-Na_2CO_3 buffer, ●.

pH 8.7 and pH 9.1 at 37°C. Therefore, the following studies were carried out
at pH 8.8 using the H_3BO_3-KCl-Na_2CO_3 buffer. The amount of GEt produced
was directly proportional to the incubation time up to 20min.

    The effect of substrate concentration on the rate of GEt formation is
shown in Fig.3. The velocity of GEt formation was increased according to
Michaelis-Menten kinetics. The apparent Km for Arg was 1.26mM in the presens
of 300mM EA, and the apparent Km for EA was 450mM in the presence of 15mM
Arg.

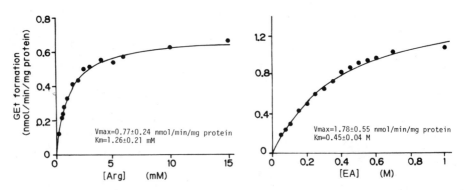

Figure 3.   The effect of Arg and ethanolamine (EA) concentration on 2-
guanidinoethanol (GEt) formation. The left panel shows the
effect of Arg concentration on GEt formation at a fixed EA
concentration (300mM). The right panel shows the effect of EA
concentration on GEt formation at a fixed Arg concentration
(15mM). The inserted Vmax and Km values are calculated from 6
independent assays (mean ± SD).

The effects of Orn, Gly, GAA, creatine and creatinine on GEt formation were shown in Fig.4-8. Orn effectively inhibited GEt formation (Fig. 4). Inhibition was of the mixed type when the Arg concentration was varied. The Ki value was 0.075mM and α value was 2.04. When the EA concentration was varied, the inhibition was competitive. The Ki value was 0.107mM. Gly also inhibited GEt formation (Fig. 5). Inhibition was also of the mixed type when the Arg concentration was varied. The Ki value was 0.59mM, and α value was 2.96. When ethanolamine was varied, the inhibition was competitive. The Ki value was 1.02mM. Fig. 6 shows the effect of GAA on GEt formation. When Arg concentration was varied, a mixed type inhibition was again observed. The Ki value was 3.2mM, and α value was 1.55. However, no inhibition was observed when the EA concentration was varied. Neither 20mM creatine (Fig. 7) nor 20mM creatinine (Fig. 8) affected GEt formation.

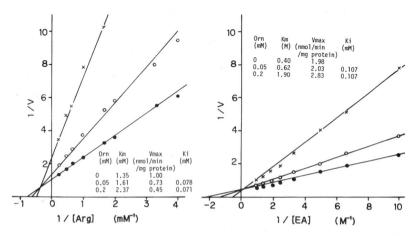

Figure 4.   Effect of Orn on 2-guanidinoethanol (GEt) formation. No Orn
            added, ●; 0.05mM Orn, ○; 0.2mM Orn, x. The left panel shows
            the Lineweaver-Burk plot obtained when the Arg concentration
            was varied at a fixed ethanolamine (EA) concentration (300mM).
            The right panel shows the Lineweaver-Burk plot obtained when
            the EA concentration was varied at a fixed Arg concentration
            (15mM). Figures 5-8 are shown in the same manner.

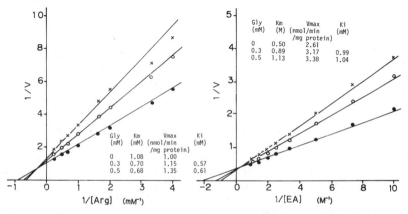

Figure 5.   Effect of Gly on 2-guanidinoethanol (GEt) formation. No Gly
            added, ●; 0.3mM Gly, ○; 0.5mM Gly, x. EA; ethanolamine.

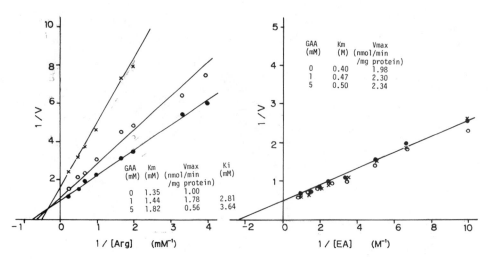

Figure 6.   Effect of guanidinoacetic acid (GAA) on 2-guanidinoethanol
            (GEt) formation. No GAA added, ●; 1mM GAA, ○; 5mM GAA, x. EA;
            ethanolamine.

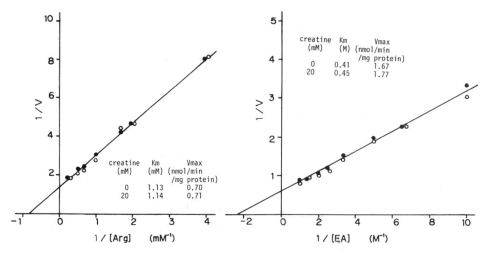

Figure 7.   Effect of creatine on 2-guanidinoethanol (GEt) formation. No
            creatine added, ●; 20mM creatine, ○. EA; ethanolamine.

DISCUSSION

The transamidination reaction in vertebrate is catalyzed by Arg:Gly
amidinotransferase [EC 2.1.4.1] (amidinotransferase)[7]. This enzyme catalyzes
the reversible transfer of amidine groups from Arg to Gly producing GAA
which is a precursor of creatine synthesis. It has been observed that other
compounds with amino groups, such as Orn, L-canaline and 4-aminobutyric acid
can act as amidine acceptors in the reaction catalyzed by the same enzyme[7].
Therefore, GEt synthesis from Arg and EA would be expected to be catalyzed by
amidinotransferase.

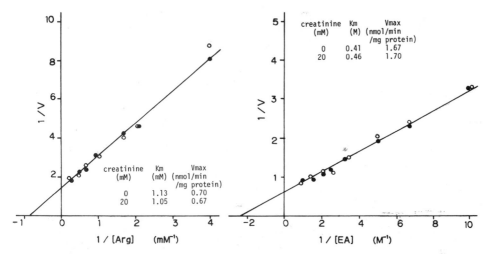

Figure 8.   Effect of creatinine on 2-guanidinoethanol (GEt) formation.
No creatinine added, ● ; 20mM creatine ○. EA; ethanolamine.

It is known that Orn and GAA inhibit the transamidination reaction from
Arg to Gly[7-8]. Further, It is reported that amidinotransferase catalyzes the
formation of GAA with a reaction mechanism involving two half reactions with
binary complexes between the substrate and enzyme[9]. Our observations on the
inhibitory effects of Orn and Gly are in agreement with this reaction mechanism. They inhibited the GEt formation in a mixed type manner when the Arg
concentration was varied at a fixed EA concentration, while they showed
competitive inhibition when the EA concentration was varied at a fixed Arg
concentration. Orn and Gly occupying the acceptor site of the enzyme appear
to inhibit the formation of the enzyme—amidine complex in a mixed type inhibitory manner, and appear to compete with EA for the catalytic site on the
enzyme—amidine complex. These observations suggest that GEt is synthesized by
a amidinotransferase catalyzing reaction.

GAA is also observed to inhibit GEt formation when the Arg concentration was varied at a fixed EA concentration. It is known that GAA donates
an amidine group to form the enzyme—amidine complex. This enzyme—amidine
complex reacts with EA resulting in GEt formation under these assay conditions. Therefore, kinetic studies for the mechanism of GAA inhibition on the
formation of the enzyme—amidine complex should be examined without the addition of EA. It is understandable that GAA does not inhibit GEt formation when
the EA concentration was varied at a fixed Arg concentration since GAA does
not compete with EA for the catalytic site on the enzyme—amidine complex.

Creatine is known to alter the levels of kidney amidinotransferase
activity, although the addition of creatine to the kidney homogenate does not
inhibit the enzyme activity[10-11]. Recently McGuire et al.[12] reported that
creatine affects amidinotransferase activity by altering its rate of synthesis at a pretranslational step. We observed that creatine did not inhibit GEt
formation in vitro. This result is comsistent with McGuire's observations.
Creatinine, which is synthesized non-enzymatically from creatine, also did
not show an inhibitory effect on GEt formation.

The optimal activity for GEt formation was observed at pH 8.7-9.1,
although it was at pH 7.5 for GAA formation. Ronca et al.[9] reported that the

acceptor amino group must be in the unprotonated form. Since the amino group of EA is more basic than that of Gly, the optimal pH for transamidination with EA as acceptor is higher than that with Gly as acceptor.

In conclusion, GEt was synthesized from Arg and EA in rat kidneys and this reaction appeares to be catelyzed by Arg:Gly amidinotransferease.

REFERENCES

1. Watanabe, Y., Shindo, S. and Mori, A., Identification of 2-guanidino-ethanol in human urine, Eur. J. Biochem. 147 (1985) 465-468.
2. Watanabe, Y., Yokoi, I. and Mori, A., The biosynthesis of 2-guanidino-ethanol in intact mice and isolated perfused rabbit Kidneys, Life Sci. 40 (1987) 293-299.
3. Lowry, O.H., Rosebrough, N.J., Farr, A.L. and Randall, R.J., Protein measurement with the Folin phenol reagent, J. Biol. Chem. 193 (1951) 263-275.
4. Higashidate, S., Maekubo, T., Saito, M., Senda, M. and Hoshino, T., New high-speed fully automated guanidino compound analyzer, in:"Guanidines," A. Mori, B.D. Cohen and A. Lowenthal Eds., Plenum Press, New York (1985) pp. 3-13.
5. Marquardt, D.W., An algorithm for least-squares estimation of nonlenear parameters, J. Soc. Indust. Appi. Math. 11 (1963) 431-441.
6. Segel, I.H., Enzymes, In: "Biochemical calculations (2nd ed)," John Wiley & Sons, New York (1976) pp. 208-323.
7. Walker, J.B., Amidinotransferase, in: "The Enzymes (3rd ed) 9," P.D. Boyer, Ed., Academic Press, New York and London (1973) pp. 497-509.
8. Ratner, S. in: "The enzymes (2nd ed) 5," P.D. Boyer, H. Lardy and K. Myrbäck, Eds., Academic Press, New York and London (1962) pp. 267-279.
9. Ronca, G., Vigi, V. and Grazi, E., Transamidinase of hog kidney, J. Biol. Chem. 241 (1966) 2589-2595.
10. Walker, J.B., Repression of arginine-glycine transamidinase activity by dietry creatine, Biochim. Biophys. Acta 36 (1959) 574-575.
11. Fitch, C.D., Hus, C. and Dinning, J.S., Some factors affecting kidney transamidinase activity in rats, J. Biol. Chem. 235 (1960) 2362-2364.
12. McGuire, D.M., Gross, M.D., Van Pilsum, J.F. and Towle, H.C., Repression of rat kidney L-arginine:glycine amidinotransferase synthesis by creatine at a pretranslational level, J. Biol. Chem. 259 (1984) 12034-12038.

THE EXISTENCE OF MULTIPLE FORMS OF RAT KIDNEY TRANSAMIDINASE

Myron D. Gross, Alexander M. Simon, Richard J. Jenny,
Ernest D. Gray, Denise M. McGuire and John F. Van Pilsum

Department of Biochemistry, Medical School, 4-225 Millard
Hall, 435 Delaware St. S.E., Minneapolis, Minnesota 55455

INTRODUCTION

Two forms of rat kidney transamidinase, called α and ß, have been purified to homogeneity[1]. No differences were found in the properties of these
forms other than their separation from each other by chromatography on DEAE
cellulose. Polyclonal antibodies made to the α-form of the enzyme reacted
with the ß-form of the enzyme and precipitated all of the enzyme activity
from a rat kidney homogenate. The low transamidinase activities in kidneys
from rats fed creatine-supplemented diets correlated well with the relative
amounts of transamidinase protein as determined by immunotitration with the
polyclonal antibodies[2]. To farther examine α and ß transamidinase regulation,
monoclonal antibodies to rat kidney transamidinase were made[3] and used to
determine the relative amounts of transamidinase protein in kidneys from normal and creatine-fed rats. The results are presented in this report. Also,
the techniques of isoelectric focusing and Western blotting with either the
monoclonal or the polyclonal antibodies were used to further characterize rat
kidney transamidinase.

MATERIALS AND METHODS

Materials

The materials listed below were purchased from the following: rats,
Holtzman, Inc.; $Na^{125}I$, New England Nuclear; microtiter plates, Dynatech
Laboratories; electrofocusing marker proteins, Pharmacia; ampholines, LKB;
nitrocellulose paper,Schleicher and Schuell; biotinylated anti-mouse Ig, and
alkaline phosphatase reagent, Vector laboratories. All other reagents were
the best grade available from commercial sources.

Animals and Diet

Weanling male rats were fed a purified diet for 18 days[4]. Subsequently,
rats were fed the control or the test diet for 10 days. Each experimental
group contained 8 rats. The rats were allowed to consume food and water, ad

libitum, during the entire period. The control and test diets were as
follows: purified diet plus a) no additive, b) 6.0% starch, c) 5.0% glycine,
d) 0.125% creatine, e) 5.0% glycine and 0.125% creatine, f) 1.0% creatine,
g) 1.0% creatine and 5.0% glycine. Rats were killed by decapitation and
kidneys removed and a 20% homogenate was made with distilled H_2O. Triton
X-100 was added to the homogenate (0.5 ml/100ml) and the homogenates stored
at -70°C. A supernatant fraction was obtained by centrifugation at 35,000 x g
for 30 min at 4°C. Transamidinase activity was determined as previously
described[5]. For all experimental groups of rats, ~85% of the transamidinase
activity in the homogenates was found in the supernatant fractions.

Purification of Rat Kidney Transamidinase

Rat kidney transamidinase was purified as described previously[3]. Both α
and ß forms of the enzymes were judged to be homogeneous by SDS polyacrylamide gel electrophoresis and specific activities.

Antibodies to Rat Kidney Transamidinase

The production and characterization of two monoclonal antibodies to rat
kidney transamidinase are described in a separate report[3]. Purified monoclonal antibody (Tran/NS-1/3) was used in this report. The IgG fraction of
mouse antiserum to transamidinase (the polyclonal antibody) was prepared as
described previously[1]. Mouse anti-transamidinase IgG was determined to be
monospecific by Ouchterlony double diffusion analysis[6].

Immunosorbent Inhibition Assay

The assay is described in a separate report[3]. Five ng of ^{125}I-labeled
purified Tran/NS-1/3 was incubated with samples of kidney supernatant overnight. The mixture was applied to microtiter plates coated with purified α rat
kidney transamidinase and allowed to incubate overnight. Plates were washed
and individual wells counted for bound ^{125}I-Trans/NS-1/3. The assay results
were expressed as the amount of radioactivity bound relative to that bound
when 5 ng of ^{125}I-Tran/NS-1/3 was incubated with transamidinase-free protein
solutions prior to application to the wells. The volumes of kidney supernatants required for a 50% inhibition of ^{125}I-Tran/NS-1/3 binding were used
to calculate the relative amounts of the monoclonal immunoreactive protein.

Isoelectric Focusing and Blots

Isoelectric focusing was by the method of Giulian et al.[7] modified to
include 0.5% agarose in the separating gel on a Hoefer "Mighty Small" apparatus. Electrofocusing marker proteins were used to determine the isoelectric
points. Proteins were detected by staining with Coomassie Blue R dye and by
blotting techniques. Proteins that had been subjected to isoelectric focusing were transferred to nitrocellulose paper by the method of Towbin et al.[8]
in 25 mM Tris/192 mM glycine/20% methanol, pH 8.3. The transfer was for 2 h,
100 mV at 10°C. The nitrocellulose membrane was incubated overnight in the
cold in PBS (phosphate buffered saline) containing 1% bovine serum albumin
(blocking buffer) to block nonspecific binding. All subsequent incubations
were at room temperature on a rocking platform. The membrane was then incubated 1 h with a dilution (in blocking buffer) of either polyclonal anti-
transamidinase or monoclonal anti-transamidinase. The nitrocellulose transfers were washed 3 times for 5 min with PBS - 0.5% Tween 20. Binding of the
anti-transamidinase antibodies was detected using a biotin-avidin amplification system (Vector Laboratories ABC alkaline phosphatase kit). The nitrocellulose transfers were incubated for 1 h with 1:100 dilution of Vector

Table 1.   The effect of dietary creatine and/or glycine on rat kidney transamidinase activities and the relative amounts of Tran/NS-1/3 immunoreactive transamidinase protein

Diet[a]	Supplement (g/100g diet)	Kidney Transamidinase Activity Supernatant (Units/ml[d] and % Control)[c]		Tran/NS-1/3 Immunoreactive Transamidinase Supernatant (% Control)[b]
a	none	10.43 ± 0.212	(100.0)	100.0
b	Starch 6.0	10.17 ± 0.080	(97.5)	105.0
c	Glycine 5.0	10.04 ± 0.311	(96.3)	95.0
d	Creatine 0.125	4.31 ± 0.245 eg	(41.3)	77.8 eg
e	Glycine 5.0 and Creatine 0.125	3.64 ± 0.147 eg	(34.9)	68.1 eg
f	Creatine 1.0	1.09 ± 0.059 ef	(10.5)	77.8 eg
g	Creatine 1.0 and Glycine 5.0	2.46 ± 0.233 ef	(23.6)	65.3 eg

a.   Diets described under methods.

b.   Tran/NS-1/3 immunoreactive transamidinase calculated from volume of supernatant required for 50% inhibition of ^{125}I-Tran/NS-1/3 bound relative to that bound when 5 ng of ^{125}I-Tran/NS-1/3 in blocking buffer was incubated in the wells coated with purified transamidinase. The volume of the supernatant from the control rat kidneys divided by the volume of the supernatant experimental rat kidneys (in both cases that resulted in a 50% inhibition of ^{125}I-Tran/NS-1/3 binding) x 100 yielded the percent of the control Tran/NS-1/3 immunoreactive protein. The linear sections of the binding curves were analyzed by the method of linear regression to determine the volume of kidney supernatants required for a 50% inhibition of ^{125}I-Tran/NS-1/3 binding.

c.   % Control = % of transamidinase activity found in kidney supernatants from rats fed diet a.

d.   Units/ml ± 1.0 standard deviation

e.   The differences from the controls (groups receiving diets a, b, and c) was statistically significant; $p < 0.01$.

f.   These groups were statistically significantly different from one another; < 0.01

g.   These groups were not statistically significantly different from one another; < 0.01

Laboratories biotinylated antimouse Ig. Following another 3 washes in PBS-
Tween the transfers were incubated for 1 h with Vector ABC alkaline phosphatase reagent (1:100). The transfers were washed 3 times with PBS-Tween and a
final time with 0.15 M Barbital (Veronal) acetate beffer, pH 9.6. Alkaline
phosphatase substrate mixture (Blake et al.)[9] was then added to the membrane
and the color allowed to develop at room temperature to a suitable intensity.
The reaction was stopped by washing the membrane in water.

High Pressure Liquid Chromatography

Purified α and β rat kidney transamidinase (10 µl, 0.5 mg/ml) were applied to a Beckman C^3(RPSC 4.6 x 75 mm) column at room temperature. Solvent A
was 0.05% trifluoroacetic acid in water. Solvent B was 0.05% trifluoroacetic acid in HPLC grade acetonitrile. Elution was achieved with a 40 min
linear gradient from 0% Buffer B to 60% Buffer B at a flow rate of 1 ml/min.
Elution peaks were detected by absorbance at 214 nm.

RESULTS

Effect of Dietary Creatine and/or Glycine on Rat Kidney Transamidinase Activities and the Amounts of Tran/NS-1/3 Immunoreactive Transamidinase Protein Expressed as % of the Control Values

The kidney transamidinase activities and relative amounts of Tran/NS-
1/3 immunoreactive protein found in rats fed various diets are shown in Table
I. The weight gains and kidney weights were similar for all experimental
groups of rats (data not shown). The addition of 6.0 g of either starch or
glycine to 100 g of diet was without any significant effect on the kidney
transamidinase activities or the relative amounts of Tran/NS-1/3 immunoreactive transamidinase protein in the kidney supernatants. The enzyme activities of kidneys from rats fed diets supplemented with 0.125 g or 1.0 g of
creatine/100 g diet were 41.3% and 10.5%, respectively, of the activities of
the control rats. The Tran/NS-1/3 immunoreactive protein found in the supernatants of the above listed rats were both 77.85% of the control values,
thus, the alterations in enzyme activities and in relative amounts of Tran/
NS-1/3 immunoreactive protein were not similar. The enzyme activities of
kidneys from rats fed diets containing 0.125 g creatine/100 g diet and 0.125
g creatine plus 5.0 g glycine/100 g diet were similar, as were the relative
amounts of Tran/NS-1/3 immunoreactive transamidinase protein. The enzyme
activities of kidneys from rats fed diets containing 1.0 g creatine/100 g
diet plus 5.0 g glycine/100 g diet were two times greater than those found in
kidneys from rats fed the diet containing 1.0 g creatine/100 g diet, with no
significant differences found in the amounts of Tran/NS-1/3 immunoreactive
transamidinase protein. The data on transamidinase activities with creatine
and/or glycine feeding were similar to those reported previously[10]. We had
previously reported a good correlation between transamidinase activities and
transamidinase protein as determined by immunotitration with polyclonal antibodies[2]. A possible explanation could be that multiple forms of transamidinase were present in kidney homogenates and that the monoclonal antibody
recognized forms of the enzyme which were not altered greatly in amounts as
the result of creatine feeding.

Evidence for Multiple Forms of Rat Kidney Transamidinase

A number of experiments were performed to investigate the possibility
that multiple forms of the transamidinase existed in the kidney supernatants
and in the purified preparations of the enzyme.

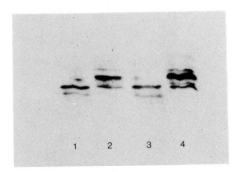

Figure 1.   Isoelectric focusing of purified α and ß rat kidney transamidinase. Samples were (1 and 3) purified α rat kidney transamidinase
(10 μg), (2 and 4) purified ß rat kidney transamidinase (10 μg).
The pH gradient was from basic (top) to acidic (bottom).

1) Isoelectric focusing of homogeneous rat kidney transamidinase.
The results of the isoelectric focusing of both the purified α- and ß-forms
of the enzyme are shown in Fig 1. Lanes 1 and 3 and lanes 2 and 4 were the
α- and ß-forms, respectively. Both the purified α- and ß-forms of the enzyme
migrated as a single band in native and in SDS polyacrylamide gel
electrophoresis. The α- and ß-forms of rat kidney transamidinase were separated by isoelectric focusing into multiple bands. The pattern of the bands
obtained with the α-form was different than with the ß-form. The isoelectric
points of the bands ranged from ~6.1 to 7.2 and from ~6.6 to 7.6 for the
α- and ß-form of the enzyme, respectively. Both the α- and ß-forms contained
a major band of similar isoelectric points. The other major band found in
the α- and in the ß-form of the enzyme had isoelectric points that were
lower or higher, respectively, than the major band of similar isoelectric
points. This observation was the first indication of a difference in the
properties of α- and ß-rat kidney transamidinase and the existence of
multiple forms of both α- and ß-transamidinase. No differences had been
found in any of the properties of the α- and ß-forms in a previous
investigation[1].

2) High pressure liquid chromatography. α and ß transamidinase were
eluted as 1 as 2 fractions, respectively. The single fraction of the
α-form coincided with the first fraction eluted with the ß-form of the
enzyme. This is additional evidence that the ß fraction was not identical
with the α fraction of the enzyme.

3) Western transfers and blotting experiments with monoclonal and
polyclonal antibodies.

a) Isoelectric focusing of the purified α- and ß-forms of
transamidinase followed by immunoblotting with a polyclonal antibody.

The results are shown in Fig.2 (A). The patterns obtained after isoelectric focusing of the α-(lane 1) and ß-(lane 2) forms of the enzyme and
immunoblotting with the polyclonal antibody resembled the bands obtained by
isoelectric focusing and staining of the gels. One of the major bands in
both the α- and ß-forms of the enzyme had similar isoelectric points. This
major band is the bottom band in the ß-form and the second from the bottom
band in the α-form. The major band in the ß-form (second from the bottom)
was barely observable in the α-form as it was in the protein stained gels

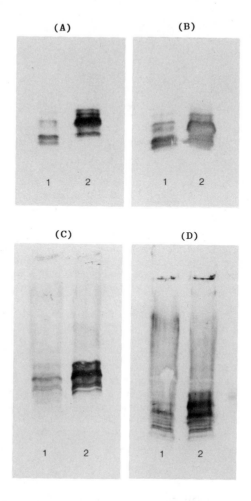

Figure 2 (A) Isoelectric focusing of the purified α- and ß-forms of rat
               kidney transamidinase followed by immunoblotting with a
               polyclonal antibody, Lanes 1 and 2 are the α- and ß-form of the
               enzyme, respectively.
          (B) Isoelectric focusing of the purified α- and B-forms of rat
               kidney transamidinase followed by immunoblotting with the mono-
               clonal antibody, Lanes 1 and 2 are the α- and ß-form of the
               enzyme, respectively.
          (C) Isoelectric focusing of supernatants of homogenates of kidneys
               from rats fed a creatine supplemented diet (lane 1) or a con-
               trol diet (lane 2) followed by immunoblotting with a monoclonal
               antibody.
          (D) Isoelectric focusing of supernatants of homogenates of kidneys
               from rats fed a creatine supplemented diet (lane 1) or a con-
               trol diet (lane 2) followed by immunoblotting with a polyclonal
               antibody.

(Fig. 1). Two additional bands at the top of the ß-form gel were also much more prominent in the ß-form than in the α-form, and also could not be seen in the α-form of the gels stained for protein. Thus, by isoelectric focusing and protein staining and by isoelectric focusing and immunoblotting with the polyclonal antibody, both α- and ß-forms of the enzyme were separated into multiple bands. In addition, the distribution of the bands in the α-form was different than in the ß-form by both protein staining and immunoblotting with the polyclonal antibody.

b) Isoelectric focusing of the purified α- and ß-forms of transamidinase followed by immunoblotting with a monoclonal antibody.

The results are shown in Fig. 2 (B). The bands observed in the α-form (lane 1) after isoelectric focusing and immunoblotting with the monoclonal antibody appeared to be similar to the bands when immunoblotted with the polyclonal antibody. The monoclonal antibody, however, appeared to recognize bands at the bottom of the ß-form (lane 2) that were not recognized by the polyclonal antibody. This is considered to be evidence for a difference in the specificity between the monoclonal and the polyclonal antibodies for recognizing the individual forms of transamidinase.

c) Immunoblotting of kidney supernatants of homogenates from rats fed a complete purified diet or a complete purified diet supplemented with creatine.

The supernatants were subjected to isoelectric focusing and immunoblotted with the monoclonal antibody (Fig. 2 (C)) and with the polyclonal antibody (Fig. 2 (D)). Multiple forms of the enzyme were present in the kidney supernatants of both the control (lane 2) and creatine-fed (lane 1) rats that were recognized by both the monoclonal and polyclonal antibodies. The range of the isoelectric points of the bands observed in the supernatants from homogenates of the control rats were similar to those in the purified α plus ß forms of the enzyme (~6-8). The relative intensities of the major bands obtained with the control rats were greater than those from the creatine-fed rats with both the monoclonal and polyclonal antibodies. This is consistent with the low amounts of enzyme protein in the creatine-fed rats when determined with either the monoclonal or polyclonal antibodies. Also, the relative intensities of the individual bands detected in supernatants from kidneys of creatine-fed rats differed from those detected in supernatants of kidneys from rats fed the control diet with either the monoclonal or the polyclonal antibodies. For example, both the upper and lower bands detected by the monoclonal antibody in the control rats could barely be seen in the creatine-fed rats (Fig. 2 (C)). With the polyclonal antibodies, the major band in the control rats could barely be seen in the creatine-fed rats (Fig. 2 (D)). Thus, an alteration of the distribution and/or the intensities of the individual bands as result of creatine feeding was indicated with both the monoclonal and polyclonal antibodies. Also, the relative intensities of the individual bands as a result of creatine feeding varied with the antibody used in the immunoblotting procedure. The relative intensities of the individual bands could be a function of the amount of the individual form of the enzyme present or it could be a function of the avidity of the individual form of the enzyme for the antibody used in the immunoblotting procedure.

DISCUSSION

Evidence for multiple forms of rat kidney transamidinase is presented in this report. A logical hypothesis to explain the lack of correlation between enzyme activities and the amount of enzyme protein determined with the immunosorbent inhibition assay with the monoclonal antibody (Table I) is that the monoclonal antibody recognized forms of the enzyme that were decreased in amounts that were not proportional to the decrease in enzyme activities with creatine feeding. The specific activities of the individual forms of the enzyme have not as yet been determined. However, the distribution of the individual forms of the enzyme appear to be altered by feeding creatine to the rats. We suggest that the specific activities of the individual forms of transamidinase found by isoelectric focusing of the supernatants of rat kidneys are not likely to be identical. If this is so, an alteration in the distribution of the individual forms could occur with creatine feeding that may have an effect on the enzyme activities of the kidney supernatants. We have previously reported that the amount of enzyme protein when determined by immunotitration with the polyclonal antibody did correlate well with the enzyme activities of control and creatine-fed rats. This suggests that the polyclonal antibody recognized forms of the enzyme that were decreased in amounts proportional to the decreases in enzyme activities. Also, we have reported previously that creatine in the diet represses the synthesis of transamidinase at a pretranslational level[2].

The multiple forms of transamidinase may arise by a number of mechanisms. The close immunological relationship between the various forms of transamidinase may be interpreted to suggest the production of the multiple forms from a common precursor. The multiple forms could arise from the covalent modification of mature transamidinase or from covalent modification and/or limited proteolysis of a precursor transamidinase. Recently a functional mRNA and its translated precursor protein have been described for transamidinase[3]. The precursor protein had a molecular weight of 58,000 daltons and presumably undergoes rapid processing to a mitochondrial membrane-bound mature transamidinase. Further study will be required to determine the relationship between the putative transamidinase precursor and the multiple forms of transamidinase which we have found.

CONCLUSION

We have established by isoelectric focusing that multiple forms of rat kidney transamidinase are present in supernatants from homogenates of kidneys from both control and creatine-fed rats. The distribution of the individual forms in the control rats when determined by immunoblotting with the monoclonal antibody differed from that obtained with the polyclonal antibody. Also, the change in the distribution of the forms as the result of creatine feeding when immunoblotted with the monoclonal antibody differed from that obtained with the polyclonal antibody. We have presented evidence that the amount of enzyme protein when determined by the immunosorbent inhibition assay with the monoclonal antibody did not correlate well with the enzyme activities in creatine feeding. We have previously reported that a good correlation was found between enzyme activities and enzyme protein when determined by immunotitration with the polyclonal antibodies. We have suggested that the specific activities of the individual forms of the enzyme are not likely to be identical and that an alteration of the distribution of these forms as a result of creatine feeding may well be a factor in the alteration of enzyme activities, in addition to the repression of transamidinase synthesis at a pretranslational level[2].

ACKNOWLEDGMENTS

This work was supported by Grant AM-26505 from the National Institutes of Health.

REFERENCES

1. McGuire, D.M., Tormanen, C.D., Segal, I.S. and Van Plsum, J.F., The effect of growth hormone and thyroxine on the amount of L-arginine: glycine amidinotransferase in kidneys of hypophysectomized rats, J. Biol. Chem. 225 (1980) 1152-1159.
2. McGuire, D.M., Gross, M.D., Van Pilsum J.F. and Towle, H.C., Repression of rat kidney L-arginine: glycine amidinotransferase synthesis by creatine at a pretranslational level, J. Biol. Chem. 259 (1984) 12034-12038.
3. Gross, M.D., McGuire, D.M. and Van Pilsum, J.F., The production and characterization of two monoclonal antibodies to rat kidney L-arginine: glycine amidinotransferase, Hybridoma 4 (1985) 257-269.
4. Van Pilsum, J.F., Taylor, D. and Boen, J.R., Evidence that creatine may be one factor in the low transamidinase activities of kidneys from protein-depleted rats, J. Nutr. 91 (1967) 383-390.
5. Van Pilsum, J.F., Taylor, D., Zakis, B. and McCormick, P., Simplified assay for transamidinase activities of rat kidney homogenates, Anal. Biochem. 35 (1970) 277-286.
6. Ouchterlony, O. and Nilsson, L.A., Immunodiffusion and immunoelectrophoresis in: "Handbook of Experimental Immunology," D.M. Weir Ed., Blackwell Scientific Publications, Oxford (1773) pp 19.1-19.39.
7. Giulian, G.C., Moss, R.L. and Greaser, M., Analytical isoelectric focusing using a high-voltage vertical slab polyacrylamide gel system, Anal. Biochem. 142 (1984) 421-436.
8. Towbin, H., Staehelin, T. and Gordon, J., Electrophoretic transfer of proteins from polyacrylamide gels to nitrocellulose sheets: procedure and some applications, Proc. Natl. Acad. Sci. 76 (1979) 4350-4354.
9. Blake, M.S., Johnson, K.H., Russell-Jones, G.J. and Gotschlich, E.C., A rapid sensitive method for detection of alkaline phosphatase-conjugated anti-antibody on Western blots, Anal. Biochem. 136 (1984) 175-179.
10. Van Pilsum, J.F., Evidence for a dual role of creatine in the regulation of kidney transamidinase activities in the rat. J. Nutr. 101 (1971) 1085-1091.

PUROMYCIN AMINONUCLEOSIDE STIMULATES THE SYNTHESIS OF METHYLGUANIDINE: A

POSSIBLE MARKER OF ACTIVE OXYGEN GENERATION IN ISOLATED RAT HEPATOCYTES

Kazumasa Aoyagi, Sohji Nagase, Masako Sakamoto, Mistuharu
Narita and Shizuo Tojo

Department of Internal Medicine, Institute of Clinical
Medicine, University of Tsukuba, Tsukuba 305, Japan

INTRODUCTION

Puromycin aminonucleoside (PA), the structure of which is similar to
that of adenosine, is known as a toxic substance which induces heavy proteinuria in rats[1]. It has also been reported that the oxygen radical plays
an important role in kidney disease[2-7]. We have reported that methylguanidine (MG), a uremic toxin[8,9], is formed by the action of oxygen radicals
reacting with creatinine in vitro[10] and in isolated rat hepatocytes[11-14]. We
have also reported that PA stimulated MG synthesis in isolated rat hepatocytes, and adenosine, adenosine analogues and its potentiators inhibited PA
stimulated MG synthesis[12-14].

In this paper, we investigate further the details of PA stimulated MG
synthesis as well as the regulatory mechanisms for oxygen radical generation
in isolated hepatocytes, by measuring the formation of MG from creatinine
reacting with active oxygen.

METHODS

Preparation of Isolated Rat Hepatocytes

Male Wistar rats weighing 300 to 350 g were used in all experiments.
The rats were allowed free access to water and laboratory chow containing
25% protein. Isolated hepatocytes were prepared essentially according to the
method of Berry and Friend[15] as described previously[16]. We calculated that
9.8×10^8 cells corresponded to 1 g of liver (wet weight)[17].

Incubation of Cells

Cells were incubated in 6 ml of Krebs-Henseleit bicarbonate buffer
containing 3% bovine serum albumin, 10 mM sodium lactate, 17.6 mM creatinine
and indicated substances with shaking at 60 cycles/min in a 30 ml conical
flask with a rubber cap under 95% oxygen and 5% carbon dioxide at 37°C for
4h. (except for the experiment on time dependence). In the experiments

perfomed to check the effect of dipyridamole, cells were preincubated with
dipyridamole for 5 min. Equilibration of the buffer was repeated every hour.
To measure the rate of non-biological conversion of creatinine to MG[11],
incubations were carried out without cells. The amount of cells used for
each experiment is indicated in the Results section. The reaction was
stopped by the addition of 0.6 ml of 100% (wt/v) trichloroacetic acid. After
sonication, the supernatant (of cells and medium) was obtained by cenrif-
ugation at 1,700 g for 15 min at 0°C and 0.25 ml of the extract was used for
MG measurement. MG was determined by high-performance liquid chromatographic
analysis using 9,10-phenanthrenequinone for post-labeling, as described
previously[11]. Aminonucleoside of puromycin, adenosine, 2-chloroadenosine
and dibutyryl cAMP were purchased from Sigma Chemical Co., (St Louis, MO).
Dipyridamole was kindly donated by Boehringer Ingelheim Ltd.

RESULTS

Stimulation of MG Synthesis in Isolated Rat Hepatocytes by PA

MG synthesis was increased in isolated rat hepatocytes incubated with
PA, as shown in Figure 1. The increase was apparent after incubation for 4 h,
and markedly increased after 8 h. The maximum rate of MG synthesis was
observed at 0.9 mM PA at 4 h, but MG synthesis was dosedependent at 8 h
(Fig.2).

Inhibition of PA Stimulated MG Synthesis by Adenosine, its Analogues and its Potentiators

MG synthesis stimulated by PA was 32 ± 2% inhibited by 200 μM adenosine,
and 54 ± 5% inhibited by 100 μM 2-chloroadenosine, as shown in Table 1.

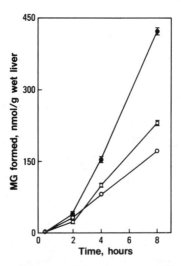

Figure 1.   Effect of puromycine aminonucleoside (PA) on methylguanidine
            (MG) synthesis in isolated hepatocytes. Cells (0.18 g wet
            weight) were incubated for 2, 4 or 8h in the absence (o) or in
            the presence of 0.2 (□) or 1.9 mM (●) PA as described in the
            Methods section. Each point represents the mean value of dupli-
            cated incubations. Bars represent the range of each determination.

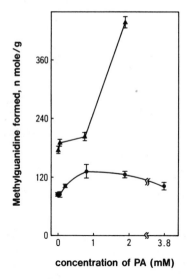

Figure 2. Dose dependence of the effect of puromycine aminonucleoside (PA) on methylguanidine (MG) synthesis. Cells (0.18 g wet weight) were incubated for 4 (o) or 8 (▲) h with various concentrations of PA as described in the Methods section. Each point represents the mean value of duplicated incubation. Bars indicate the range of each determination.

Table 1. Effect of adenosine and 2-chloroadenosine on MG synthesis

reagents	concentration (mM)	MG formed	
		nmol/g/6 h	(%)
none		110.9 ± 0	(100 ± 0)
adenosine	0.2	90.3 ± 2.6	(81.4 ± 2.3)
PA	1.9	162.5 ± 2.5	(146.5 ± 2.3)
PA + adenosine	1.9 0.2	114.8 ± 1.3	(103.5 ± 1.1)
PA + 2-chloroadenosine	1.9 0.1	74.8 ± 8.1	(67.4 ± 10)

Cells (0.21 g wet weight) were incubated for 6 h as described in the Methods section. Values are expressed as the mean of duplicated incubations ± range of each incubation. MG; methylguanidine PA; puromycin aminonucleoside.

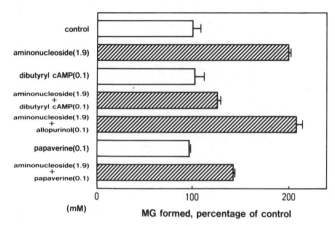

Figure 3.   Effect of dibutyryl cAMP, papaverine and allopurinol on methyl-
            guanidine (MG) synthesis. Cells (0.1 g wet weight) were incubated
            for 6 h as described in the Methods section. Each column repre-
            sents the mean of duplicated incubations expressed as a percent-
            age of the control value. Bars express the range of each incuba-
            tion. The control value was 57.8 ± 4 nmol/g/6 h.

Effect of Dibutyryl-cyclic-AMP, Papaverine and Allopurinol on MG Synthesis

Many of the effects of adenosine are attributed to its ability to
stimulate or inhibit adenylate cyclase, and these effects are mediated by
distinct binding sites for the nucleoside[18]. The existence of stimulatory
(A2) sites and of high-affinity inhibitory (A1) sites for the nucleoside has
been reported in liver and isolated rat hepatocytes[19,20].

Addition of 100 μM dibutyryl cAMP or 100 μM papaverine (which inhibits
phosphodiesterase) inhibited MG synthesis stimulated by PA. However, MG
synthesis in isolated hepatocytes incubated without PA was not inhibited by
these reagents (Fig. 3).

In tissues, the hypoxanthine oxidase reaction is one of the metabolic
pathways of oxygen radical generation[21]. To examine the role of this reaction in MG synthesis in isolated rat hepatocytes under our conditions, we
tested the effect of allopurinol, which inhibits xanthine oxidase[22].
Allopurinol had no effect on MG synthesis stimulated by PA, as shown in
Figure 3, and also had no effect on MG synthesis in isolated hepatocytes
incubated without PA[14].

DISCUSSION

Activated oxygen radicals are thought to induce deleterious effects
such as lipid peroxidation, inflammation, carcinogenesis, cataracts and
atherosclerosis. Recently, it has been reported that active oxygen plays an
important role in the pathogenesis of acute renal failure[2,6] and glomerular
injury[3,6]. We have reported that MG, a uremic toxin, is formed from creatinine by the action of active oxygen in vitro[10] and in isolated rat hepatocytes[12,14]. We have found that adenosine, its analogues and its potentiators
inhibit MG synthesis in the presence and absence of PA. Further, it has been
reported that adenosine inhibits superoxide anion generation in human neu-

trophils at physiological concentrations[23], and the injury to endothelial cell caused by neutrophils is inhibited by adenosine and its agonist[24]. Most of the reagents that inhibited MG synthesis in our experiments[12-14] such as adenosine[25], N^6-monomethyl adenosine[26], dipyridamole[27,28], dilazep[29] and saikosaponins[30], have been reported to block nephrosis induced by PA. Moreover, dipyridamole is effective in ameliorating experimental nephritis[31] and in decreasing proteinuria in patients with glomerulonephritis[32,33]. From these results and the existence of adenosine receptors in various tissues, we propose that the inhibition of active oxygen generation may explain, in part, the favorable effect of adenosine potentiators on renal disease and also that generation of excess active oxygen by PA might explain the pathogenesis of heavy proteinusia induced by PA[12-14]. Recently, Diamond et al. reported that PA-induced nephrosis was inhibited by superoxide dismutase and allopurinol[34]. They suggested that the increase of hypoxanthine (which is a substrate for xanthine oxidase) by the degradation of PA and/or the change of the enzyme form from NAD-reducing dehydrogenase (Type D) to superoxide-producing oxidase (Type O) led to an excess oxygen radical generation.

Concerning the regulatory mechanism for active oxygen generation by PA, cAMP may play important role. MG synthesis stimulated by PA is inhibited by the addition of dibutyryl cyclic AMP and papaverine, both of which increase the cyclic AMP content of hepatocytes. Also the addition of $10 \mu M$ 2-chloro-adenosine to hepatocytes increases the cyclic AMP level in isolated rat hepatocytes[21]. In addition to these results, it is reported that kidneys transplanted 15 min after PA injection developed heavy proteinuria[35]. These results suggested irreversible binding of PA to the adenosine receptors which in turn decreases the intracellular cAMP concentration rather than increasing hypoxanthine derived from PA degradation. It is reported that in isolated glomeruli from PA-treated rats, even though histamine is increased, the cAMP level is not changed and the increase of cAMP accumulation in response to added histamine, serotonin, and carbamylcholine is markedly reduced, whereas elevation of cAMP content by nonhormonal agents (cholera toxin, nitroprusside) is not diminished[36]. These results also suggest that the receptors or the linkage of the receptors to adenyl cyclase modulating protein are damaged by PA.

The result that allopurinol did not inhibit MG synthesis stimulated by PA within 4 h suggests that the hypoxanthine oxidase reaction is not responsible for the generation of active oxygen in isolated hepatocytes, since this is an initial phase phenomenon. However, in vivo, the supply of adenosine or hypoxanthine from other organs and the increase in permeability to these substances caused by PA could lead to an increase of oxygen radical generation. Moreover, in PA nephrosis, modification of the biosynthesis of glycoproteins following upon this initial phenomenon may also be an important factor[37].

ACKNOWLEDGMENTS

These research findings were presented, in part, at the IV international congress on nutrition and metabolism in renal disease in October, 1985, in Williamsburg, Virginia. This study was supported in part by research grants from the University of Tsukuba Project Research Fund and the Intractable Disease Division, Public Health Bureau, Ministry of Health and Welfare, Japan. We are indebted to Mrs. Satomi Kawamura for her valuable assistance.

REFERENCES

1. Lamnigan, R., Kark, R. and Pollak, VE., The effect of a single intravenous injection of aminonucleoside of puromycin on the rat Kidney,
   J. Pathol. Bacteriol. 83 (1962) 357-362.
2. Paller, M.S., Hoidal, J.R. and Ferries, T.F., Oxygen free radical in
   ischemic acute renal failure in the rat, J. Clin. Invest. 74 (1984)
   1156-1164.
3. Rehan, A., Johnson, K.J., Wiggs, R.C., Kunkel, R.G. and Ward, P.A.,
   Evidence for the role of oxygen radicals in acute nephrotoxic nephritis,
   Lab. Invest. 51 (1984) 396-403.
4. Rehan, A., Johnson, K.J., Kunkel, R.G. and Wiggs, R.C., Role of oxygen
   radicals in phorbol myristate acetate-induced glomerular injury, Kidney
   Int. 27 (1985) 503-511.
5. Adachi, T., Fukuta, M., Ito, Y., Hirano, K., Sugiura, M. and Sugiura,
   K., Effect of superoxide dismutase on glomerular nephritis, Biochem.
   Pharamcol. 35 (1986) 341-345.
6. Dobyan, D.C., Bull, J.M. Strebel, F.R., Sunderland, B.A. and Bulger,
   R.E., Protective effects of 0-(ß-hydroxyethyl)-rutoside on cis-platinum-
   induced acute renal failure in the rat, Lab. Invest. 55 (1986) 557-563.
7. Holdsworth, S.R. and Boyce, N.W., Hydroxyl radical mediation of immune
   renal injury by desferrioxamine, Kidney Int. 30 (1986) 813-817.
8. Giovannetti, S., Biagini, M., Balestri, P.L., Navasesi, R., Giagnoni,
   P., deMatteis, A., Ferro-Milone, P. and Perfetti, C., Uremia-like
   syndrome in dogs chronically intoxicated with methylguanidine and
   creatinine, Clin. Sci. 36 (1969) 445-452.
9. Barsotti, G., Bevilacqua, G., Morelli, E., Cappelli, P., Balestri, P.L.
   and Giovannetti, S., Toxicity arising from guanidine compounds: Role of
   methylguanidine as a uremic toxin, Kidney Int. 7 (1975) s299-s301.
10. Nagase, S., Aoyagi, K., Narita, M. and Tojo, S., Active oxygen in methyl-
    guanidine synthesis, Nephron 44 (1986) 299-303.
11. Nagase, S., Aoyagi, K., Karita, M. and Tojo, S., Biosynthesis of methyl-
    guanidine in isolated rat hepatocytes and in vivo, Nephron 40 (1985)
    470-475.
12. Aoyagi, K., Nagase, S., Ohba, S., Miyazaki, M., Sakamoto, M., Narita,
    M. and Tojo, S., Significance of active oxygen in methylguanidine
    synthesis in isolated hepatocytes, Jpn. J. Nephrol. 28 (1986) 536
    (abstract of Japanese Nephrology Congress in 1985).
13. Aoyagi, K., Narita, M. and Tojo, S., Uremic toxin-recent progress of our
    study, Medical Practice 3 (1986) 256-259.
14. Aoyagi, K., Nagase, S., Narita, M. and Tojo, S., Role of active oxygen
    on methylguanidine synthesis in isolated rat hepatocytes, Kidney Int.
    (in press)
15. Berry, M.N. and Friend, D.S., High-yield preparation of isolated liver
    cells, J. Cell Biol. 43 (1969) 506-520.
16. Aoyagi, K., Ohba, S., Narita, M. and Tojo, S., Regulation of biosynthesis
    of guanidinosuccinic acid in isolated rat hepatocytes and in vivo, Kidney
    Int. 24 (1983) s224-s228.
17. Zahlten, R.N., Stratman, F.W. and Lardy, H.A., Regulation of glucose
    synthesis in hormone-sensitive isolated rat hepatocytes, Proc. Natl.
    Acad. Sci. USA 70 (1973) 3213-3218.
18. Sattin, A. and Roll, T.W., The effect of adenosine and adenosine
    uncleotides on the cyclic adenosine 3', 5'-phosphate content of guinea
    pig cerebral cortex slices, Mol. Pharmacol. 6 (1973) 13-24.
19. Cooper, D.M.F. and Londos, C., Evaluation of the effects of adenosine on
    hepatic and adipocyte adenylate cyclase under conditions where adenosine
    is not generated endogenously, J. Cyclic Nucleotide Res. 5 (1979) 289-
    302.

20. Claus, T.H., Anand-Srivastava, M.B. and Johnson, R.A., Regulation of
    hepatocytes cAMP and pyruvate kinase by site-specific analogs of
    adenosine, Mol. Cell Endocrinol. 26 (1982) 269-279.
21. Parks, D.A. and Granger, D.N., Ischemia-induced vascular changes:Role
    of xanthine oxidase and hydroxyl radicals, Am. J. Physiol. 245 (1983)
    G285-G289.
22. Bulkley, G.B., The role of oxygen free radicals in human disease
    processes, Surgery 94 (1983) 407-411.
23. Bruce, N., Cronstein, B.N., Karmer, S.B., Weissmann, G.D. and
    Hirschorn, R., Adenosine: A physiological modulator of superoxide anion
    generation by human neutrophils, J. Exp. Med. 158 (1983) 1160-1177.
24. Cronstein, B.N., Levin, R.I., Belanoff, J., Weissmann, G. and Hirschorn,
    R., Adenosine: A endogenous inhibitor of neutrophil-mediated injury to
    endothelian cells, J. Clin. Invest. 78 (1986) 760-770.
25. Alexander, C.S. and Hunt, V.R., Inhibition of aminonucleoside nephrosis
    in rat: II Effect of nucleic acid precursors and 2-triiodothyronine,
    Proc. Soc. Exp. Biol. 108 (1961) 706-709.
26. Deir, R.F., Loechler, D.K., Alxander, C.S. and Nagasawa, C.S., Inhibition
    of aminonucleoside nephrosis in rats: IV Prevention by N^6-methyl-
    adenosine, J. Lab. Clin. Med. 72 (1968) 363-369.
27. Kimura, K., Endo, H. and Sakai, F., Suppressive effect of dipyridamole
    on the proteinuria of aminonucleoside nephrosis in rat, J. Toxicol. Sci.
    4 (1978) 1-10.
28. Nagase, M., Kumagi, H. and Honda, N., Suppression of proteinuria by
    dipyridamole in rats with aminonucleoside nephropathy, Renal Physiol. 7
    (1984) 218-226.
29. Nagase, M., Kobayashi, S., Sakakibara, K. and Honda, N., Amelioration of
    albuminuria in aminonucleoside nephrosis of rats, J. Nephrol. 27 (1985)
    385-391.
30. Abe, H., Orita, M., Konishi, H. and Arichi, S., Effects of saikosaponin-d
    on aminonucleoside nephrosis, Jpn. J. Pharm. 38 (1985) 221p.
31. Suzaki, Y. and Ito, M., Studies on antinephritic action of dipyridamole.
    (I) The effect of dipyridamole on anti-GBM induced nephritis in rats,
    Jpn. J. Nephrol. 23 (1981) 323-332.
32. Kan, K., Wada, T., J. Kitamoto, K., Konishi, K., Ozawa, Y., Kato, E. and
    Matsuki, S., Dipyridamole for proteinuria supression: Use in a patient
    with proliferative glomerulonephritis, J A M A 229 (1974) 557-558.
33. Tojo, S., Narita, M., Koyama, A., Sano, M., Suzuki, H., Tsuchiya, T.,
    Tsuchida, H., Yamamoto, S. and Shishido, H., Dipyridamole therapy in the
    nephrotic syndrome, Contrib. Nephrol. 9 (1978) 111-127.
34. Diamond, J.R., Bonventre, J.V. and Karnovsky, M.J., A role for oxygen
    free radicals in aminonucleoside nephrosis, Kidney Int. 29 (1986) 478-
    483.
35. Hoyer, J.R., Ratte, J., Potter, A.H. and Michael, A.F., Transfer of
    aminonucleoside nephrosis by renal transplantation, J. Clin. Invest. 51
    (1972) 2777-2780.
36. Shan, S.V., Abboud, H.E., Velosa, J.A. and Douso, T.P., Responsiveness
    of glomerular cAMP and cGMP to hormonal agents in aminonucleoside
    nephrosis, Kidney Int. 16 (1980) 788.
37. Kerjaski, D., Vernillo, A.T. and Forquhar, M.G., Reduced sialylation of
    podocalyxin - The major sialoprotein of the rat kidney glomerulus - in
    aminonucleoside nephrosis, Am. J. Pathol. 118 (1985) 343-349.

ACTIVE OXYGEN IN METHYLGUANIDINE SYNTHESIS BY ISOLATED RAT HEPATOCYTES

Kazumasa Aoyagi, Sohji Nagase, Masako Sakamoto, Mitsuharu
Narita and Shizuo Tojo

Department of Internal Medicine, Institute of Clinical
Medicine, University of Tsukuba, Tskuba, 305, Japan

INTRODUCTION

Methylguanidine (MG) has been implicated as a potent uremic toxin[1-3].
Furthermore, the carcinogenecity of the nitrosated compound, MG, has been
reported[4]. Chemically, MG is an oxidative product which results from the
exposure of creatinine (CRN) to silver, mercury, copper, Fe^{3+} or charcoal[5].
However the factors affecting the oxidation of CRN to MG are far different
from the internal environment of uremics, and the mechanism of its synthesis
in vivo still remains unclear. We have reported MG synthesis in various
tissues of rats and also have shown MG synthesis in liver using isolated
hepatocytes[6].

We have also reported that rat liver homogenates had no capacity to form
MG from CRN, although MG synthetic activity appeared in the homogenate after
incubation at 60°C for 15 min[7]. MG synthesis by the treated homogenate is
almost proportional to the amount added to the reaction mixture[7]. Furthermore, the activity disappeared after incubation at 100°C, or by the addition
of glutathione[7]. The activity of MG synthesis both in isolated hepatocytes
and in heat treated liver homogenates increased almost in proportion to the
concentration of CRN[6,7]. These results suggest that the mechanism of MG
synthesis depends on both biological factors and chemical reactions. Consequently, we considered the role of active oxygen in MG formation from CRN
in vivo. We have already reported that hydrogen peroxide, superoxide radicals and hydroxyl radicals stimulate MG synthesis, and their scavengers
inhibit MG synthesis in vitro[8]. Among the active oxygen species tested,
hydroxyl radicals formed by the Fenton's reaction with hydrogen peroxide and
Fe^{2+} have the strongest effect of MG synthesis from CRN in vitro[8].

In this study, we investigate the role of active oxygen on MG synthesis
in vivo, at the cellular level using isolated rat hepatocytes.

METHODS

Preparation of Isolated Rat Hepatocytes

Male Wistar rats weighing 300 to 350 g were used in all experiments. The rats were allowed free access to water and laboratory chow containing 25% protein. Isolated hepatocytes were prepared essentially according to the method of Berry and Friend[9] as described previously[10]. We calculated that 9.8×10^7 cells correspond to 1 g (liver wet weight)[11].

Incubation of Cells

Cells were incubated in 6 ml of Kreb-Henseleit bicarbonate buffer containing 3% bovine serum albumin, 10 mM sodium lactate, 200 mg/dl of CRN and indicated substances. The incubation mixture was shaken at 60 cycles/min in 30 ml conical flask with a rubber cap, under 95% oxygen and 5% carbon dioxide at 37°C for 4 h (except for the time dependence experiment). Equilibration of the buffer was repeated every hour. To measure the rate of nonbiological conversion of CRN to MG[6], incubations were carried out without cells. The amount of cells used for each experment is indicated in the Results section. Incubation was arrested by the addition of 0.6 ml of 100% (wt/v) trichloroacetic acid. After sonication, the supernatant of cells and medicine was obtained by centrifugation at 1,700 x g for 15 min at 0°C, and 0.25 ml of the extract was used for MG measurements. MG was determined by high-performance liquid chromatographic analysis using 9,10-phenanthrenequione for the post-labeling method as described previously[6]. Dimethyl formamide for fluorometrical use was purchased from Wako Co.[6](Japan).

RESULTS

Effects of Dimethyl Sulfoxide

Dimethyl sulfoxide (DMSO) is known to be a specific hydroxyl radical scavenger[12], and it easily enters into cels. MG synthesis from CRN in isolated rat hepatocytes was inhibited almost completely by 100 mM DMSO as shown in Fig. 1. At this concentration of DMSO, synthesis of urea, a major

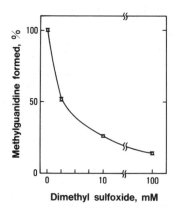

Figure 1.   Effect of dimethyl sulfoxide on methylguanidine synthesis in iso-
            lated rat hepatocytes. Cells (0.16 g liver wet weight) were in-
            cubated with indicated concentrations of dimethyl solfoxide. Each
            point represents the mean of duplicated incubations. Bars indicat
            the range of each determination.

product of hepatocytes from ammonium chloride, was not affected.

Effect of Ethanol, Lactulose and Sorbitol

Other hydroxyl radical scavengers, ethanol, lactulose and sorbitol[13] were tested for activity in the isolated rat hepatocyte system. Among these hydroxyl radical scavengers, ethanol inhibited MG synthesis in hepatocytes most effectively. Lactulose and sorbitol inhibited MG synthesis in hepatocytes to almost the same extent, as shown in Fig. 2.

Effect of Acetaminophenol, Indomethacin and Piroxicam

Acetaminophenol is a known anti-inflammatory drug. This reagent easily reacts with hydroxyl radicals and makes an organic hydrocompound[14]. Acetaminophenol at the concentrations of 0.1, 0.5 and 2.5 mM inhibited MG synthesis by 0, 30 and 76%, respectively, as shown in Fig. 3, whereas other non-steroidal anti-inflamatory drugs, both indomethacin and piroxicam, had little effect.

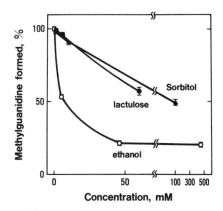

Figure 2. Effect of ethanol, lactulose and sorbitol. Cells (0.16 g of liver wet weight) were incubated with the indicated concentration of ethano (o——o), lactulose (●——●) or sorbitol (▲——▲), each point represents the mean of duplicated incubations. Bars represent the range of each determination.

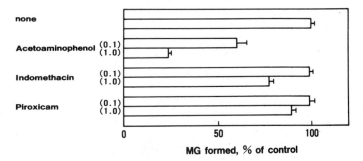

Figure 3. Effect of acetoaminophenol on methylguanidine (MG) synthesis in isolated rat hepatocytes. Cells (0.1 g of liver wet weight) were incubated with 0.1 or 1.0 mM acetoaminophenol, indomethacin or piroxicam for 4 h. Each column represents the mean of duplicated incubations. Bars indicate the range of each determination.

Effect of Catalase and Allopurinol

Catalase degrades hydrogen peroxide. The addition of 1,700 units of catalase to the incubation medium decreased synthesis in isolated hepatocyte by 25%. Because catalase can not enter cells, a proportion of hydrogen peroxide formed by isolated cells might leak from cells and then undergo degradation by catalase. Allopurinol is an inhibitor of xanthine and hypoxan tine oxidase and inhibits the formation of superoxide radicals. However, allopurinol had no effect on MG synthesis in isolated hepatocytes as shown i Fig. 4.

Effect of Anti-Oxidants

Non specific anti-oxidants, butylated hydroxytoluene, nordihydroguaiare tic acid (NDGA), and glutathione were investigated. These reagents also inhibited MG synthesis in isolated rat hepatocytes. However, degradation of glutathione to cysteine and glycine and the reutilization of these amino acids to form glutathione is still a possibility. We, therefore, tested the effect of methionine or cysteine plus glycine on the formation of glutathione[14]. These amino acids had no significant effect on MG synthesis in isolated rat hepatocytes under the conditions shown in Fig. 5.

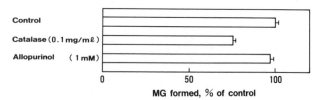

Figure 4.   Effect of catalase and allopurinol on methylguanidine (MG) synthe sis in isolated rat hepatocytes. Cells (0.16 g of liver wet weight were incubated with indicated reagents. Each column represents th the mean of duplicated incubations. Vertical bars indicate the range of each determination.

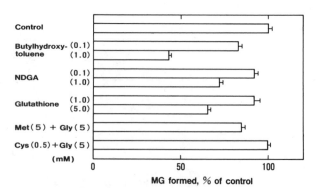

Figure 5.   Effect of antioxidant. Cells (0.2 g of liver wet weight) were incubated with indicated reagents for 4 h. Each column represents the mean of duplicated incubations. Vertical bars indicate the range of each determination. NDGA; nordihydroguaiaretic acid and MG; methylguanidine.

Effect of Ascorbic Acid

The Fenton's reaction which forms hydroxyl radicals requires Fe^{2+}, while ascorbic acid reduced Fe^{3+} to Fe^{2+} [15]. Ascorbic acid stimulates MG synthesis in isolated hepatocytes as shown in Fig. 6.

DISCUSSION

The role of active oxygen in MG synthesis in isolated hepatocytes was investigated using free radical scavengers and reagents which affect active oxygen generation. DMSO, which is a specific hydroxyl radical scavenger, inhibited MG synthesis by 90%. This finding suggests that MG synthesis from CRN in isolated rat hepatocytes mainly depends on the hydroxyl radical. Other data also support the significance of hydroxyl radicals as observed <u>in vitro</u> experiments[7]. A proposed pathway for the synthesis of MG from CRN is shown in Fig. 7.

Figure 6.   Effect of ascorbic acid. Cells (0.16 g of liver wet weight) were incubated with indicated concentration of ascorbic acid for 4h. Each point represents the mean of duplicated incubations. Bars indicate the range of each determination.

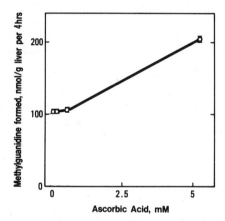

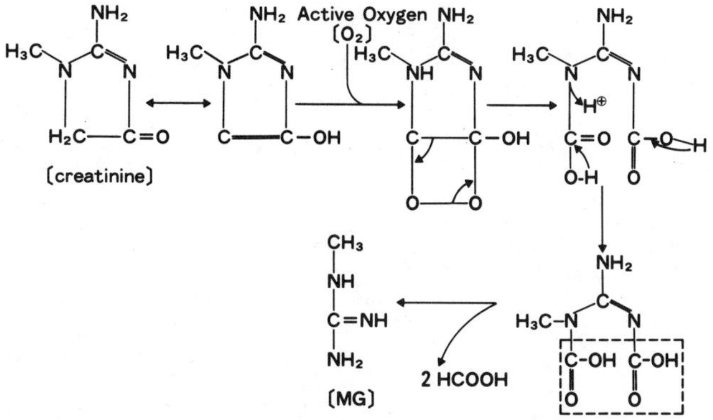

Figure 7.   Hypothetical pathway of methylguanidine (MG) synthesis from creatinine.

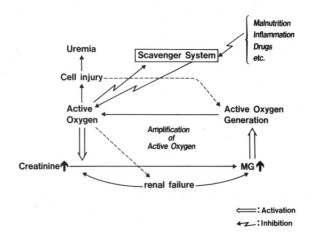

Figure 8. Hypothetical amplification of active oxygen by the creatinine-
           methylguanidine (MG)-active oxygen generation system.

We have proposed that the role of MG in the pathogenesis of uremia is
shown in Fig. 8[16-18]. CRN, a substrate of MG, increases and the clearance o
MG decreases in patients with renal failure. MG stimulates active oxygen
generation which, in turn, stimulates MG synthesis. Thus, active oxygen migl
be amplified by a CRN-MG-active oxygen generation system. We propose this
hypothesis based on the report that MG inhibited oxidative phosphorylation[1]
Recently, it was reported that MG itself generates active oxygen as was
demonstrated using ESR and radical aduct[20].

Based on these observations, we suggest that the amount of MG synthesized might be a good marker of hydroxyl radical generation *in vivo*.

ACKNOWLEDGMENTS

These research findings were presented, in part, at the IV Internation.
Congress on Nutrition and Metabolism in Renal Disease in October, 1985, in
Williamsburg, Virginia.

This study was supported in part by a research grant from University o
Tsukuba Project Research and Intractable Disease Division, Public Health
Bureau, Ministry of Health and Welfare, Japan. We are indepted to Mrs. Sato
Kawamura for her valuable technical assistance.

REFERENCES

1. Giovannetti, S., Biagini, M., Balestri, P.L., Navasesi, R., Giagnoni,
   P., deMatteis, A., Ferro-Milone, P. and Perfetti, C., Uremia-like
   syndrome in dogs chronically intoxicated with methylguanidine and
   creatinine, Clin. Sci. 36 (1969) 445-452.
2. Giovannetti, S., Balestri, P.L. and Barsotti, G., Methylguanidine in
   uremia, Arch. Intern. Med. 131 (1973) 709-713
3. Barsotti, G., Bevilacqua, G., Morelli, E., Cappelli, P., Balestri, P.L.
   and Giovannetti, S., Toxicity arising from guanidine compounds: Role of
   methylguanidine as a uremic toxin, Kidney Int. 7 (1975) S299-S301.

4. Endo, G. and Takahashi, K., Identification and property of the muta-
   genic principle formed from a food-component, methylguanidine, after
   nitrosation in simulated gastric juice, Biochem. Biophys. Res. Commun.
   54 (1973) 1384-1392.
5. Baumann, L. and Ingvalden, T., An oxidation product of creatinine, J.
   Biol. Chem. 36 (1918) 277-280.
6. Nagase, S., Aoyagi, K., Narita, M. and Tojo, S., Biosynthesis of methylguanidine in isolated rat hepatocytes and in vivo, Nephron 40 (1985)
   470-475.
7. Nagase, S., Aoyagi, K., Narita, M. and Tojo, S., Stimulatory and inhibitory factors of methylguanidine synthesis in rat organs, Jpn. J.
   Nephrol. 27 (1985) 1141-1147.
8. Nagase, S., Aoyagi, K., Narita, M. and Tojo, S., Synthesis of methylguanidine by active oxygen, with special reference to the signifcant
   role of hydroxyl radical, Jpn. J. Nephro. 27 (1985) 1149-1159.
9. Berry, M.N. and Friend, D.S., High-yield preparation of isolated liver
   cells, J. Cell Biol. 43 (1969) 506-520.
10. Aoyagi, K., Ohba, S., Narita, M. and Tojo, S., Regulation of biosynthe-
    sis of guanidinosuccinic acid in isolated rat hepatocytes and in vivo,
    Kidney Int. 24 (1983) S224-S228.
11. Zahlten, R.N., Stratman, F.W. and Lardy, H.A., Regulation of glucose
    synthesis in hormone-sensitive isolated rat hepatocytes, Proc. Natl.
    Acad, Sci. USA 70 (1973) 3213-3218.
12. Paller, M.S., Free radical scanvengers in mercuric chloride-induced
    acute renal failure in the rat, J. Lab. Clin. Med. 28 (1985) 459-463.
13. Halliwll, B. and Gutteridge, J.M.C., Oxygen toxicity, oxygen radicals,
    transition metals and disease, Biochem. J. 219 (1984) 1-14.
14. Orrenius, S., Thor, H., Belloma, G. and Moldéus, P., Glutathione and
    tissue toxicity, in: "Proceedings of Iuphar 9th international congress
    of pharmacology. volume 2," S.W. Paton, J. Mitchell and P. Turner Eds.,
    The Macmillan Press Ltd, Houndmills, Hampshire (1984) pp. 57.
15. Halliwell, B. and Foyer, C.H., Ascorbic acid, metal ions and the super-
    oxide radical, Biochem. J. 155 (1976) 697-700.
16. Aoyagi, K., Ohba, S. and Nagase, S., Mechanism of synthesis of guanidino
    compound, Jpn. J. Nephrolo. 27 (1985) 1034-1036.
17. Aoyagi, K., Narita, M. and Tojo, S., Uremic toxin: recent progress of
    our study, Medical Practice vol. 3 No2. (1986) 256-259.
18. Aoyagi, K., Nagase, S., Narita, M. and Tojo, S., Role of active oxygen
    on methylguanidine synthesis in isolated rat hepatocytes, Kidney Int.
    (in press)
19. Hollunger, G., Guanidines and oxidative phosphorylations, Acta pharmacol.
    Toxicol. 11 Supl. (1955) 1-84.
20. Hiramatsu, M., Edamatsu R., Kohno, M. and Mori, A., Proc. Annual Meeting
    Jpn. Guanidino Compound Res. Assoc. 9 (1986) 44-45.

EFFECT OF ACTIVE OXYGEN ON GUANIDINE SYNTHESIS IN VITRO

Masako Sakamoto, Kazumasa Aoyagi, Sohji Nagase, Shoji Ohba, Mitsuhiro Miyazaki, Mitsuharu Narita and Shizuo Tojo

Department of Internal Medicine, Institute of Clinical Medicine, University of Tsukuba, Tsukuba, 305, Japan

INTRODUCTION

Recently there have been many reports describing the toxicity of oxygen in various disease states[1-4]. Some investigators have reported that patients with chronic renal failure are in a stronger peroxidative state than normal persons[5-9]. We have reported that methylguanidine (MG) is a peroxidative product of creatinine (CRN) and that the hydroxyl radical plays an important role in MG synthesis both in vitro and in isolated rat hepatocytes. We also suggest that MG is a useful indicator of peroxidation in uremic serum[10-13].

There are numerous studies of various guanidino compounds in the serum and urine of patients with renal disease including guanidine (Gua) which is reported increased in uremia[14-15]. These reports prompted us to suggest a possible role for active oxygen in Gua synthesis. In the present study, we investigate the effect of active oxygen on Gua synthesis from various precursors.

MATERIALS AND METHODS

The reaction mixture consisted of 1 ml of 50mM phosphate buffer (pH 7.4 at 37°C) containing various possible precursors of Gua, i.e. guanidino-succinic acid (GSA), guanidinoacetic acid (GAA), creatine, CRN, arginine, MG and canavanine at a final concentration of 1mM.

As a generator of superoxide radicals (O_2^-), 3mM hypoxanthine and 0.1 units xanthine oxidase were added to the mixture. As a scavenger of O_2^-, 300 units of superoxide dismutase (SOD) was added to the mixture.

We investigated the amount of Gua synthesized by H_2O_2, $FeCl_2$ or $FeCl_3$ and the effect of ascorbic acid on Gua synthesis by Fe.

Hydrogen peroxide (H_2O_2) and $FeCl_2$ were added to the mixture to a final concentration of 10mM as a generator of the hydroxyl radical (·OH) (Fenton's reaction). The effect of the concentration of each factor of the ·OH generating system was examined. The effect of ascorbic acid on Gua synthesis by ·OH was also examined. As scavengers of ·OH, dimethylsulfoxide (DMSO), lactulose, sorbitol or ethanol was added to the mixture at a final concentration of 1mM.

The reaction was terminated by the addition of trichloroacetic acid (TCA) to a final concentration of 10%. When H_2O_2 was present in the mixture, 2,800 units of catalase were added and the mixture was incubated for 30s at 37°C before the addition of TCA. All samples were stored in an ice bath and centrifuged at 1,700 x g for 20min. Gua in the supernatant was determined by high pressure liquid chromatography as previously reported[16].

RESULTS

Guanidine Synthesis from Arginine

Gua was synthesized from arginine by O_2^-. And Gua synthesis by O_2^- was inhibited by the addition of SOD (Fig. 1). Gua was not synthesized in the absence of hypoxanthine, xanthine oxidase or both. H_2O_2 formed Gua depending on the concentration and the period of incubation (Fig. 2).

Gua production by Fe reached a maximum level within 15min. But in the case of $FeCl_2$, a little more Gua production was observed after 15min (Fig.3-a). The addition of ascorbic acid to $FeCl_2$ served to inhibit Gua formation. When ascorbic acid was added to $FeCl_3$, Gua formation was potentiated until the concentration of ascorbic acid reached 10mM. Higher concentration of

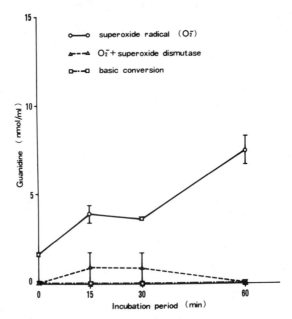

Figure 1. Time course of guanidine synthesis from arginine by the superoxide radical (O_2^-). Incubation mixture consisted of 1ml of 50mM phosphate buffer (pH 7.4 at 37°C), containing 1mM arginine (□-·-□) 1mM arginine, 3mM hypoxanthine (HX) and 0.1 units xanthine oxidase (XO) (O—O), 1mM arginine, 3mM HX, 0.1 units XO and 300 units superoxide dismutase (SOD) (Δ·····Δ). Each point represents the mean of duplicated incubations. Bars indicate the range of each determination.

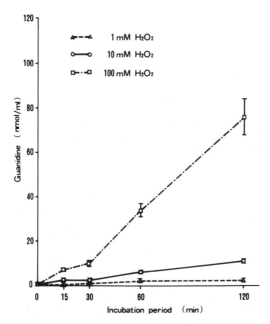

Figure 2. Time course of guanidine synthesis from arginine by hydrogen
         peroxide (H_2O_2). Incubation mixture consisted of 1ml of 50mM
         phosphate buffer (pH 7.4 at 37°C), containing 1mM arginine and 1mM
         H_2O_2 (Δ····Δ) or 10mM H_2O_2(0—0) or 100mM H_2O_2 (□-·-□).

ascorbic acid resulted in decreased Gua formation by $FeCl_3$ (Fig. 3-b). Gua
formation by $FeCl_2$ depended on the arginine concentration. Gua formation by
$FeCl_2$ or $FeCl_3$ increased depending on concentration and reached a maximum
level at 5 and 10mM of $FeCl_2$ and $FeCl_3$, respectively (data not shown).

   The hydroxyl radical achieved a maximum Gua formation within 5s (Fig.4-
a). The mixture of $FeCl_3$ and H_2O_2 formed much less Gua than $FeCl_2$ and H_2O_2
mixture. In the ·OH generating system Gua synthesis depended on the concentration of Fe and arginine. Gua synthesis increased depending on the concentration of H_2O_2 up to 5mM. Concentrations of H_2O_2 higher than 5mM resulted in
decreased Gua formation (Fig. 4-b). The addition of ascorbic acid to the ·OH
generating system served to inhibit Gua formation. The potentiation of Gua
formation by the addition of ascorbic acid to the mixture of $FeCl_3$ and H_2O_2
was observed until the concentration of ascorbic acid was at least 20mM.
Ascorbic acid at 50mM was not sufficient to inhibit Gua formation by $FeCl_3$
and H_2O_2 (Fig. 5). The addition of the ·OH scavengers, i.e. DMSO, lactulose,
sorbitol and ethanol inhibited Gua formation by 31.6, 30.2, 26.5 and 29.4%,
respectively (Fig. 6-a). The inhibitory effect of ·OH scavengers depends on
their concentrations (data not shown).

Guanidine Synthesis from GAA

   Gua formation by O_2^- was inhibited by the addition of SOD. The same
tendency was observed in the effect of H_2O_2 or Fe on Gua formation from GAA.
However, the amount produced from GAA was smaller than that from arginine
except in the case of H_2O_2 (Table 1). Gua formation by ·OH did not reach
maximum within 5s but progressed slowly after 5s. The addition of the ·OH
scavenger i.e. DMSO, lactulose, sorbitol and ethanol inhibited Gua formation
by 85.6, 82.5, 71.4 and 70.0%, respectively (Fig. 6-b). These inhibitory

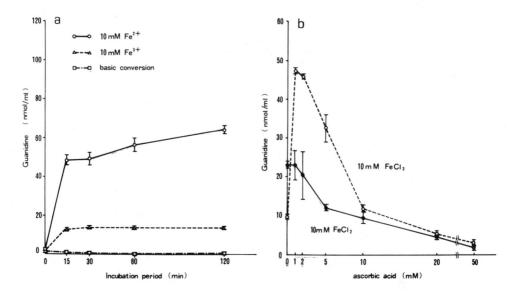

Figure 3. Time course of guanidine synthesis from arginine by Fe and the effect of the concentration of ascorbic acid. Incubation mixture consisted of 1ml of 50mM phosphate buffer (pH 7.4 at 37°C), containing (a) 1mM arginine (□-·-□), 1mM arginine and 10mM $FeCl_2$ (0—0) or $FeCl_3$ (Δ·····Δ), (b) 1mM arginine, 10mM $FeCl_2$ (●—●) or $FeCl_3$ (0—0) and the indicated amount of ascorbic acid. Incubation period was 120min.

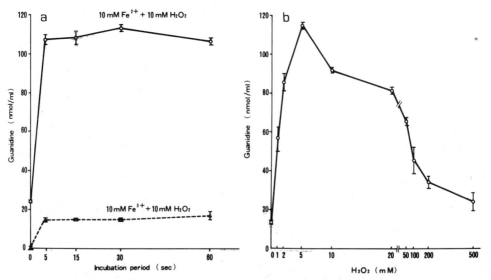

Figure 4. Time course of guanidine synthesis from arginine by Fe and H_2O_2. And the effect of the concentration of H_2O_2. Incubation mixture consisted of 1ml of 50mM phosphate buffer (pH 7.4 at 37°C), containing (a) 1mM arginine, 10mM H_2O_2 and 10mM $FeCl_3$ (Δ·····Δ) or $FeCl_2$ (0—0), (b) 1mM arginine, 10mM $FeCl_3$ and the indicated amount of H_2O_2. Incubation period was 60s.

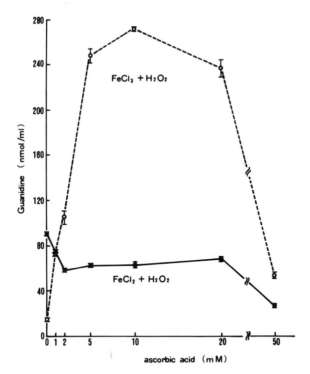

Figure 5. Effect of ascorbic acid on guanidine synthesis from arginine by the
          hydroxyl radical. Incubation mixture consisted of 1ml of 50mM
          phosphate buffer (pH 7.4 at 37°C), containing the indicated amount
          of ascorbic acid, 1mM arginine, 10mM H_2O_2 and 10mM $FeCl_2$ (●—●) or
          10mM $FeCl_3$ (0····0). Incubation period was 60s.

effects were stronger than those observed with arginine.

Guanidine Synthesis from Various Guanidino Compounds by the Hydroxyl Radical

     Gua was formed from GSA, GAA, creatine, CRN, arginine, MG and canavanine.
The amount of Gua formed occurred in the following order: canavanine, arginine, GAA, CRN, creatine, MG and GSA. In addition, MG was formed from CRN and
creatine (Fig. 7).

DISCUSSION

     Some investigators have reported that Gua is increased in uremia[14,15]
and in rats with acute or chronic renal failure[17] and its elevation is in
proportion to the elevation of blood urea nitrogen or CRN[18]. Other investigators have reported that Gua inhibits oxygen uptake in mitochondria of
rabbit kidney or rat liver[19] and also inhibits various enzymes including
xanthine oxidase[20]. Gua has been reported to be formed from canavanine[21,22],
arginine[23-25], GAA[11,25] and CRN[10]. But the mechanism of its formation has
not been identified yet. In this study, we investigate the effect of active
oxygen on Gua synthesis in vitro.

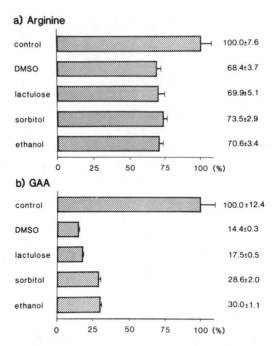

Figure 6. Effects of the scavengers of hydroxyl radical on guanidine synthesis from arginine or guanidinoacetic acid (GAA) by the hydroxyl
radical. Incubation mixture consisted of 1ml of 50mM phosphate
buffer (pH 7.4 at 37°C), containing 1mM arginine (6-a) or GAA
(6-b), 10mM H_2O_2, 10mM $FeCl_2$ and 1mM scavengers of the hydroxyl
radical. Each column represents the mean of duplicated incubations.
Bars indicate the range of each determination. Control values are
(a) 113.4 ± 8.4 and (b) 65.5 ± 8.2 nmol/ml/min.

Table 1. Comparison of guanidine synthesis from arginine or GAA by various
species of active oxygen.

precursor	sources of active oxygen				
	HX + XO	Fe^{2+}	Fe^{3+}	H_2O_2	$Fe^{2+}+H_2O_2$
arginine	0.26 ± 0.03	3.25 ± 0.17	0.89 ± 0.03	0.16 ± 0.03	113.4 ± 8.4
GAA	0.16 ± 0.04	1.38 ± 0.13	0.54 ± 0.03	0.23 ± 0.00	65.5 ± 8.2

The incubation mixture consisted of 1ml of 50mM phosphate buffer (pH 7.4 at
37°C), containing 1mM arginine or guanidinoacetic acid (GAA) and one of the
following: 3mM hypoxanthine (HX) and 0.1 units xanthine oxidase (XO), 10mM
$FeCl_2$, 10mM $FeCl_3$, 10mM H_2O_2, 10mM $FeCl_2$ and 10mM H_2O_2. The incubation period
was 60s. The amount of guanidine synthesized by each active oxygen except for
the hydroxyl radical was calculated from the amount of guanidine synthesized
during a 15 min incubation. Each value represents the mean of duplicated
incubations ± the range of each determination.

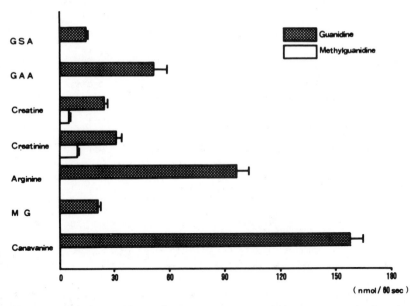

Figure 7. Guanidine synthesis from various guanidino compounds by the
hydroxyl radical. Incubation mixture consisted of 1ml of 50mM
phosphate buffer (pH 7.4 at 37°C), containing 1mM of the indicated
guanidino compounds, 10mM $FeCl_2$ and 10mM H_2O_2. Incubation period
was 60s.

We show that Gua is formed from arginine and GAA by every kind of
active oxygen, i.e. O_2^-, H_2O_2, Fe and ·OH. Gua synthesis was observed from
various guanidino compounds by ·OH. As summarized in Table 1, the effect of
O_2^-, H_2O_2 or Fe seems less important than ·OH and the effect of ·OH was the
strongest. Gua synthesis was greater from arginine than from GAA except for
the case of H_2O_2.

The addition of ascorbic acid resulted in two opposite effects. One is
to supply substrate for Fenton's reaction via the reduction of Fe^{3+}, and the
other is a scavenging effect of active oxygen as previously reported[26,27].
The difference in the amount of ascorbic acid necessary for the inhibition of
Gua formation by Fe versus Fenton's reaction suggests a high reactivity for
Fenton's reaction.

The rate of Gua formation at various H_2O_2 concentratins in Fenton's
reaction suggests that arginine degrades to a smaller molecule than Gua when
the concentration of H_2O_2 is higher than 5mM because the quantity of ·OH
produced is too much to preserve Gua base. Gua may be just one of the degradation products formed by the destructive activity of active oxygen. Gua
formation from MG was also explainable by this hypothesis. The difference in
the inhibitory effects of scavengers on Gua formation from arginine and GAA
suggests that arginine is more fragile to the attack of ·OH or arginine has a
more suitable structure than GAA for degradation to Gua base.

CONCLUSION

Gua was formed from arginine and GAA by active oxygen generated from multiple sources, i.e. superoxide radicals, hydrogen peroxide and hydroxyl radicals. Among the active oxygen species, the hydroxyl radical formed the largest amount of Gua. Gua was also formed from various kinds of guanidino compounds, i.e. arginine, GSA, GAA, creatine, CRN, MG and canavanine, by the hydroxyl radical. These results suggest that Gua is one of the degradation products of guanidino compounds via the destructive activity of active oxygen. Gua may also be a useful marker of peroxidation.

ACKNOWLEDGMENT

These research findings were presented at the IXth Annual Meeting of the Japan Guanidino Compounds Research Association in October 1986, in Tokyo and at the XXIXth Annual Meeting of Japanese Congress of Nephrology in November 1986, in Tokyo. We thank Ms. Satomi Kawamura for her skilled technical assistance.

REFERENCES

1. Paller, M.S., Hoidal. J.R. and Ferris, T.F., Oxygen free radicals in ischemic acute renal failure in the rat, J. Clin. Invest. 74 (1984) 1156-1164.
2. Koyama, I., Bulkley, G.B., Williams, G.M. and Im, M. J., The role of oxygen free radicals in mediating the re-perfusion injury of cold-preserved ischemic kidneys, Transplantation 40 (1985) 590-595.
3. Kobayashi, Y., Okahata, S., Tanabe, K., Tanaka, Y., Ueda, K. and Usui, T., Erythrocyte superoxide dismutase activity in hemolytic uremic syndrome, Hiroshima J. Med. Sci. 27 (1978) 181-183.
4. Carrel, R.W., Winterbourn, C.C. and Rachmilewitz, E.A., Activated oxygen and hemolysis, Brit. J. Haematol. 30 (1975) 259-264.
5. Giardini, O., Taccone-Gallucci, M., Lubrano, R., Ricciardi-Tenore, G., Bandino, D., Silvi, I., Ruberto, U. and Casciani, C.U., Evidence of red blood cell membrane lipid peroxidation in haemodialysis patients, Nephron 36 (1984) 235-237.
6. Giardini, O., Taccone-Gallucci, M., Lubrano, R., Ricciardi-Tenore, G., Bandino, D., Silvi, I., Paradisi, C., Mannarino, O., Citti G., Elli, M. and Casciani, C.U., Effects of alpha-tocopherol administration on red blood cell membrane lipid peroxidation in hemodialysis patients, Clin. Nephrol. 21 (1984) 174-177.
7. Kuroda, M., Asaka, S., Tofuku, Y. and Takeda, R., Serum anti oxidant activity in uremic patients, Nephron 41 (1985) 293-298.
8. Chauhan, D.P., Gupta, P.H., Nampoothiri, M.R.N., Singhal, P.C., Chugh, K.C. and Nair, C.R., Determination of erythrocyte superoxide dismutase, catalase, glucose-6-phosphate dehydrogenase, reduced glutathione and malondialdehyde in uremia, Clin. Chimi. Acta 123 (1982) 153-159.
9. Fillit, H., Elion, E., Sullivan, R., Sherman, R. and Zabriskie, J.B., Thiobarbituric acid reactive material in uremic blood, Nephron 29 (1981) 40-43.
10. Nagase, S., Aoyagi, K. and Tojo, S., On methylguanidine synthesizing organs - estimation from short term effect of creatinine -, Jpn. J. Nephrol. 26 (1984) 1099-1104.
11. Nagase, S., Aoyagi, K., Narita, M. and Tojo, S., Biosynthesis of methylguanidine in isolated rat hepatocytes and in vitro, Nephron 40 (1985) 470-475.
12. Nagase, S., Aoyagi. K., Narita, M. and Tojo, S., Active oxygen in methyguanidine synthesis, Nephron 44 (1986) 299-303.
13. Aoyagi, K., Nagase, S., Ohba, S., Miyazaki, M., Sakamoto, M., Narita, M. and Tojo, S., Significance of active oxygen in methylguanidine synthesis in isolated rat hepatocytes, Jpn. J. Nephro. 28 (1986) 536. (abstract of

Japanese Nephrology Congress in 1985)

14. Menichini, G.C. and Giovannetti, S., A new method for measuring guanidine in uremia, Experientia 29 (1973) 506–507.

15. Carr, M.H. and Schloerb, P.R., Analysis for guanidine and methylguanidine in uremic plasma, Anal. Biochem. 1 (1960) 221–227.

16. Yamamoto, Y., Manji, T., Saito, A., Maeda, K. and Ohta, K., Ion-exchange chromatography separation and fluorometric detection of guanidino compounds in physiologic fluids, J. Chromatography 162 (1979) 327–340.

17. Koide, H. and Azushima, C., Metabolic profiles of guanidino compounds in various tissues of uremic rats, in: "Guanidines," A. Mori, B.D. Cohen and A. Lowenthal Eds., Plenum Press, New York (1985) pp. 365–372.

18. Andes, J.E., Linegar, C.R. and Myers, V.C., Guanidine-like substances in blood. II. Blood guanidine in nitrogen retention and hypertention, J. Lab. Clin. Med. 22 (1937) 1209–1216.

19. Hollunger, G., Guanidines and oxidative phosphorylations, Acta Pharmacol. Toxicol. 111 (suppl 7) (1955) 7–83.

20. Rajagopalan, K.V., Fridovich, I. and Handler, P., Inhibition of enzyme activity by urea, Fed. Proc. 19 (1969) 49.

21. Kihara, H., Prescott, J.M. and Snell, E., The bacterial cleavage of canavanine to homoserine and guanidine, J. Biol. Chem. 217 (1955) 497–503.

22. Takahara, K., Nakanishi, S. and Natelson, S., Studies on the reductive cleavage of canavanine and canavaninosuccinic acid, Arch. Biochem. Biophys. 145 (1971) 85–95.

23. Walker, J.B. and Walker, M.S., The enzymatic reduction of hydroxyguanidine, J. Biol. Chem. 134 (1959) 1481–1484.

24. Reiter, A.J. and Horner, W.H., Studies on the metabolism of guanidino compounds in mammals. Formation of guanidine and hydroxyguanidine in the rat, Arch. Biochem. Biophys. 197 (1979) 126–131.

25. Mikami, H., Orita, Y., Ando, A., Fujii, M., Kikuchi, T., Yoshihara, K., Okada, A. and Abe, H., Metabolic pathway of guanidino compounds in chronic renal failure, Adv. Exp. Med. Biol. 153 (1982) 449–458.

26. Nishikimi, M., Oxidation of ascorbic acid with superoxide anion generated by the xanthine-xanthine oxidase system, Biochm. Biophys. Res. Commun. 63 (1975) 463–468.

27. Halliwell, B. and Gutteridge, J.M.C., Oxygen free radicals and iron in relation to biology and medicine: Some problems and concepts, Arch. Biochem. Biophys. 246 (1986) 501–514.

THE REACTIVITY OF GUANIDINO COMPOUNDS WITH HYDROXYL RADICALS

Midori Hiramatsu, Rei Edamatsu, Masahiro Kohno* and
Akitane Mori

Department of Neurochemistry, Institute for Neurobiology
Okayama University Medical School, 2-5-1 Shikata-cho
Okayama 700 and *JEOL Ltd., 1-2 3choo-me, Musashino
Akishima, Tokyo 196, Japan

INTRODUCTION

Guanidinoethanesulfonic acid[1], guanidinoacetic acid (GAA)[2], N-acetyl-arginine (NAA)[3], γ-guanidinobutyric acid (GBA)[4], methylguanidine (MGua)[5], α-guanidinoglutaric acid[6] and homoarginine[7], which are detected in the mammalian brain, induce epileptic discharges and/or induce convulsions after intracisternal injection or local administration into the rodent cortex. The seizure mechanisms are still unclear.

Willmore et al. have reported that the administration of vitamin E and selenium[8] or administration of vitamin E and dimethyl sulfoxide[9], which is a scavenger of hydroxyl radicals, inhibited the appearance of iron-induced epileptic discharges in rats. Free radicals[10] and active oxygen mediated radicals[11] were also increased in the rat cortex 5 min after injection into the sensory motor cortex. These results show that hydroxyl radicals and other active oxygen radicals generated by iron may be a trigger for the initiation of iron-induced epileptic discharges. Recently, we found that levels of MGua and GAA were increased in the rat brain after iron injection into the sensory motor cortex. In the present study, the effect of MGua and guanidine (Gua) on hydroxyl radicals and superoxide anion radicals were examined using electron spin resonance (ESR) spectrometry. In addition, the effect of the Fenton reagent[12], which generates hydroxyl radicals from ferric ion and hydrogen peroxide, on levels on Gua, MGua and other guanidino compounds in the homogenates of rat brains were studied.

MATERIALS AND METHODS

Source of Compounds

5,5-Dimethyl-1-pyrroline-1-oxide (DMPO), hypoxanthine, xanthine oxidase and dimethylenetriaminepentaacetic acid (DETAPAC) were obtained from Sigma Chemical Company (St. Louis, Mo). All other chemicals and reagents were of the highest grade available from commercial suppliers.

Hydroxyl Radical Analysis

Fifty microliters of 6.8 µM ascorbic acid, 35 µl of 1.37 M hydrogen per oxide, 50 µl of 0.1 mM $FeCl_3$, 50 µl of guanidino compounds dissolved in solution and 15 µl of DMPO were put into a test tube and mixed by an automatic mixer. The solution was then placed in a special flat cell and the presence of hydroxyl radicals was analyzed by an ESR spectrometer (JEOL JES-FE1XG). Manganese oxide was used as an internal standard.

The conditions of ESR spectrometry for the measurement of hydroxyl radicals were as follows : magnetic field, 335 ± 5 mT, power, 8.0 mW; response, 0.3 s; modulation, 0.2 mT; temperature, room temperature; sweep time 0.5 s.

Superoxide Anion (O_2^-) Radical Analysis

Fifty microliters of 2 mM hypoxanthine, 35 µl of 5.5 mM DETAPAC, 50 µl of 0.1 M sodium phosphate buffer at pH 7.8, 15 µl of DMPO and 50 µl of xanthine oxidase (0.272 unit/ml) were put into a test tube and mixed by the automatic mixer. The solution was then placed in a special flat cell and the $DMPO-O_2^-$ spin adduct was analyzed by an ESR spectrometer (JEOL JES-FE1XG). Manganese oxide was used as an internal standard.

The conditions of ESR spectrometry for measurement of $DMPO-O_2^-$ spin adduct were as follows : magnetic field, 335 ± 5mT; power, 8.0 mW; response, 0.1 s; modulation, 0.2 mT; temperature, room temperature; sweep time, 2 min.

Guanidino Compound Analysis

Rat brain tissue was homogenized with 5 vol. of 0.03 M Tris-HCl / 0.1 M KCl (1 : 19, pH 7.4). Five microliters of 0.1 M $FeCl_3$ and 50 µl of 3.5% hydrogen peroxide were added to 0.5 ml of the homogenate and the sample was incubated at 37°C for 0-90 min. Then the sample was centrifuged at 3,000 rpm for 15 min with 0.5 ml of 30% trichloroacetic acid. Guanidino compounds in the resultant supernatnat fluid were analyzed by a high performance liquid chromatogrph (Jasco, Tokyo) with the phenanthrenequinone reaction[13,14]. The conditions of analysis for guanidino compounds were as follows : column, Guanidino pak (6.0 x 35 mm, Jasco); column temperature, 70°C; flow rate of eluent, 1.1 ml/min; flow rate of 0.025% 9,10-phenanthrenequinone in dimethylformamide), 0.5 ml/min; flow rate of 1 M sodium hydroxide, 0.5 ml/min; reaction coil, 0.5 mm I.D. x 5 m; detector, fluorometer (Jasco FP-110, Ex 365 nm, Em 495 nm); eluent, (1) 0.4 M sodium citrate buffer at pH 3.00 (0.2 min), (2) 0.4 M sodium citrate buffer at pH 3.50 (4.0 min), (3) 0.4 M sodium citrate buffer at pH 5.25 (4.2 min), (4) 0.4 M sodium citrate buffer at pH 10.00 (9.4 min), (5) 1 M sodium hydroxide (5.2 min); fluorescence reagent, (1) 2 M sodium hydroxide, (2) 0.025% 9,10-phenanthrenequinone in dimethylformamide. Other detailed conditions have been written elsewhere[13,14].

Statistics

Statistical analysis was performed using the Student's t-test.

RESULTS

Effects of Guanidine and Methylguanidine on Active Oxygen Radicals

Fig. 1 shows the spectrum of the quartet signal of the hydroxyl radical. Both 25 mM Gua and MGua accelerated the generation of hydroxyl radicals by the Fenton reagent. However, both Gua and Mgua in the concentration range of 0.0025 to 2.5 mM did not affect the generation of hydroxyl radicals (Figs 2 and 3).

The generation of superoxide anion radicals from the hypoxanthine-xanthine oxidase system (Fig. 4) were not affected by the addition of Gua and MGua in the concentration range of 0.0025 to 25 mM (Figs. 5 and 6).

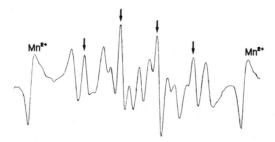

Figure 1. Hydroxyl radical generation by the Fenton reagent.

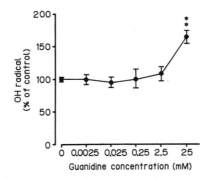

Figure 2. Effect of guanidine on hydroxyl radical generation by the Fenton reagnet. Each value represents the mean ± SD of 5 determinations. **$p < 0.005$ vs control.

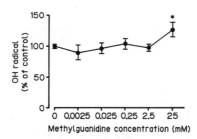

Figure 3. Effect of methylguanidine on hydroxyl radical generation by the Fenton reagent. Each value represents the mean ± SD of 5 to 6 determinations. *$p < 0.01$ vs control.

Figure 4. Superoxide anion radicals generated by the hypoxanthine-xanthine oxidase system.

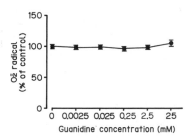

Figure 5. Effect of guanidine on superoxide anion radical generation by the hypoxanthine-xanthine oxidase system. Each value represents the mean ± SD of 3 to 5 determinations.

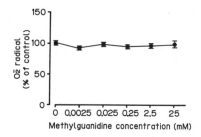

Figure 6. Effect of methylguanidine on superoxide anion radical generation by the hypoxanthine-xanthine oxidase system. Each value represents the mean ± SD of 3 to 6 determinations.

Biosynthesis of Guanidino Compounds in Brain Homogenates by the Fenton Reagent

Gua, MGua, GAA and arginine (Arg) were generated immediatly after the addition of Fenton reagent to the homogenates of rat brin. Gua levels increased remarkably though Gua was not detected in the control homogenates of rat brain (Fig. 7). MGua levels increased about 12 times that of controls (Fig. 8), and levels of GAA and Arg were increased about 38% and 21% respectively (Figs. 9 and 10). Levels of Gua and MGua decreased 5 min after incubation and became stable 15 min after the incubation. Levels of GAA and Arg were slightly increased up to 90 min after incubation. NAA started to increase 30 min after incubation and then increased linearly until 90 min after incubation (Fig 11). Creatinine levels decreased only 30 min after incubation through the incubation period of 90 min (Fig. 12). There was no significant change in GBA levels induced by the Fenton reagent (Fig. 13).

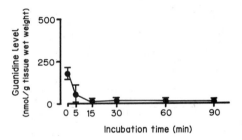

Figure 7. Effect of the Fenton reagent on guanidine levels in the homogenates of rat brin. Each value represents the mean ± SD of 5 to 7 determinations.

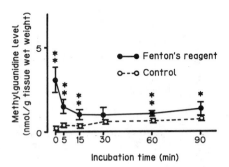

Figure 8. Effect of the Fenton reagent on methylguanidine levels in the homogenates of rat brain. Each value represents the mean ± SD of 5 to 8 determinations. *p < 0.01, **p < 0.005 vs control.

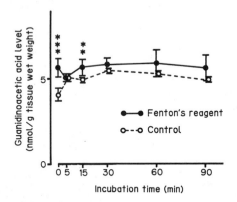

Figure 9. Effect of the Fenton reagent on guanidinoacetic acid levels in the homogenates of rat brain. Each value represents the mean ± SD of 5 to 7 determinations. **p < 0.005, ***p < 0.001 vs control.

DISCUSSION

In the present study it is found that Gua and MGua accelerated the generation of hydroxyl radicals by the Fenton reagent and that the Fenton reagent increased the biosynthesis of Gua and MGua in homogenates of rat brain.

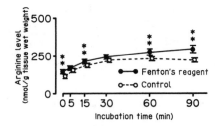

Figure 10. Effect of the Fenton reagent on arginine levels in the homogenates
of rat brain. Each value represents the mean ± SD of 5 to 7 determinations. **p < 0.005 vs control.

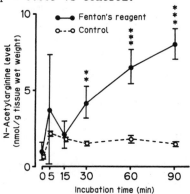

Figure 11. Effect of the Fenton reagent on N-acetylarginine levels in the
homogenates of rat brain. Each value represents the mean ± SD of
4 to 7 determinations. **p < 0.005, ***p < 0.001 vs control.

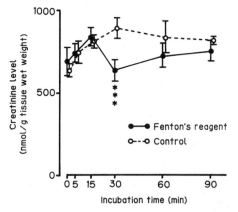

Figure 12. Effect of the Fenton reagent on creatinine levels in the homogenates of rat brain. Each value represents the mean ± SD of 5
to 8 determinations. ***p < 0.001 vs control.

        Recently we found that MGua and GAA levels were remarkably increased in
brain at the initiation of epileptic discharges after the injection of iron
into the sensory motor cortex and that the level decreased 60 min after the
injection. It was also increased markedly 2 months after injection, when the
epileptic focus had become chronic. In mammalian tissues, the hydroxyl radi-

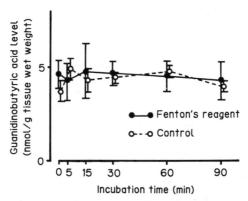

Figure 13. Effect of the Fenton reagent on guanidinobutyric acid levels
        in the homogenates of rat brain. Each value represents the
        mean ± SD of 5 to 8 determinations.

cal is thought to be generated in a part from the Fenton reaction and the
Haber-Weiss reaction[15-16]. From these facts, our finding may be explained as
follows : iron-generated active oxygen radicals, specially hydroxyl radical,
may accelerate the biosynthesis of MGua and GAA. MGua and GAA could induce
epileptic discharges after intracisternal injection into rodents[2-5]. MGua
has been reported to inhibit the Na^+, K^+-ATPase in the microsomes of rabbit
brains[17]. In consideration of these facts, excess MGua biosynthesized in the
presence of ferric ion may act on neuronal membranes directly and the epileptic neuronal disfunction follows.

    In our experiment it was found that excess Gua and MGua increased the
hydroxyl radicals generated by the Fenton reagent. On the other hand, these
guanidino compounds did not affect the superoxide anion radical. Hydroxyl
radicals are unstable and its half life is much shorter than other active
oxygen radicals.

    Gua is an extremely strong base. Gua acquires a proton readily at any
pH below 14 or thereabout, and it converted into the guanidium ion. This ion
is completely symmetrical with respect to the three $-NH_2$ groups. All three
C-N distances are equal and intermediate in character between double and
single bonds. Owing to the complete equivalence of the three forms, the resonance energy is high, and the ion is very stable (pK 13.65). Removal of a
proton diminishes the symmetry and stability, that is, the guanidium ion is
in an extremely weak acid (pK ca. 14)[18]. In our experiment it was found that
the extent of the increase in the hydroxyl radicals induced by guanidine was
much larger than that induced by MGua. This fact is supported by the above
explanation.

    On the other hand, Gua and MGua levels in the homogenates of rat brain
were increased immediatly after the addition of the Fenton reagent. The
extent of increase in the Gua level was much larger than that of the MGua
level. This fact also supports the hypothesis that removal of a proton
diminishes symmetry and stability. Other guanidino compounds were slightly
or hardly affected by the Fenton reagent. The Fenton reagent generates
hydroxyl radicals. From our results it is found that hydroxyl radicals are
easily reactive with Gua and MGua.

SUMMARY

(1) The effect of guanidino compounds on hydroxyl radical generation by the Fenton reagent and superoxide anion radical generation by the hypoxanthine-xanthine oxidase system was examined using electron spin resonance (ESR) spectrometry. Excess guanidine and methylguanidine increased hydroxyl radical concentrations and there was no significant change in superoxide anion radicals produced by guanidine and methylguanidine. The extent of increase of hydroxyl radicals by guanidine was much larger than that seen with methylguanidine.

(2) The effect of the Fenton reagent on the biosynthesis of guanidino compounds in the homogenates of rat brain was examined. The levels of guanidine and methylguanidine were remarkably increased immediately after the addition of the Fenton reagent to the homogenates of rat brain. The extent of increase in the guanidine levels were much larger than that of methylguanidine levels. Levels of arginine and guanidinoacetic acid were increased slightly and levels of creatinine and guanidinobutyric acid were not increased. N-Acetylarginine levels were increased 30 min after incubation.

These results show that hydroxyl radicals are responsible for guanidine and methylguanidine synthesis in brain tissue.

REFERENCES

1. Mizuno, A., Mukawa, J., Kobayashi, K. and Mori, A., Convulsive activity of taurocyamine in cats and rabbits, IRCS Med. Sci. 3 (1975) 385.
2. Jinnai, D., Mori, A., Mukawa, J., Ohkusu, H., Hosotani, M., Mizuno, A. and Tye, L.C., Biological and physiological studies on guanidino compounds induced convulsion, Jpn. J. Brain Physiol. 106 (1969) 3668-3673.
3. Ohkusu, H., Isolation of α-N-acetyl-L-arginine from calf brain and convulsive seizure induced by this substance, Osaka-Igakkai-Zasshi 21 (1970) 49-55.
4. Jinnai, D., Sawai, A. and Mori, A., α-Guanidinobutyric acid as a convulsive substance, Nature 212 (1966) 617.
5. Matsumoto, M., Kobayashi, K., Kishikawa, H. and Mori, A., Convulsive activity of methylguanidine in cats and rabbits, IRCS Med. Sci. 4 (1976) 65.
6. Shiraga, H., Hiramatsu, M. and Mori, A., Convulsive activity of α-guanidinoglutaric acid and the possible involvement of 5-hydroxytryptamine in the α-guanidinoglutaric acid induced seizure mechanism, J. Neurochem. 47 (1986) 1832-1836.
7. Yokoi, I., Toma, J. and Mori, A., The effect of homoarginine on the EEG of rats, Neurochem. Pathol. 2 (1984) 295-300.
8. Willmore, L.J. and Rubin, J.J., Antiperoxidant pretreatment and iron-induced epileptiform discharges in the rat : EEG and histopathologic studies, Neurology 31 (1981) 63-69.
9. Willmore, L.J. and Rubin, J.J., The effect of tocopherol and dimethylsulfoxide on focal edema and lipid peroxidation induced by isocortical injection of ferrous chloride, Brain Res. 296 (1984) 389-392.
10. Willmore, L.J., Hiramatsu, M., Kochi, H. and Mori, A., Formation of superoxide radicals after FeCl$_3$ injection into rat isocortex, Brain Res. 277 (1983) 293-296.
11. Hiramatsu, M., Mori, A. and Kohno, M., Formation of peroxyl radical after FeCl$_3$ injection into rat isocortex, Neurosciences 10 (1984) 281-284.
12. Walling, C., Fenton's reagent revisited, Accounts Chem. Res. 8 (1975) 125-131.

13. Mori, A., Katayama, Y., Higashidate, S. and Kimura, S., Fluorometrical analysis of guanidino compounds in mouse brain, J. Neurochem. 32 (1979) 643-644.
14. Mori, A., Watanabe, Y. and Fujimoto, N., Fluorometrical analysis of guanidino compounds in human cerebrospinal fluid, J. Neurochem. 38 (1982) 448-450.
15. Czapski, G. and Ilan, Y.A., On the generation of the hydroxylation agent from superoxide radical. Can the Harber-Weiss reaction be the source of OH radicals ? Photochem. Photobiol. 28 (1978) 651-653.
16. Koppenol, W.H., Butler, J. and Lecuwen, J.W., The Harber-Weiss cycle, Photochem. Photobiol. 28 (1978) 655-660.
17. Matsumoto, M. and Mori, A., Effects of guanidino compounds on rabbit brain microsomal Na^+, K^+ -ATPase activity, J. Neurochem. 27 (1976) 635-636.
18. Cohn, E.J. and Edsall, J.T. in : "Proteins, Amino acids and Peptides as Ions and Dipolar Ions," Reinhold Publishing Corporation, New York (1943).

III. PHYSIOLOGICAL, PHARMACOLOGICAL AND TOXICOLOGICAL ASPECTS
   OF GUANIDINO COMPOUNDS

GUANIDINES AS DRUGS

Burton D. Cohen

Bronx-Lebanon Hospital
Bronx, N.Y.

In 1983, at the first international meeting of this group, Drs. Robin and Marescau treated us to a survey of naturally occurring guanidino compounds that was detailed and authoratative. It was the highlight of the congress and ultimately became the most significant contribution to the text which followed.

It occurred to me that the second volume of Guanidines also needs a survey of some general interest and, while this was touched on briefly 4 years ago, the subject of synthetic guanidines and their pharmacologic properties might be of interest. This will hardly be definitive, however, since there is no single source available which classifies drugs in accordance with shared chemical radicals or ligands. We are, moreover, saddled with a vast array of copyrighted trade names and even generic terms which are not universal. Since the use of standard chemical nomenclature in describing these synthetics is useless in this context, I shall present generic names in current use and offer trade names parenthetically. Similarly, I will draw largely from my own experience with these agents and cannot, therefore, hope to achieve the exhaustive survey that would enhance the value of Guanidines II in the manner in which Drs. Robin and Marescau contributed to Guanidines I[1].

You will recall from that summary that many of the pharmacologic properties of synthetic guanidines are inherent and, indeed, predicted by the biologic actions or toxic effects of naturally occurring guanides. The biologic role of these substances can be reduced to two major areas:

1. They are the principle source of phosphagens such as phosphocreatine, phosphoarginine, phosphoglycocyamine and several others. The resonances created by the opposition of the phosphoryl and guanidine groups is particularly suited to maintaining a pool of labile phosphate available to a variety of tissues in addition to muscle.

2. They form an economically efficient nitrogen sink from which organisms can derive purines (such as guanidine which gave the group its name), pyrimidines, various excretory products and several defensive poisons. In this last role, they are obviously cytotoxic and a wide variety of antibiotics are derivatives of guanidines. Many are complex molecules such as the

aminoglycosides of which streptomycin, the earliest of these, is an example. Some of the simplest linear guanidines, such as homarginine, have antibacterial activity and several guanidino amides have antifungal, antiviral and, even, chemotherapeutic properties.

Although they are certainly important guanidino drugs, I will not review the guanidines in use as antimicrobials since they are mostly naturally occurring and are, in fact, discussed in detail in the earlier volume[2]. I would call attention, however, to their use by primitive forms as protective toxins. Here they function primarily on the cellular level as sodium or chloride channel blockers inhibiting passage across cell membranes in both neuronal and myocardial tissues. Both gammaaminobutyric acid (GABA), a guanidine analogue, and glycine, an important guanidine precursor, function in the central nervous system to open such channels forming the theoretic basis for competition by many of the naturally occurring and synthetic guanidines.

The toxicities, therefore, of the naturally occurring guanidines are consequently also of interest with regard to their role as pharmacologic agents. These toxicities can be organized into four headings:

1. They disrupt muscular histoanatomy. Pathologists studying the linkages of large proteins and their subunits, such as actin and myosin in muscle, have long known and made use of the fact that high concentratations of guanidine produces dissociation into smaller units[2]. This has been a useful tool in studies of muscular histopathology and serves as rationale for some of the earliest clinical applications of guanidine, the prototypical guanide.

2. They are convulsants which forms the basis for much of our interest at these meetings. The consensus, as can be judged from numerous references throughout this text, is that the convulsant effect results from competitive inhibition of central neurotransmitters and, indeed, the effect increases with increasing structural similarity to GABA.

3. They are hypotensive and this is, in fact, another prominent feature shared by guanidine itself.

4. They are hypoglycemic and this was the principle stimulus (along with the obvious structural analogy linking guanidine to urea) for the initial explorations into the role of guanidines in uremia.

Keeping in mind these biologic actions and effects, we will look into a variety of guanidines as drugs.

Guanidine hydrochloride (Guanadyne) was briefly marketed 2-3 decades ago for use in amyotrophic lateral sclerosis and a variety of idiopathic myelopathies. Its use was largely empiric, though based upon its known interaction with muscle in vitro. Dosage was titrated to an effect and highly individualized. Its use was complicated by tetany, convulsions and nephrotoxicity and, to my knowledge, it is largely discontinued. I have been unable to contact the manufacturer, a subsidiary of a larger drug company which no longer exists.

The biguanides begin with synthalin which was never released for general use but gave rise to phenformin (DBI) and several analogues. They produce hypoglycemia by a mechanism totally distinct from insulin or the sufonylureas. They do not alter the beta-cell nor increase circulating immunoreactive insulin levels. They disrupt oxidative phosphorylation leading to two

separate and complementary effects: a compensatory increase in anaerobic
glycolysis and a reduction in energy-dependent gastrointestinal glucose
absorption. This led to the popularity of phenformin in the treatment of
obese diabetics, as a supplement to both insulin and sufonylurea, and as a
therapy for alimentary hypoglycemia or the "flaccid pylorus syndrome".
Increase in glucose fermentation, however, led to lactic acidosis and the
ultimate abandonment of phenformin and the other biguanides.

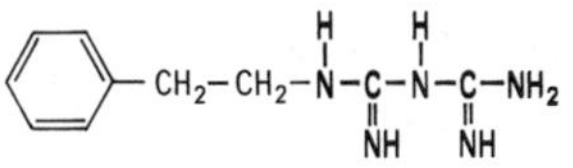

Phenformin

Chloroguanide (Proguanil, Paludrine) and similar antimalarials seems to
have a related action. They inhibit glucose oxidation by parasitized erythrocytes meaning that they block oxidative phosphorylation by the parasite. The
biguanides are thus, antimitochondrial in cells of the GI tract and anti-
plasmodial in parasitized red cells. It is of further interest that two side
effects of these antimalarials (including chloroproguanil and cycloguanil)
which made them unpopular are a decrease in gastric acidity in response to
histamine and an antiepinephrine effect on the myocardium.

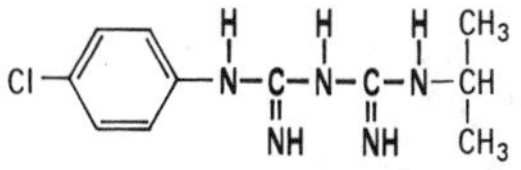

Chloroguanide

Cimetidine (Tagamet) is a histamine H_2 receptor antagonist which means
it blocks gastric acid production in response to most secretogogues (it does
not function as a peripheral antihistamine all of which are H_1 receptor mediated). It is the widest selling drug in the United States and generating
numerous analogues. Ranitidine (Zantac) is not a guanide. Like histamine
itself, both are ethyl amines (as is chloroguanide, the antimalarial which
inhibits gastric acidity). Ethyl amine rather than guanidine is, therefore,
probably the essential radical. The effect of cimetidine on the central nervous system which results in increased release of prolactin is independent of
ranitidine and shared by other guanides.

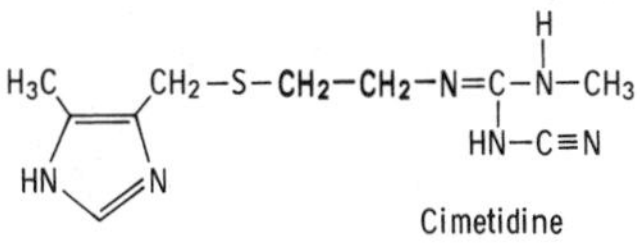

Cimetidine

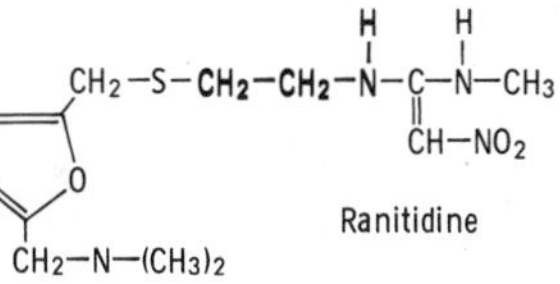

Ranitidine

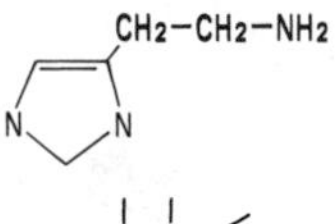

Histamine

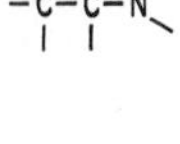

Ethyl amine

Amiloride (Midamor) is an antikaliuretic natriuretic agent which is not
an aldosterone antagonist. It also reduces calciuria and, in conjunction with
thiazide, is the drug of choice for managing hypercalciuric stoneformers. It
is a pure natriuretic acting at an as yet unknown locus. It may function
through inhibition of oxidative ATP production in the distal nephron or may
serve as a sodium or chloride channel blocker.

Amiloride

Finally, of the currently marketed guanidines the most important are a
series of sympatholytics. These are agents which block norepinephrine release at neuronal junctions. They differ in terms of their sites of action.
Guanethidine (Ismelin) and guanadrel (Hyrolel) are peripheral or ganglionic
blockers which also act to reduce cardiac output (as did chloroguanide, the
antimalarial). Guanabenz (Wytensin) and guanfacine (Tenex) act centrally
and, therefore, have fewer reflex effects such as orthostatic hypotension
and impotence but more generalized antiadrenergic activity such as sedation,
depression and dryness.

Three other antihypertensives fall into this chemically related class.
Clonidine (Catapres) is an aminoimidazole and like guanabenz is an alpha-2
adrenergic antagonist giving primarily central blockade. Prazosin (Minipress)
is an aminopteridine. It is an alpha-1 or peripheral antagonist having no
beta-1 or beta-2 inhibitory effect such as is seen with guanethedine. Mino-
xidil (Loniten), an aminopyrimidine (of considerable fame now as a treatment
for baldness) is thought to bypass the sympathetic nervous system and produce
vasodilatation by a direct effect on smooth muscle cells. You will recall
that guanidine is thought to be an actinomycin antagonist and guanidinosuccinic acid is known to block microtubular function in platelets[3].

Perhaps the most important guanidine to serve clinically is yet on the
horizon[4]. Aminoguanidine is currently available only in an experimental
format. It functions, however, to inhibit the non-enzymatic glycosylation
and protein cross-linking which occur with aging and are so disastrously

accelerated in diabetes. It may relate to the disruption by guanidine of non-covalent bonding in protein alluded to earlier and serves currently as a means for determining the importance of glycosylation in the complications of diabetes.

In summary, the synthetic guanidines seem to have the following physiologic effects in common differing primarily in the site of activity.

1. They disrupt protein bonds and linkages
2. They interfere with oxidative phosphorylation
3. They antagonize norepinephrine
4. They block cellular ion channels

REFERENCES

1. Robin, Y. and Marescau, B., Natural guanidino compounds, in: "Guanidines," A. Mori, B.D. Cohen and A. Lowenthal Eds., Plenum Press, New York (1985) pp 383-438.
2. Dreizen, H.I., Hartshorne, D.J. and Stracher, A., The subunit structure of myosin I: polydispersity in 5M guanidine, J. Biol. Chem. 241 (1966) 443-448.
3. Horowitz, H.I., Stein, I.M., Cohen, B.D. and White, J.G., Further studies on the platelet-inhibitory effect of guanidinosuccinic acid and its role in uremic bleeding, Am. J. Med. 49 (1970) 336-345.
4. Brownlee, M., Vlassara, H., Kooney, A., Ulrich, P. and Cerami, A., Aminoguanidine prevents diabetes-induced arterial wall protein cross-linking, Science 232 (1986) 1629-1632.

DIMETHYLARGININE HAS DIGITALIS-LIKE ACTIVITY

Katsuya Nagai, Akiko Hashida and Hachiro Nakagawa

Division of Protein Metabolism, Institute for Protein
Research, Osaka University, Suita, Osaka 565, Japan

INTRODUCTION

Recently it has been suggested that a digitalis-like substance, having cross-reactivity with an anti-digoxin antibody, natriuretic activity and Na^+,K^+-ATPase (ATP phosphohydrolase, EC 3.6.1.3)-inhibiting activity, plays an important role in the pathogenesis of essential hypertension[1,2]. It was also reported that in human urine there were two natriuretic substances with the molecular weight of greater than 30,000 and smaller than 3,000[3], and that the latter was ultrafiltrable through a membrance of Amicon UMO5 which cuts off the compounds with molecular weight less than 500[4], and was relatively resistant to heating for 1 h at 100°C at a pH of 10 or for 90 h at 110°C in 6N hydrochloric acid under anaerobic condition[4]. On the other hand, it was found that methylamino acids such as the dimethylarginines and monomethylarginine were isolated from human urine, and that the dimethylarginines were stable under acid hydrolysis[5]. In the course of our effort to isolate a digitalis-like subtances in rat urine with inhibition of ouabain-sensitive Na^+,K^+-ATPase as a marker, we found a low molecular weight (ultarfiltrable through Amicon YCO5) substance which inhibited Na^+,K^+-ATPase was stable to acid hydrolysis (K. Nagai et al.; unpublished observation). Thus we examined whether or not methylamino acids such as the dimethylarginines not only inhibit brain Na^+,K^+-ATPase activity, but also react with anti-digoxin antibody. In this paper we report that the dimethyl-L-arginines and monomethyl-L-arginine had such properties.

MATERIALS AND METHODS

Preparation and Assay of Na^+,K^+ ATPase

Na^+,K^+-ATPase was prepared from brains of Wistar rats, weighting 300-350 g, by the following procedure. The brains were homogenized with 10 vol of 1 mM sodium phosphate buffer (pH 7.3), containing 0.32 M sucrose and 0.1 mM EDTA, and the homogenate was centrifuged at 17,000 x g for 12 min. The supernatant was centrifuged at 100,000 x g for 60 min and the resultant precipitate was suspended in 2 vol of 20 mM imidazole buffer (pH 7.25), containing 0.32 M sucrose and 1 mM EDTA. The suspension was used as a partially

115

purified preparation of Na^+,K^+-ATPase. Na^+,K^+-ATPase activity was assayed by the method of Matsuda and Cooper[6]. Briefly, the enzyme preparation was incubated for 20 min at 37°C in 25 mM imidazole buffer (pH 7.25), containing 3 mM $MgCl_2$, 3 mM ATP, 140 Mm NaCl, and 20 mM KCl. The reaction was stopped b the addition of trichloracetic acid, and the concentration of Pi liberated in the supernatant was determind by the method of Baginski et al.[7]. Mg^{2+}-ATPase activity was measured in the presence of 1 mM ouabain. Net Na^+,K^+-ATPase activity was calculated by subtracting Mg^{2+}-ATPase activity from total activity. The dimethylarginines and other compounds examined in this experiment were added to the incubation medium, and the inhibitory effect was expressed as the percent decrease in activity of which the value with incubation without test compounds was defined at 100 %. The specific activit of brain Na^+,K^+-ATPase in the suspension was 257 units (nmoles/min)/mg protein, and 1.285 unit of the enzyme/0.1 ml of incubation medium was used i this experiment.

Preparation of Anti-Digoxin Antibody-Sepharose 4B Affinity Column and Affinity Chromatography

Anti-digoxin/BSA antibody was obtained from Cappel Laboratories Inc., (U.S.A.) and coupled with Sepharose 4B using CNBr-activated Sepharose 4B (Pharmacia, Uppsala) in 0.5 M NaCl containing 0.1 M $NaHCO_3$ (pH 8.3).

Treatments with Heat and Prolidase

The solution of N^G,N'^G-dimethyl-L-arginine was adjusted to pH 10 by the addition of potassium hydroxide and was boiled for 2 h. The solution was neutralized with acetic acid and applied on anti-digoxin antibody-Sepharose 4B affinity column. The column was then eluted with 1 N acetic acid, and the eluate was evaporated and dissolved in water. The solution was adjusted to pH 7.3 and was studied for Na^+,K^+-ATPase inhibiting activity. A control solution of N^G,N'^G-dimethyl-L-arginine was treated in the same way except for boiling.

To examine the effect of prolidase, N^G,N'^G-dimethyl-L-arginine was incubated for 16 h at 37°C with 0.63 units of prolidase (from porcine kidney, Sigma Chemical Co., St. Louis)/ml, 1.3 mM Tris-HCl buffer (pH 8.0), containing 14.3 mM $MnCl_2$ and 33 µM dithiothreitol. The reaction was stopped by boiling for 10 min. After deproteinization by centrifugation the supernatant was adjusted to pH 7 and subjected to anti-digoxin antibody affinity chromatography. Its Na^+,K^+-ATPase-inhibiting actitity was measured as mentioned above. For control experiment the solution of Na^+,K^+-ATPase-inhibiting activity was measured as mentioned above. For control experiment the solution of N^G,N'^G-dimethyl-L-arginine was incubated for 16 h at 37°C without prolidase, then, prolidase was added to the incubation medium and treated using the same procedure as described above.

RESULTS

Inhibitory Actions of Methylamino Acids and Other Agents

In order to solve the problem of whether methylamino acids elicit inhibitory effects on rat Na^+,K^+-ATPase or not, we first examined their inhibitory actions in vitro. As shown in Fig. 1, 1-methyl-L-histidine, 3-methyl-L-histidine and γ-guanidinobutyric acid did not affect the Na^+,K^+-ATPase activity at 1 mM. However, N^ε-methyl-L-lysine, N^G-monomethyl-L-arginine, N^G,N^G-dimethyl-L-arginine, N^G,N'^G-dimethyl-L-arginine and guanidinosuccinic acid (GSA) showed inhibitory activity at 1 mM. It should be noted that guanidino compounds such as N^G-monomethyl-L-arginine, N^G,N^G-dimethyl-L-arginine and N^G,N'^G-dimethyl-L-arginine showed especially remarkable inhibition, among them the inhibitory action of N^G,N'^G-dimethyl-L-arginine being strongest (Fig. 1).

Fig. 2 shows the dose-dependent inhibition of the Na^+,K^+-ATPase activity by N^G,N'^G-dimethyl-L-arginine at concentrations between 0.25 mM and 2.26 mM.

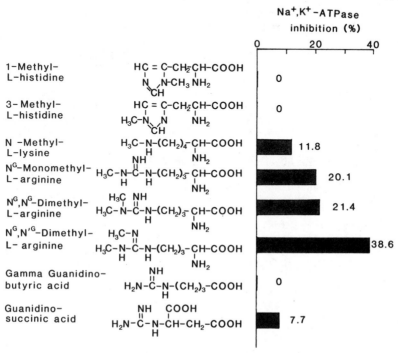

Figure 1. Effects of methylamino acids and their structually related compounds on activity of brain Na^+,K^+-ATPase. Each amino acid was added to a final concentration of 1 mM. The values are shown as means of duplicate determinations.

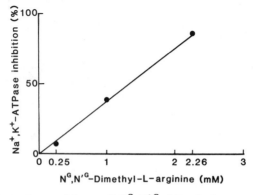

Figure 2. Dose-dependent effect of N^G,N'^G-dimethyl-L-arginene on brain Na^+,K^+-ATPase.

Inhibitory Activity of N^G,N'^G-Dimethyl-L-Arginine Retained in the Eluate by
Acetic Acid from Anti-Digoxin Antibody-Sepharose 4B Affinity Chromatography

In order to see the cross-reactivity of N^G,N'^G-dimethyl-L-arginine
with anti-digoxin antibody, we examined whether or not the inhibitory activity of the compounds was bound to an affinity column of anti-digoxin antibody-Sepharose 4B, and whether or not it was eluted with acetic acid from the column, if it bound to the latter. N^G,N'^G-Dimethyl-L-arginine was applied on the affinity column. The column was washed with 20 mM Tris-HCl buffer (pH 7.4) containing 0.15 M NaCl, 20 mM Tris-HCl buffer (pH 7.4) containing 0.5 M NaCl and water, and eluted with 1 N acetic acid. The eluate was evaporated and the residue was neutralized with Tris. Then the neutralized fraction was examine for Na^+,K^+-ATPase-inhibiting activity. The activity recovered was compared with that of the compound without affinity chromatography. As shown in Fig. 3, almost all inhibitory activity (98 % of original activity) was retained in the eluate. It was also found that the Na^+,K^+-ATPase-inhibiting activity of N^G-monomethyl-L-arginine was bound to the anti-digoxin affinity column and eluted with 1 N acetic acid (data is not shown here). These findings suggest that eigher N^G,N'^G-dimethyl-L-arginine or N^G-monomethyl-L-arginine cross-reacts with anti-digoxin antibody.

Effects of Heating at pH 10 and Treatment with Prolidase

Since it was shown that natriuretic activity in the urine of normal man was partially labile but relatively resistent to heating at a pH of 10^4, we examined the effect of heating on Na^+,K^+-ATPase-inhibiting activity of N^G,N'^G-dimethyl-L-arginine. After boiling for 2 h at pH 10 (by KOH), the inhibitory activity of N^G,N'^G-dimethyl-L-arginine (750 nmole) was assayed. For control, the pH of N^G,N'^G-dimethyl-L-arginine (750 nmole) solution was adjusted to 10, the solution was stood for 2 h at room temperature, and was neutralized with acetic acid. Then the solution was applied on an anti-digoxin affinity column, eluted with acetic acid, and subjected to determination of brain Na^+,K^+-ATPase-inhibiting activity. As shown in Fig. 4A, the inhibiting activity was not decreased after boiling, but rather tended to increase.

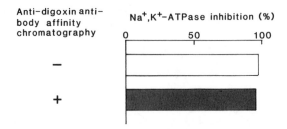

Figure 3. Binding of Na^+,K^+-ATPase inhibiting activity of N^G,N'^G-dimethyl-L-arginine to anti-digoxin antibody-Sepharose 4B affinity column. Na^+,K^+-ATPase-inhibiting activity of N^G,N'^G-dimethyl-L-arginine(260 nmole) before and after the anti-digoxin affinity chromatography were determined. The inhibition before the affinity chromatography was 97.3 % of ouabain-sensitive Na^+,K^+-ATPase at the concentration of 2.6 mM (260 nmol/0.1 ml of incubation medium). The values are means of the duplicted results.

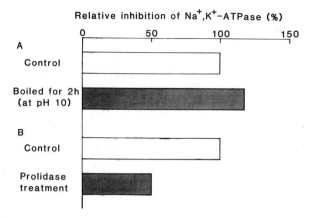

Figure 4. Effects of heating at a pH of 10 (A) and treatment with prolidase
(B). A) A solution containing N^G,N$^{'G}$-dimethyl-L-arginine (75
nmole) was adjusted to pH 10 by KOH, boiled for 2 h, neutralized
with acetic acid, and applied to an anti-digoxin affinity column.
The column was eluted with 1 N acetic acid and the eluate was
neutralized with Tris. Then its Na$^+$,K$^+$-ATPase-inhibiting activity
was determined. For control, N^G,N$^{'G}$-dimethyl-L-arginine (75 nmole)
was treated as the same way except for the boiling treatment. The
inhibition was expressed as the percentage of control (100 %).
N^G,N$^{'G}$-dimethyl-L-arginine without boiling (control) inhibited 25.3
% activity of ouabain-sensitive Na$^+$,K$^+$-ATPase at a concentration of
0.75 mM (75 nmole/0.1 ml). B) N^G,N$^{'G}$-Dimethyl-L-arginine (80 nmole)
was incubated with porcine kidney prolidase (0.63 U/ml) for 16 h
at 37°C, and then the mixture was deproteinized by boiling. The
supernatant was bound to an anti-digoxin affinity column. The
column was eluted with 1 N acetic acid, and the eluate was evaporated and neutralized with Tris. Then its Na$^+$,K$^+$-ATPase-inhibiting
activity was determined. For control, the compound (80 nmole) was
incubated with prolidase-free medium for 16 h at 37°C, and then
porlidase was added to the medium. The mixture was deproteinized
by boiling, and its Na$^+$,K$^+$-ATPase-inhibiting activity was determined by the same procedure mentioned above. The inhibiting activity was expressed as the percentage of control (100 %). N^G,N$^{'G}$-
Dimethyl-L-arginine (0.8 mM, 80 nmole/0.1 ml) inhibited 32.1 %
activity of the Na$^+$,K$^+$-ATPase (control).

DISCUSSIONS

In the present experiments, we obtained the following results; 1) Methylamino acids such as N$^\varepsilon$-methyl-L-lysine, N^G-monomethyl-L-arginine, N^G,N^G-
dimethyl-L-arginine, N^G,N$^{'G}$-dimethyl-L-arginine elicited strong inhibiting
effect on brain Na$^+$,K$^+$-ATPase at concentrations of 1 mM. Among the methylamino acids tested N^G,N$^{'G}$-dimethyl-L-arginine had the highest inhibiting
activity followed by di- and mono-methyl-L-arginine. 2) N^G,N$^{'G}$-dimethyl-L-
arginine showed dose-dependent inhibition on the Na$^+$,K$^+$-ATPase. 3) Na$^+$,K$^+$-
ATPase-inhibiting activities of N^G,N$^{'G}$-dimethyl-L-arginine and N^G-monometh-
yl-L-arginine bound to the anti-digoxin antibody-Sepharose 4B affinity column
and was eluted with 1 N acetic acid. 4) The inhibitory activity of N^G,N$^{'G}$-
dimethyl-L-arginine was not decreased by heating at a pH of 10, but was
reduced by the treatment with prolidase. From these findings it became clear
that mono- and di-methyl-L-arginines had not only inhibiting activity on

brain Na^+,K^+-ATPase, but also cross-reactivity with anti-digoxin antibody.

As previously mentioned, the natriuretic substance in human urine is relatively resistant to heating for 1 h at 100°C at a pH of 10, or to heating for 90 h at 110°C in 6 N hydrochloric acid under anaerobic condition[4]. Coincident with these facts, the Na^+,K^+-ATPase-inhibiting activity of N^G,N'^G dimethyl-L-arginine is not reduced by heating for 2h at 100°C at a pH of 10, but it rather tends to increase. N^G-Monomethyl-L-arginine showed the inhibitory activity, but its inhibitory activity does not exceed that of N^G,N'^G-dimethyl-L-arginine. This tempts us to speculate that by heating at a pH of 10 dimethylarginine might not be degraded to monomethylarginine, but to ureido-N-methylcitrulline which was shown to be produced from N^G,N'^G-dimethyl-L-arginine by alkaline hydrolysis[5], and that the latter might cause greater inhibition of Na^+,K^+-ATPase. In this study we did not examine the problem of whether the inhibitory action of the dimethylarginine is affected by acid hydrolysis. It is shown, however, that N^G,N'^G-dimethyl-L-arginine is stable to acid hydrolysis of 6 N HCl for 2 h at 105°C[5]. Thus it seems quite possible that the Na^+,K^+ATPase-inhibiting activity of dimethylarginine is stable to acid hydrolysis.

It is reported that the natriuretic substance from human urine is totally destroyed by treatment with prolidase[4]. The present experiment shows that Na^+,K^+-ATPase-inhibiting activity of N^G,N'^G-dimethyl-L-arginine is reduced to about half of the original activity after incubating with porcine kidney prolidase. Althought we do not know yet whether the prolidase itself is responsible for the disappearance of the inhibitory activity, (some other enzyme(s) or substance(s) with which prolidae preparation is contaminated could be responsible for the activity), these findings indicate the possibility that this compound might be the natriuretic substance mentioned previously[4]. The fact that N^G,N'^G-dimethyl-L-arginine exists in human urine supports this possibility.

As seen in Fig 2, N^G,N'^G-dimethyl-L-arginine inhibited brain Na^+,K^+-ATPase dose-dependently. The Na^+,K^+-ATPase-inhibiting activity of N^G,N'^G-dimethyl-L-arginine bound to an anti-digoxin affinity column and was eluted with 1 N acetic acid (Fig. 3). These findings suggest that the dimethylarginine cross-reacted with anti-digoxin antibody. In addition, 75 nmole (0.75 mM) and 80 nmole (0.8 mM) of N^G,N'^G-dimethyl-L-arginine after anti-digoxin affinity chromatography showed 25.3 and 32.1 % inhibition of the Na^+,K^+-ATPase activity in the control experiments (Fig. 4A and 4B), respectively. These values also fit in well with the dose-dependent inhibition curve without the affinity chromatography shown in Fig 2.

Since it was recently found that the inhibitory substance existed in bovine[8] and rat[9,10] hypothalami and rat[11] and pig[12] brains, and that in rats the hypothalamus contained the highest inhibiting activity among the brain areas tested. On the other hand, it was shown that methylamino acids such as N^G-monomethyl-L-arginine and N^G,N'^G-dimethyl-L-arginine also existed in human urine, bovine brain[5,13] and human hypothalamus[14], and that the enzyme catalizing methylation of arginine residue in proteins existed in rat and bovine brain[15,16]. Thus it seems quite possible that these methylarginines are endogenous inhibitors of Na^+,K^+-ATPase. Furthermore, it has been shown that in the uremic state guanidino compounds such as methylguanidine and GSA are increased[17]. In the present study we demonstrated that GSA elicited slight inhibitory action on brain Na^+,K^+-ATPase (7.7 % in Fig. 1) at a concentration of 1 mM. This finding suggests that guanidino compounds might be the entities inducing brain symptoms in the uremic state by their inhibitory action on brain Na^+,K^+-ATPase.

ACKNOWLEDGEMENTS

    Authors would like to express their deep gratitude to Professor Yasuo
Kakimoto and Dr. Masaharu Miyake, Ehime University School of Medicine, for
their kind gift of methylamino acids such as N^G-monomethyl-L-arginine, N^G,N^G-
dimethyl-L-arginine, N^G,N'^G-dimethyl-L-arginine and others. They also wish to
thank Dr. S. Sakakibara and Dr. Y. Kishida, Institute for Peptide Research,
and Dr. S. Tsunazawa, Institute for Protein Research, Osaka University for
their valuable discussions. This study is supported by a Grant-in-Aid for
General Research "C" (Grant No 62570132) from the Ministry of Education,
Science and Culture, Japan.

REFERENCES

1. De Wardener, H.E. and Clarkson, E.M., Concept of natriuretic hormone,
   Physiol. Rev. 65 (1985) 658–759.
2. De Wardener, H.E., The relation of circulating sodium transport inhibitor (the natriuretic hormone?) to hypertension, Medicine 62 (1983)
   310–326.
3. Clarkson, E.M., Raw, S.M. and De Wardener, H.E., Two natriuretic substances extracts of urine from normal man when salt-depleted and salt-
   loaded, Kidney Int. 10 (1976) 381–394.
4. Clarkson, E.M., Raw, S.M. and De Wardener, H.E., Further observations
   on a low-molecular-weight natriuretic substance in the urine of normal
   man, Kidney, Int. 16 (1979) 710–721.
5. Kakimoto, Y. and Akazawa, S., Isolation and identification of N^G,N^G-
   and N^G,N'^G-dimethylarginine, N^ε-mono-, di-, and trimethyllysine, and
   glucosyl-galactosyl- and galactosyl-8-hydroxylysine from human urine,
   J. Biol. Chem. 245 (1970) 5751–5758.
6. Matsuda, T. and Cooper, J.R., Inhibition of neuronal sodium and potassium ion activated adenosinetriphosphatase by pyrithiamin, Biochem. 22
   (1983) 2209–2213.
7. Baginski, E.S., Foa, P.P. and Zak, B., Determination of phosphate:Study
   of labile organic phosphate interference, Clin. Chim. Acta 15 (1967)
   155–158.
8. Haupert, G.T. and Sancho, J.M., Sodium transport inhibitor from bovine
   hypothalamus, Proc. Natl. Acad. Sci. USA 76 (1979) 4658–4660.
9. Lichstein, D. and Samuelow, S., Endogenous 'ouabain like' activity in
   rat brain, Biochem. Biophys. Res. Comm. 96 (1980) 1518–1523.
10. Alaghband-Zadeh, J., Fenton, S., Hancock, K., Millett, J. and De
    Wardener, H.E., Evidence that the hypothalamus may be a source of a
    circulating Na+,K+-ATPase inhibitor, J. Endocr. 98 (1983) 221–226.
11. Whitmer, K.R., Wallick, E.T., Epps, D.E., Lane, L.K., Collins, J.H. and
    Schwarts, A., Effects of extracts of rat brain on the digitalis receptor,
    Life Sci. 30 (1982) 2261–2275.
12. Fishman, M.C., Endogenous digitalis-like activity in mammalian brain,
    Proc, Natl. Acad. Sci. USA 76 (1979) 4661–4663.
13. Nakajima, T., Matsuoka, Y. and Kakimoto, Y., Isolation and identification
    of N^G-monomethyl-, N^G,N^G-dimethyl- and N^G,N'^G-dimethylarginine from the
    hydrolysate of proteins of bovine brain, biochim. Biophys. Acta 230
    (1971) 212–222.
14. Kakimoto, Y. and Miyake, M., Methylation of protein, Taisha 11 (1974)
    1103–1319 (in Japanese).
15. Paik, W.K. and Kim, S., Protein methylation in rat brain in vitro, J.
    Neurochem. 16 (1969) 1257–1261.

16. Miyake, M. and Kakimoto, Y., Protein methylation by cerebral tissue, J.
    Neurochem. 20 (1973) 859-871,
17. Stein, I.M., Cohen, B.D. and Kornhauser, R.S., Guanidinosuccinic acid
    in renal failure, experimental azotemia inborn errors of the urea cycle,
    New Engl. J. Med. 280 (1969) 926-930.

ADENOSINE, ADENOSINE ANALOGUES AND THEIR POTENTIATORS INHIBIT METHYLGUANIDINE SYNTHESIS, A POSSIBLE MARKER OF ACTIVE OXYGEN IN ISOLATED RAT HEPATOCYTES

Kazumasa Aoyagi, Sohji Nagase, Masako Sakamoto, Mitsuharu Narita and Shizuo Tojo

Department of Internal Medicine, Institute of Clinical Medicine, University of Tsukuba, Tsukuba 305, Japan

INTRODUCTION

Recently it has been reported that the oxygen radical plays an important role in kidney disease[1~6] and epilepsy[7]. We have reported that methylguanidine (MG), which is a uremic toxin[8~9], is formed by the reaction of oxygen radicals with creatinine both in vitro[10] and in isolated rat hepatocytes[11-14]. In addition, we have reported that MG synthesis in isolated rat hepatocytes is inhibited by adenosine, adenosine potentiators (i.e. dipyridamole and dilazep), and adenosine analogues[12-14]. Moreover it has been reported that diphenylhydantoin, an anti-convulsant drug is also an adenosine potentiator[15].

In this paper, we investigate further details of the action of adenosine and its potentiators including diphenylhydantoin and discuss the role of adenosine potentiators in kidney disease and epilepsy.

METHODS

Preparation of Isolated Rat Hepatocytes

Male Wistar rats weighing 300 to 350 g were used in all experiments. The rats were allowed free access to water and laboratory chow containing 25% protein. Isolated hepatocytes were prepared essentially according to the method of Berry and Friend[16] as described previously[17]. We caluculated that 9.8×10^7 cells corresponded to 1 g of liver (wet weight)[18].

Incubation of Cells

Cells were incubated in 6 ml of Krebs-Henseleit bicarbonate buffer containing 3% bovine serum albumin, 10 mM sodium lactate, 17.6 mM creatinine and the test substances with shaking at 60 cycles/min in a 30 ml conical flask with a rubber cap under 95% oxygen and 5% carbon dioxide at 37°C for 4 h (except for the experiment on time dependence). In the experiments to check the effect of dipyridamole, cells were preincubated with dipyridamole for 5 min. Equilibration of the buffer was repeated every hour. To measure

the rate of non-biological conversion of creatinine to MG[11], incubations were carried out without cells. The amount of cells used for each
experiment is indicated in the Results section. The reaction was stopped by
the addition of 0.6 ml of 100% (wt/v) trichloroacetic acid. After sonication,
the supernatant (of cells and medium) was obtained by centrifugation at
1,700 g for 15 min at 0°C and 0.25 ml of the extract was used for MG measurement. MG was determined by high-performance liquid chromatographic analysis
using 9,10-phenanthrenequinone for post-labeling, as described previously[11].
Adenosine, 2-chloroadenosine and dibutyryl cAMP were purchased from Sigma
Chemical Co., (St Louis, MO). Dipyridamole was kindly donated by Boehringer
Ingelheim Ltd.

RESULTS

Effect of Adenosine and Dipyridamole on MG Synthesis in Isolated Rat
Hepatocytes

Incubation of isolated rat hepatocytes with adenosine at concentrations
ranging from 0.1 μM to 100 μM resulted in inhibition of MG synthesis. The
inhibition was dose-dependent and was maximum at 100 μM adenosine (60 ± 2%)
(Fig. 1). The addition of 1 mM dipyridomole, which inhibits the uptake of
adenosine into the cell, enhanced the effect of adenosine. The co-operative
effect of dipyridamole and adenosine is shown in Fig. 2. In the absence of
adenosine in the medium, dipyridamole inhibited MG synthesis. This result
suggests the existence of endogenous adenosine in the medium. The effect of
dipyridamole was dose dependent in the concentration range from 20 μM to
1000 μM.

Specificity of Adenosine Analogues for the Inhibition of MG Synthesis

2-Chloroadenosine, which is not metabolized by adenosine deaminase,
inhibited MG synthesis by 50% at the concentration of 100 μM (Table 1). In
contrast, 2-deoxyadenosine, a naturally occurring anologue modiffied in the
ribose moiety, did not inhibit MG synthesis. Adenosine is degraded to inosine
and hypoxanthine[19]. Hypoxanthine and inosine at 100 μM had little effect on
MG synthesis under the present conditions.

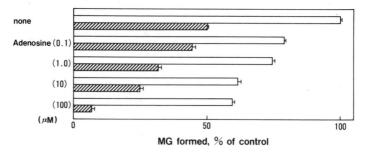

Figure 1.   Effect of adenosine and dipyridamole on methylguanidine (MG) syn-
            thesis in isolated rat hepatocytes. Hepatocytes (76 mg of wet
            weight) were incubated for 4 h with adenosine in the absence
            (☐) or in the presence of 1 mM dipyridamole (▨) as
            described in the Methods section. Various concentrations of
            adenosine were added to the incubation medium 5 min after the
            addition of dipyridamole. Each column represets the mean value of
            duplicate incubations. Bars represent the range of each deter-
            mination.

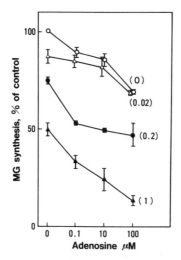

Figure 2. Cooperative effect of adenosine and dipyridamole on methylguanidine (MG) synthesis in isolated rat hepatocytes. Hepatocytes
(85 mg of wet weight) were incubated for 4 with various concentrations of adenosine and dipyridamole as described in the Methods
section. Adenosine was added to the incubation medium 5 min after
the addition of dipyridamole. The concentration of dipyridamole
is given (as mM) in parenthesis. Each point represents the mean
value of duplicate incubations. Bars represent the range of each
determination.

Effect of Diphenylhydantoin on MG Synthesis in Isolated Hepatocytes

MG synthesis in isolated hepatocytes was inhibited by diphenylhydantoin
in the presence and absence of puromycin which stimulated MG synthesis[12-14]
as shown in Table 2.

Table 1. Effect of adenosine analogues on methylguanidine (MG) synthesis

reagent	(mM)	MG formed	
		μmole/g/4 h	(%)
control		218 ± 2	(100 ± 0.9)
adenosine	(0.1)	130 ± 4	(60 ± 2)
2-chloroadenosine	(0.1)	118 ± 18	(54 ± 8)
2'-deoxyadenosine	(0.1)	179 ± 8	(82 ± 4)
inosine	(0.1)	198 ± 10	(90 ± 5)
hypoxanthine	(0.1)	241 ± 16	(110 ± 7)

Hepatocytes (0.14 g wet weight) were incubated with various reagents for 4 h.
Each value represents the mean value of duplicate incubations ± range of each
determination.

Table 2. Effect of diphenylhydantoin on methylguanidine synthesis in isolated rat hepatocytes

Diphenylhydantoin (mM)	Methylguanidine formed (nmol/g wet liver/4 h)			
	-Puromycin Aminonucleoside		+Puromycin Aminonucleoside	
0	58 ± 0.8	(100 ± 1.3)	79.5 ± 0	(100 ± 0)
0.02	62.3 ± 3.2	(107 ± 5.5)	80.2 ± 18.3	(101 ± 23)
0.20	66.7 ± 0	(115 ± 0)	54.8 ± 0.8	(68.9 ± 1.0)
1.0	42.9 ± 0.3	(73.9 ± 0.5)	49.3 ± 3.2	(62.0 ± 4.0)

Cells (0.11 g of wet liver) were incubated for 4 h with the indicated substances in the presence and absence of 1.9 mM puromycin aminonucleoside as
described in the method. Each value represents the mean of duplicate incubations ± range of each determination.

DISCUSSION

Activated oxygen radicals are thought to induce deleterious effects such
as lipid peroxidation, inflammation, carcinogenesis, cataracts and atherosclerosis. Recently, it has been reported that active oxygen plays an important role in the pathogenesis of acute renal failure[1-5] and glomerular injury[2-5]. We have reported that MG, a uremic toxin, was formed from creatinine
by the action of active oxygen both in vitro[10] and in isolated rat hepatocytes[12-14]. In our system, MG synthesis was inhibited by low concentrations
of adenosine and dipyridamole (which inhibits the uptake of adenosine into
cells). Further, it has been reported that adenosine inhibited superoxide
anion generation in human neutrophils at physiological concentrations[20], and
the injury to endothelial cells caused by neutrophils was inhibited by
adenosine and its agonist[21].

Increase of extracellular adenosine levels within the physiological
range by adenosine potentiators, such as dipyridamole, inhibited oxygen
radical generation not only in neutrophils but also in isolated hepatocytes.
These results suggest that the inhibition of oxygen radical generation in
neutrophils and glomerular epithelial cells by adenosine potentiators may
explain in part their proteinuria-reducing effect in human glomerulonephritis
as summarized in Fig. 3[22,23].

In addition, it has been reported that convulsions were induced by
active oxygen generation[7] and diphenylhydantoin, an adenosine potentiator
depresses the firing of cortical neurons[15]. Furthermore, diphenylhydantoin
reduces the 24 hour urinary protein excretion in patients with IgA
nephritis[24].

Production of MG in patients with chronic renal failure who had clinical symptoms and needed hemodialysis, was about twice that in patients without uremic symptoms even though the serum creatinine levels are almost the
same[25]. The administration of adenosine potentiators to patients with chronic renal failure might be an effective way to reduce the production of MG and
also prevent uremic symptoms.

ACKNOWLEDGMENTS

These research findings were presented, in part, at the IV international congress on nutrition and metabolism in renal disease in October, 1985, in Williamsburg, Virginia. This study was supported in part by research grants from the University of Tsukuba Project Research Fund and the Intractable Disease Division, Public Health Bureau, Ministry of Health and Welfare, Japan. We are indebted to Mrs. Satomi Kawamura for her valuable assistance.

REFERENCES

1. Paller, M.S., Hoidal, J.R. and Ferris, T.E., Oxygen free radical in ischemic acute renal failure in the rat, J. Clin. Invest. 74 (1984) 1156–1164.
2. Rehan, A., Johnson, K.J., Wiggs, R.C, Kunkel, R.G. and Ward, P.A., Evidence for the role of oxygen radicals in acute nephrotoxic nephritis, Lab. Invest. 51 (1984) 396–403.
3. Rehan, A., Johnson, K.J., Kunkel, R.G. and Wiggs, R.C., Role of oxygen radicals in phorbol myristate acetate-induced glomerular injury, Kidney Int. 27 (1985) 503–511.
4. Adachi, T., Fukuta, M., Ito, Y., Hirano, K., Sugiura, M., and Sugiura, K., Effect of superoxide dismutase on glomerular nephritis, Biochem. Pharamcol. 35 (1986) 341–345.
5. Dobyan, D.C., Bull, J.M., Strebel, F.R. Sunderland, B.A. and Bulger, R.E., Protective effects of 0-(ß-hydroxyethyl)-rutoside on cis-platinum-induced acute renal failure in the rat, Lab. Invest. 55 (1986) 557–563.
6. Holdsworth, S.R. and Boyce, N.W., Hydroxyl radical mediation of immune renal injury by desferrioxamine, Kidney Int. 30 (1986) 813–817.
7. Hiramatsu, M., Khochi, Y. and Mori, A., Active oxygen and free radical reaction in iron(3+)-induced epileptogenic focus of rat, Folia psychiatr. Neurol. Jpn. 37 (1983) 295.
8. Giovannetti, S., Biagini, M., Balestri, P.L., Navasesi, R., Giagnoni, P., deMatteis, A., Ferro-Milone, P. and Perfetti, C., Uremia-like syndrome in dogs chronically intoxicated with methylguanidine and creatinine, Clin. Sci. 36 (1969) 445–452.
9. Barsotti, G., Bevilacqua, G., Morelli, E., Cappelli, P., Balestri, P.L. and Giovannetti, S., Toxicity arising from guanidine compounds: Role of methylguanidine as a uremic toxin, Kidney Int. 7 (1975) s299–s301.
10. Nagase, S., Aoyagi, K., Narita, M. and Tojo, S., Active oxygen in methyl-guanidine synthesis, Nephron 44 (1986) 299–303.
11. Nagase, S., Aoyagi, K., Narita, M. and Tojo, S., Biosynthesis of methyl-guanidine in isolated rat hepatocytes and in vivo, Nephron 40 (1985) 470–475.
12. Aoyagi, K., Nagase, S., Ohba, S., Miyazaki, M., Sakamoto, M., Narita, M. and Tojo, S., Significance of active oxygen in methylguanidine synthesis in isolated hepatocytes, Jpn. J. Nephrol. 28 (1986) 536 (abstract of Japanese Nephrology Congress in 1985).
13. Aoyagi, K., Narita, M. and Tojo, S., Uremic toxin - recent progress of our study, Medical Practice 3 (1986) 256–259.
14. Aoyagi, K., Nagase, S., Narita, M. and Tojo, S., Role of active oxygen on methylguanidine synthesis in isolated rat hepatocytes, Kidney Int. 32 (1987) s229–s233.
15. Phillis, J.W., Interactions of the anticonvulsants diphenylhydantoin and carbamazepine with adenosine on cerebral cortical neurons, Epilepsia 25 (1984) 765.

16. Berry, M.N. and Friend, D.S., High-yield preparation of isolated liver cells, J. Cell Biol. 43 (1969) 506-520.
17. Aoyagi, K, Ohba, S., Narita, M. and Tojo, S., Regulation of biosynthesis of guanidinosuccinic acid in isolated rat hepatocytes and in vivo, kidney Int. 24 (1983) s224-s228.
18. Zahlten, R.N., Stratman, F.W. and Lardy, H.A., Regulation of glucose synthesis in hormone-sensitive isolated rat hepatocytes, Proc. Natl. Acad. Sci. USA 70 (1973) 3213-3218.
19. Granger, D.N., Rutil, G. and McCord, J., Superoxide radicals in feline intestinal ischemia, Gastroenterology 81 (1981) 22-29.
20. Bruce, N., Cronstein, B.N., Karmer, S.B., Weissmann, G.D. and Hirschorn, R., Adenosine: A physiological modulator of superoxide anion generation by human neutrophils, J. Exp. Med. 158 (1983) 1160-1177.
21. Cronstein, B.N., Levin, R.I., Belanoff, J., Weissmann, G. and Hirschorn, R., Adenosine: An endogenous inhibitor of neutrophil-mediated injury to endothelial cells, J. Clin. Invest. 78 (1986) 760-770.
22. Kan, K., Wada, T., Kitamoto, K., Konishi, K., Ozawa, Y., Kato, E. and Matsuki, S., Dipyridamole for proteinuria supression: Use in a patient with proliferative glomerulonephritis, J A M A 229 (1974) 557-558.
23. Tojo, S., Narita, M., Koyama, A., Sano, M., Suzuki, H., Tsuchiya, T., Tsuchida, H., Yamamoto, S. and Shishido, H., Dipyridamole therapy in the nephrotic syndrome, Contrib. Nephrol. 9 (1978) 111-127.
24. Clarkson, A.R., Seymour, A.E., Woodroffe, A.J., Mckenzie, P.E., Chan. Y.-L. and Wootton, A.M., Controlled trial of phenytoin therapy in a IgA nephropathy, Clin. Neph. 13 (1980) 215.
25. Kitamura, H, Ishizaki, M, Kitamoto, Y., Niki, M., Ueda, H., Tasaki, Y., Monma, H. and Takahashi, H., Proc. Annual Meeting Jpn. Guanidino Compounds Res. Assoc. 7 (1984) 40-41. (in Japanese).

THE ALTERATION OF GUANIDINO COMPOUNDS IN THE LIVER, KIDNEY AND BLOOD

FOLLOWING HALOTHANE INHALATION IN RATS

Michio Kitaura, Yoji Ochiai, Seiji Sugimoto, Koji Kabutan,
Yoshinori Kosogabe, Chiaki Tsuji and Futami Kosaka

Department of Anesthesiology and Resuscitology, Okayama
University Medical School, Okayama 700, Japan

INTRODUCTION

Analysis of serum guanidino compounds in patients with acute renal or hepatic insufficiency has revealed certain variations among several guanidino compounds relevant to the condition of the liver or kidney[1,2]. As for guanidinoacetic acid (GAA), while patients with acute renal insufficiency, had a low serum level, those with acute hepatic insufficiency showed a remarkably increased serum level[3,4].

It has been reported that halothane, general anesthetic, exerts various effects on individual organs, especially on the liver[5,6].

In this study, we investigated the effect of halothane on the metabolism of guanidino compounds in the liver and kidney.

MATERIALS AND METHODS

Male Wistar rats (200-250 g body weight) were made to inhale halothane, followed by an immediate and 24th hour collection of blood samples. Concurrently their livers and kidneys were excised while perfusing the veins with physiological saline. Rats were classified into 4 groups according to periods of halothane inhalation in addition to a control group.

Group 1:control group for oxygen inhalation alone
Group 2:(1 MAC*4 h--0 h)
Group 3:(1 MAC*4 h--after 24 h)
Group 4:(1 MAC*2 h--0 h)
Group 5:(1 MAC*2 h--after 24 h)

Ketamine (250 mg/kg) was given to the rats as the anesthetic. In order to make the respiratory conditions constant, tracheostomy was performed and connected to a respirator to keep $PaCO_2$ level constant. Blood samples were taken from the femoral artery. The liver and homogenized individually with distilled water as much as twice the wet weight of each organ, were centrif-

129

uged at 10,000 x g for 30 min., 4°C, followed by deproteinization with 50% trichloroacetic acid (TCA), and it was kept in the icebath (at 0°C) for 1hr. The concentration of TCA was set to make the final level 10%. Serum was also deproteinized by the same procedure.

These samples were centrifuged at 1,000 x g for 10 min. The supernatant were filtered through a microfilter with 0.45 μm pores and then analyzed by an HPLC ninhydrin method employing LC-3A (Shimadzu, Japan). The guanidino compounds were separated by the stepwise gradient system using a strong acidic cation exchanger ISC-05/S0504, at a reaction temperature of 55°C and a flow rate of 0.8 ml/min. Protein was determined by the Biuret method, and the contents of guanidino compounds were expressed in nmol/mg of protein in the tissues. As for an index of hepatic dysfunction, serum levels of ornithine carbamyl transferase (OCT) activities were determined. Student t-test was used for statistical calculations.

RESULTS

No guanidinosuccinic acid (GSA), guanidine (Gua), or methylguanidine (MG) were detected as definite peaks in the liver, and no GSA, guanidino-propionic acid (GPA), Gua, and MG were detected as definite peaks in the kidney. In the blood, only creatine (CTN), guanidinoacetic acid (GAA), guanidinobutyric acid (GBA) and arginine (ARG) were detected.

(1) No significant difference in serum OCT activities was observed among observed among all groups

(2) Results of analysis of guanidino compounds in the liver, kidney and serum are shown in Table 1,2. In the liver, concentrations of CTN and GBA significantly increased immediately after inhalation of halothane, without any change in GBA, GAA and ARG. GSA decreased significantly. In the 24th-hour samples, no significant differences in CTN, GPA, GBA and ARG were observed as compared with the control group. In the kidney, no change was

Table 1.   The concentrations of guanidino compounds in the liver and kidney

	Control	1Mx4h-0	1Mx4h-24	1Mx2h-0	1Mx2h-24
(Liver)					
CTN	1.23 ± 0.27	2.23 ± 0.54**	0.88 ± 0.19**	1.87 ± 0.36**	1.58 ± 0.16*
GAA	0.04 ± 0.01	0.03 ± 0.01	0.04 ± 0.01	0.04 ± 0.01	0.05 ± 0.01
GBA	0.05 ± 0.20	0.70 ± 0.16	0.34 ± 0.09**	0.62 ± 0.27	0.44 ± 0.08
ARG	0.48 ± 0.13	0.39 ± 0.07	0.37 ± 0.07	0.38 ± 0.09	0.32 ± 0.08
(Kidney)					
CTN	8.61 ± 1.47	12.03 ± 5.25	6.86 ± 2.19	8.25 ± 1.63	5.95 ± 0.93**
GAA	15.55 ± 2.92	3.36 ± 0.99**	8.02 ± 2.13**	6.27 ± 1.37**	2.44 ± 0.14**
GBA	0.17 ± 0.04	0.18 ± 0.04	0.17 ± 0.04	0.13 ± 0.04	0.18 ± 0.01
ARG	32.52 ± 4.00	5.66 ± 3.39**	21.14 ± 4.38**	17.83 ± 2.88**	10.42 ± 0.77**

**Significantly different ($p < 0.01$) from the control Values are expressed a nmol/mg protein (Mean ± SD, n=6)
CTN; creatine, GAA; guanidinoacetic acid, GBA; guanidinobutyric acid and ARG arginine.

Table 2. The concentrations of guanidino compounds in serum

(Serum)	Control	1Mx2h-0
CTN	46.7 ± 6.34	70.59 ± 15.1**
GAA	0.52 ± 0.19	0.73 ± 0.09**
GBA	Not Detected	0.12 ± 0.01
ARG	32.3 ± 9.3	51.8 ± 5.1**

**Significantly different (p < 0.01) from the control Values are expressed as nmol/mg protein. The abbreviations of gvanidino compounds are the same as shown in Table 1. (Mean ± SD, n=6)

Table 3. Intra and inter organ distribution of guanidino compounds

		Control	1MACx2hrs.-0hr
Intra-Organ Distribution			
ARG/GAA	Liver	10.1 ± 3.0	10.6 ± 3.45
	Kidney	2.12 ± 0.25	2.88 ± 0.32**
	Blood	63.6 ± 5.0	71.2 ± 2.1
CTN/GAA	Liver	32.1 ± 4.5	51.6 ± 15.0*
	Kidney	0.56 ± 0.02	1.34 ± 0.23**
Inter-Organ Distribution			
ARG	Kidney/Blood	1.01 ± 0.12	0.34 ± 0.06**
GAA	Kidney/Blood	29.9 ± 5.6	8.43 ± 1.88**
	Liver/Blood	0.07 ± 0.02	0.05 ± 0.02**

* Significantly different (p < 0.05) from the control
** Significantly different (p < 0.01) from the control
Values are expressed as nmol/nmol (Mean ± SD, n=6)

observed in CTN, GBA, and N-acetyl arginine (NAA), with significant decreases occurring in ARG and GAA. In the 24th-hour samples, no significant change was observed in CTN, GBA, and ARG. As for the serum concentrations of guanidino compounds, a comparison between the control and the halothane treated (1MAC*2 h--0 h) group revealed significant increases in CTN, GAA and ARG. GBA was detectable in the halothane-treated group, while it was not in the control group.

(3) Concentration ratios of ARG, GAA and CTN in organs are shown in Table 3. Inhalation of halothane caused the concentration ratio of renal ARG/blood ARG to diminish, the concentration ratio of renal ARG/renal GAA to elevate, the concentration ratio of renal CTN/renal GAA to elevate, the concentration ratio of hepatic CTN/hepatic GAA to elevate and the ratio of renal GAA/blood GAA to diminish.

DISCUSSION

The urea cycle, is localized in the liver because ornithine trans-carbamylase is only present in the mitochondria of liver cells. The activity of argininosuccinate synthetase and argininosuccinate lyase are highest in the liver[7], while in the kidney citrulline is incorporated from blood to synthesize ARG. ARG synthesized in the liver is immediately converted into urea with the help of arginase. ARG synthesized in the kidney consists of two types: one to be released in the blood and utilized in various tissues and the other to be metabolized in the presence of glycine amidinotrans-ferase to GAA and ornithine. GAA produced in the kidney consists of two types: one to be excreted in urine and the other to be released in the blood and taken up by liver where it is metabolized with GAA methyltransferase to CTN. CTN, released in the blood, will be metabolized in skeletal muscle to creatinine and them excreted in the urine. Based on the above, it is suggested that GAA in the kidney is synthesized as an alternate modality for the excretion of nitrogen and might also be useful as an index of liver func-tion. CTN concentration in the blood would influence the course of synthesis and excretion of GAA. While putting a negative feedback on glycine amidino-transferase, CTN might promote GAA to be excreted. Analysis of guanidino compounds in blood specimens from clinical cases disclosed that blood levels of GAA generally tended to be low in patients with acute renal insufficiency and extremely high in those with acute hepatic insufficiency (Fig.1).

These clinical findings, together with laboratory findings on changes in guanidino compounds in rats with hepatic and renal insufficiency, were taken into consideration in designing the present study to investigate the effects of halothane on metabolism in the liver and kidney. Results from analysis of guanidino compounds in halothane-treated rats included the following. Though a high concentration was observed in the kidney, ARG decreased significantly in Group 4 (1 MAC *2 h--0 h sampling) as compared with Group 1 (control). Serum levels of ARG, on the other hand, rose signifi-cantly. In order to determine a releasing rate of ARG synthesized in the kidney, the ratio of renal ARG/serum ARG was calculated: it was significant-ly low (Table 3), suggesting that the releasing rate of ARG in the kidney in halothane-treated rats increases. Among the metabolites of ARG, urea and GAA have been known to originate in the liver and kidney, respectively. In the present study, a high concentration of GAA was observed in the kidney in

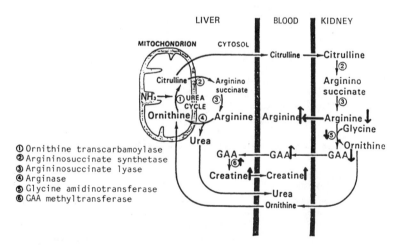

① Ornithine transcarbamoylase
② Argininosuccinate synthetase
③ Argininosuccinate lyase
④ Arginase
⑤ Glycine amidinotransferase
⑥ GAA methyltransferase

Figure 1. The alteration of urea cycle by halothane inhalation.

every group. Renal GAA was significantly lower in the halothane-treated group than in the controls. Serum levels of GAA were slightly higher in the halothane-treated group than in the control. The present findings indicated a significant decrease in the concentration ratio of renal GAA/ blood GAA, and a significant increase in the concentration of renal ARG/ renal GAA, suggesting that the decrease in renal glycine amidinotransferase activity might be cause by inhalation of halothane. There might be some increase in GAA released from the kidney to the blood. This could have resulted from a diminished uptake of GAA by the liver, making a judgment solely based on the concentration ratios difficult.

As for CTN, hepatic CTN increased in Group 2 and 4 as compared with Group 1. The concentration ratio of hepatic CTN/hepatic GAA increased significantly as compared with Group 1. These findings suggest that the elevation fo GAA methyltransferase activity are caused by halothane.

Examination of the liver and kidney obtained 24 after inhalation of halothane disclosed that there were no significant changes in concentrations of ARG and GAA between 0 hour and 24th hour specimens, while hepatic and renal CTRs which increased immediately after inhalation went back to the control levels. This suggests that after inhalation, the effect of halothane on various enzymes and concentrations of metabolites returns to normal as time passes.

In the present study, among other guanidino compounds, GBA, was observed in the liver and kidney at a high concentration. Inhalation of halothane seemed generally to cause GBA to appear in the blood. It has been reported that hepatic dysfunction occurring during acute hepatic insufficiency could result in the transmigration of GBA into blood: this might be a finding indirectly suggestive of the hepatotoxicity of halothane.

SUMMARY

It is suggested that inhalation of halothane changes the concentrations of GAA, ARG and CTN in the liver and kidney, and also alters the activities of their converting enzymes.

It is also suggested that inhalation of halothane causes GBA to appear in the blood, thus indicating a possible association of halothane with hepatotoxicity.

REFERENCES

1. Funahashi, M., Kato, H., Shiosaka, S. and Nakagawa, H., Formation of arginine and guanidinoacetic acid in the kidney in vivo. Their relations with the liver and their regulation, J. Biochem. 89 (1981) 1347-1356.
2. Cohen, B.D., Guanidinosuccinic acid in uremia, Arch. Intern. Med. 126 (1970) 846-850.
3. Kosogabe, Y., Ochiai, Y., Matsuda, R., Nishitani, K., Abe, S., Itano, Y., Yamada, T. and Kosaka, F., Guanidino compounds in patients with acute renal failure, In: "Guanidines," A. Mori, B.D. Cohen and A. Lowenthal, Eds. Plenum Press, New York (1985) pp. 287-294.
4. Nishitani, K., Metabolic role of the kidney in acute hepatic failure, Okayama Igakkai Zasshi 98 (1986) 135-143.

5. Pohl, L.R. and Gillete, J.R., A perspective on halothane-induced hepatotoxicity, Anesth. Analg. 61 (1982) 809-811.
6. Van Dyke, R.A., Effect of fasting on anesthetic-associated liver toxicity, Anesthesiol 55 (1981) A 181.
7. Kato, H., Oyamada, I., Funahashi, M. and Nakagawa, H., New radio
   isotopic assays of arginosuccinate synthetase and argininosuccinase,
   J. Biochem. 79 (1976) 945-953.

AMELIORATION OF GUANIDINOACETIC ACID METABOLISM IN STREPTOZOTOCIN-INDUCED DIABETIC RATS BY INSULIN TREATMENT

Masayoshi Hirata, Shuichi Hatakeyama, Katsuo Haruki,
Hiroaki Muramoto, Yohei Tofuku and Ryoyu Takeda

The 2nd Department of Internal Medicine, School of Medicine
Kanazawa University, 13-1, Takaramachi, Kanazawa 920, Japan

We have already reported[1] that in diabetic patients serum guanidino-acetic acid (GAA) is lower than in normal subjects in spite of the absence of renal dysfunction and the pancreatic glycine amidinotransferase (GAT) activity decreases in the experimental diabetic rats, although the effect of insulin on pancreatic GAT activity was yet under investigation. Funahashi et al.[2] reported that in the perfusion experiments on isolated kidney, GAT activity was significantly diminished in diabetic rats, and was not restored to the control level by insulin administration. From this point of view, we measured serum concentration of GAA and the renal and pancreatic GAT activities in controls, streptozotocin (STZ)-induced diabetic, and insulin-treated diabetic rats.

Male Wistar rats (300-500g body weight) were divided into three groups.

Diabetic rats were obtained by injecting STZ (30 or 40 mg/kg body weight) dissolved in 0.01M citrate buffer, pH 4.5. Five days later, half the diabetic rats were treated with 10 units of insulin (Novo Lente) per kg body weight twice a day for 2 days. Twenty rats were injected only citrate buffer instead of STZ as the controls. Seven days after injection, the rats were killed after blood sampling from the abdominal aorta. The kidney and the pancreas were removed, and GAT activity of these two organs were measured according to Van Pilsum et al.[3].

Serum urea-nitrogen (BUN) levels in the untreated rats were significantly higher than those in the control rats. Creatinine (CRN) levels were almost similar among three groups. Blood glucose (BG) levels in the untreated diabetic rats were significantly higher (624 ± 27 mg/dl) than those in other two groups. BG levels in the insulin-treated diabetic rats (133 ± 41 mg/dl) were not higher than those in the controls. However, the serum GAA concentrations were depressed in the untreated rats (20.8 ± 3.1 μg/dl) compared to the levels in the controls, and were restored to control levels after insulin treatment (79.6 ± 13.1 μg/dl). Renal GAT activity was depressed in the untreated diabetic rats (459 ± 41 μg/g tissue/h) compared to the level in the controls (907 ± 102 μg/g tissue/h), and was not restored in spite of insulin treatment. Pancreatic GAT activity was also depressed in

the untreated diabetic rats (329 ± 87 μg/g tissue/h), however, the GAT
activity was restored to control level (759 ± 76 μg/g tissue/h) after
insulin treatment, showing a contrast to renal GAT activity. Histological
examination in the untreated rats revealed no abnormalities in the glomeruli
and the renal tubular cells in the kidney, and atrophic change of the islet
of Langerhans without remarkable changes of the acinar cells around the
islet in the pancreas. In the insulin-treated rats, there was no remarkable
difference from the histology in the untreated rats. In the present experiment, STZ-induced diabetic rats showed higher BG levels, and lower levels of
serum GAA, and GAT activity in the kidney and pancreas than normal controls.
Higher BUN levels in these rats are presumably caused by dehydration due to
extreme hyperglycemia, because serum CRN levels were not increased.
Histological examination revealed no abnormalities in the kidney and only a
little change in the pancreas. So it is conceivable that STZ injection to
the rat might bring about only insulin deficiency and no renal dysfunction.
Lower concentration of serum GAA is presumed to be a result of the decreased
activity of renal and pancreatic GAT. In order to study whether insulin
deficiency decreases GAT activity of the kidney and pancreas and if
exogenous insulin administration restores the decreased GAT activity, we
examined the effect of insulin on renal and pancreatic GAT activities.
Extreme hyperglycemia was improved and serum GAA concentration was also
restored to control level in the insulin-treated rats. Renal GAT activity was
not restored, however, pancreatic GAT activity was restored to control level.

Judging from these results, it might be considered that in STZ-induced
diabetic rats depressed pancreatic GAT activity is ameliorated by insulin
administration, consequently the level of serum GAA is restored. In the
state of decreased GAT activity in the kidney, the pancreas might play a
compensatory role to produce GAA in order to maintain the energy supply of
muscle tissue.

REFERENCE

1. Hirata, M., Muramoto, H., Haruki, K., Miyazaki, F., Tofuku, Y. and
   Takeda, R., Impared metabolism of guanidinoacetic acid in uremia with
   special reference to diabetic nephropathy, in: "3rd ASIAN-PACIFIC
   CONGRESS OF NEPHROLOGY," Singapore, in 1986.
2. Funahashi, M., Kato, H., Shiosaka, S. and Nakagawa, H., Formation of
   arginine and guanidinoacetic acid in the kidney in vivo. Their relations
   with the liver and their regulation, J. Biochem. 89 (1981) 1347-1356.
3. Van Pilsum, J.F., Berman, D.A. and Wol, E.A., Assay and some properties
   of kidney transamidinase, Proc. Soc. Exp. Biol. Med. 95 (1957) 96-100.

URINARY EXCRETION RATE OF GUANIDINOACETIC ACID IN ESSENTIAL HYPERTENSION

Yoshiyuki Takano, Fumitake Gejyo, Yoshio Shirokane*,
Moto-o Nakajima* and Masaaki Arakawa

Department Medicine (II), Niigata University School of
Medicine, Niigata, and* Bioscience Research Laboratory of
Kikkoman Corporation, Noda, Japan

INTRODUCTION

The kidney is one of the major target organs in hypertension. Longterm persistent hypertension causes nephrosclerosis pathologicaly in 70-90% of patients[1,2]. Concerning the laboratory findings correlated with this pathological change, urinary micro-albumin (mAlb), ß-2-microglobuline (BMG) and N-acetyl-D-glucosaminidase (NAG) have been reported as early markers of renal damage[3,4]. Moreover, the production of guanidinoacetic acid (GAA) has been reported to be decreased in renal disease[5].

To determine the most sensitive marker of renal involvement in hypertensive patients without clinical evidence of renal damage, we measured the urinary excretion of mAlb, BMG, NAG and GAA in essential hypertensive patients and normal controls.

MATERIAL AND METHOD

Twenty-nine hypertensive patients (one female and 28 males), aged 35-67 y (mean 49), and 6 borderline hypertensive patients (all males), aged 32-55 y (mean 45), were selected for this study. None had proteinuria by Albustix. No other clinical evidence of renal damage was apparent. They were withdrawn from all medication and had eaten the same diet for two weeks prior to this study. After taking water for diuresis, all subjects rest for 30 min in the sitting position following which urine was collected for 1 h in the same sitting position for measurment of GAA, NAG, mAlb and prostagrandin-E2 (PG-E2). Venous blood was collected at this time in a cold tube with EDTA-2Na and centrifuged immediately at 4°C. Plasma, serum and urine were stored deep-frozen until analysis. The blood pressure was calculated as the average of at least two determinations at different times during the urine-sampling period.

Plasma renin activity (PRA), plasma aldosterone concentration (PAC), plasma angiotensin-II (Ang. II) and urinary mAlb were measured by radioimmunoassay. Serum angiotensin converting enzyme activity (ACE) was measured by a colorimetric method[6]. BMG was measured with automated photometry of Latex agglutination with the use of an integrated sphere turbidity assay system (Kyowa Medics, Tokyo, Japan). Urinary NAG was determined by a fluorometric method[7]. Urinary GAA was measured by an enzymatic determination[8].

Twelve age-matched normal subjects (one female and 11 males), aged 37–70 y (mean 47 y), were studied as controls, all of whom were without organic disease. The examination was performed according to the same protocol as tha of hypertensive patients.

All result were expressed as mean ± SD and were analysed statistically using the unpaired t-test.

RESULTS

There is no significant difference in age, Ccr, blood urea nitrogen (BUN) and creatinine (Cr) among the three groups. The serum level of electrolytes, such as Na, K, Cl and Ca, also showed no significant difference (Table 1).

The urinary excretion rate of of electrolytes and PG-E2 are shown in Table 2. In hypertensive subjects, the urinary excretion rate of Na, K and C were significantly lower than in the controls (p < 0.01). In borderline hypertensive subjects, these values were lower than in controls, but they were not significant. On the other hand, the urinary excretion rate of Ca and Cr showed no significant difference in these three groups. In borderline hypertensive patients the urinary excretion rate of PG-E2 was significantly higher than in controls (P < 0.01), but not in hypertensive patients. The ratio of urinary electrolytes to urinary Cr and that of urinary Na to K are presented in Table 3. In the hypertensive group the ratio of Na, K and Cl to Cr was significantly lower than that in the controls (U-Na/U-Cr; p < 0.05, U-K/U-Cr; p < 0.01, U-Cl/U-Cr: p < 0.05).

The paramaters of the renin-angiotensin-aldosterone system including PRA, PAC, Ang. II and ACE are presented in Table 4. There was no significant

Table 1. Clinical and laboratory findings in the three groups

	Age (years)	Ccr (ml/min.)	BUN (mg/dl)	Cr (mg/dl)	Na (mEq/L)	K (mEq/L)	Cl (mEq/L)	Ca (mEq/L)
Control N=12 (M/F=11/1)	46.5 ±8.8	113.3 ±28.6	11.6 ±1.6	0.8 ±0.1	140.3 ±2.1	4.9 ±0.5	101.4 ±2.2	4.4 ±0.1
Borderline N=6 (M/F=6/0)	45.2 ±8.1	89.1 ±20.3	12.1 ±2.1	0.9 ±0.1	139.8 ±1.5	4.9 ±0.6	100.3 ±1.6	4.3 ±0.1
Hypertension N=29 (M/F=28/1)	49.6 ±8.0	94.4 ±29.4	11.9 ±2.9	0.9 ±0.1	140.2 ±2.1	4.8 ±0.5	101.4 ±2.9	4.3 ±0.2

BUN; blood urea nitrogen
Ccr; creatininine clearance and Cr; creatinine. Mean ± SD

Table 2. Urinary findings in the three groups

	U-Na (mEq/min)	U-K (mEq/min)	U-Cl (mEq/min)	U-Ca (μEq/min)	U-Cr (mg/min)	U-PGE2 (pg/min)
Control N=12 (M/F=11/1)	0.40 ±0.19	0.06 ±0.02	0.42 ±0.21	9.48 ±9.53	1.02 ±0.25	587 ±258
Borderline N=6 (M/F=6/0)	0.35 ±0.09	0.06 ±0.02	0.39 ±0.10	7.93 ±7.64	0.92 ±0.22	1076 ±313***
Hypertension N=29 (M/F=28/1)	0.24 ±0.13 ***	0.04 ±0.02 ***	0.26 ±0.12 **	6.63 ±5.27	0.83 ±0.28	666 ±308

Cr; creatinine and PGE2; prostagrandin-E2.
Mean ± SD, p. value (vs. control); **:p < 0.05, ***:p < 0.01

Table 3. The ratios of urinary electrolytes to urinary creatinine in the three groups

	U-Na/U-Cr	U-K/U-Cr	U-Cl/U-Cr	U-Ca/U-Cr x100	U-Na/U-K
Control N=12 (M/F=11/1)	0.39 ±0.16	0.07 ±0.02	0.39 ±0.14	0.87 ±0.76	5.61 ±1.32
Borderline N=6 (M/F=6/0)	0.40 ±0.09	0.06 ±0.02	0.43 ±0.09	0.88 ±0.79	7.01 ±2.68
Hypertension N=29 (M/F=28/1)	0.28 ±0.14 **	0.05 ±0.02 ***	0.29 ±0.14 **	0.74 ±0.57	5.71 ±2.28

Cr; creatinine, Mean ± SD, p. value (vs. control); **:p < 0.05, ***:p < 0.01

difference in these factors among the three groups who showed wide variations.

The ratio of urinary GAA, BMG, NAG and mAlb to urinary Cr are shown in Table 5. The ratio of GAA to Cr was markedly lower in the borderline hypertensive and hypertensive patients than in controls (borderline group; p < 0.05, hypertensive group; p < 0.01). The ratio of mAlb to Cr was significantly higher than controls only in the hypertensive group (p < 0.05). The ratio of of NAG and BMG to Cr showed no significant difference in the three groups. Moreover the urinary excretion rate of GAA had a significant positive correlation with Ccr (r = 0.57, p < 0.01) (Fig. 1). However the urinary excretion rate of mAlb, NAG and BMG showed no correlation with Ccr (Fig. 2,3 and 4).

Table 4. Renin-angiotensin-aldosterone system in the three groups

	sBP (mmHg)	dBP (mmHg)	PRA (ng/ml/hr)	PAC (ng/dl)	Ang. II (pg/ml)	ACE (mU/L)
Control N=12 (M/F=11/1)	123 ±10	76.8 ±7.5	1.52 ±1.49	6.04 ±4.37	19.3 ±6.9	19.8 ±5.5
Borderline N=6 (M/F=6/0)	144 ±9***	87.8 ±4.8***	1.15 ±0.79	6.75 ±4.37	23.3 ±9.9	24.4 ±6.5
Hypertension N=29 (M/F=28/1)	17 ±22***	107.9 ±5.7***	1.19 ±1.08	6.88 ±3.99	26.0 ±14.7	24.9 ±6.7

sBP; systric blood pressure, dBP; diastric blood pressure, PRA: plasma renin activity, PAC; plasma aldosterne concentration, Ang. II; angiotensin − II an ACE; angiotensin conventing enzyme activity. Mean ± SD, P. value (vs. control); ***:$p < 0.01$

Table 5. Urinary excretion rates of GAA, BMG, NAG and mAlb in the three groups

	U-GAA/U-Cr x100	U-BMG/U-Cr x10000	U-NAG/U-Cr	U-mAlb/U-Cr x100
Control N=12 (M/F=11/1)	7.39 ±1.99	0.23 ±0.18	7.04 ±5.13	1.64 ±1.09
Borderline N=6 (M/F=6/0)	5.13 ±1.61**	0.22 ±0.19	8.50 ±4.24	1.46 ±1.36
Hypertension N=29 (M/F=28/1)	4.93 ±1.76***	0.19 ±0.21	7.66 ±4.50	3.15 ±3.47**

Cr; creatinine, BMG; beta-2-microglobuline, NAG; N-acetyl-D-glucosamindase and mAlb; micro-albumin. Mean ± SD, p. value (vs.control); **: $p < 0.05$, ***: $p < 0.01$

Meanwhile there were negative correlations between the urinary excretion rate of GAA and NAG ($r = -0.24$, $p < 0.05$) (Fig. 5), and between urinary excretion rate of GAA and the serum BMG ($r = -0.31$, $p < 0.05$) (Fig. 6).

DISCUSSION

All subjects in this study had been on a high Na diet containing about 25 g/day Na for at least two weeks. The chronic administration of a high Na

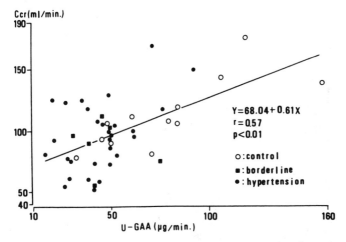

Figure 1. Relationship between creatinine clearance (Ccr) and urinary excretion rate of guanidino acetic acid (U-GAA).

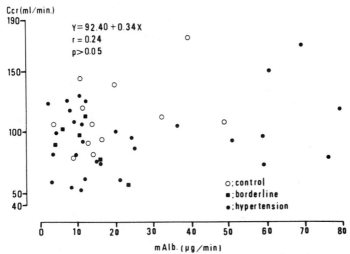

Figure 2. Relationship between creatinine clearance (Ccr) and urinary excretion rate of micro-albumin (mAlb).

diet is known to cause a state of extracellular fluid (ECF) expansion, which may suppress the renin-angiotensin-aldosterone system. The reason that the findings of this system showed no difference in hypertensive, borderline hypertensive and control groups seemed to be due to ECF expansion (Table 4).

Aldosterone, which is stimulated by Ang. II or hyperpotassemia, acts to enhance Na reabsorption and K secretion in the distal nephron. In this study PAC had a significant positive correlation with Ang. II ($r = 0.38$, $p < 0.01$), and tended to have a negative correlation with the urinary excretion of Na ($r = -0.14$, $p < 0.05$). Vasodilation of the kidney, which may be due to bradykinin, acetylcholine and prostaglandin, is usually associated with natriuresis. This study also showed that urinary PG-E2 tended to effect natriuresis ($r = 0.22$, $p < 0.05$).

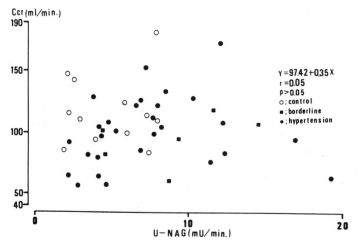

Figure 3. Relationship between creatinine clearance (Ccr) and the urinary excretion rate of N-acetyl-D-glucosaminidese (NAG).

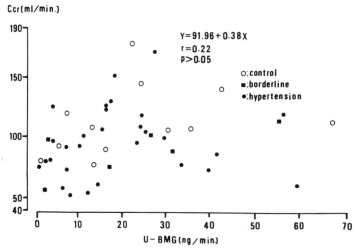

Figure 4. Relationship between creatinine clearance (Ccr) and urinary excretion rate of beta-2 microglobuline (BGM).

Elevation of the arterial pressure has a direct negative effect on the tubular Na reabsorption by means of increased peritubular hydrostatic pressure. This study showed that the urinary excretion of Na in the hypertensive group was significantly lower than that in the control (Table 2 and 3), and that it had a significant negative correlation with mean blood pressure ($r = -0.49$, $p < 0.01$) (Fig. 7). The change in glomerular filtration rate (GFR) could also induce parallel changes in urinary Na excretion. A significant positive correlation was noticed between Ccr and urinary excretion of Na in this study ($r = 0.57$, $p < 0.01$) (Fig. 8).

These findings suggest that glomerulotubular feedback may be distorted in the hypertensive state before a decreased in GFR occurs. The hypertensive

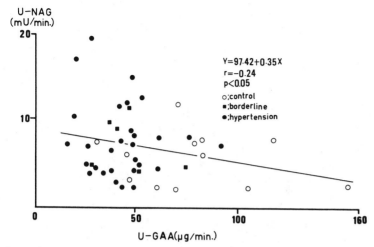

Figure 5. Relationship between urinary excretion rates of guanidinoacetic acid (GAA) and N-acetly-D-glucosaminidase (NAG).

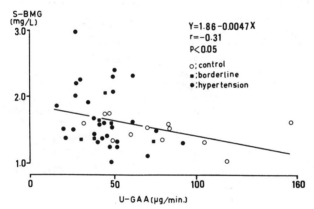

Figure 6. Relationship between urinary excretion rate of guanidinoacetic acid (GAA) and serum beta-2 microglobuline (BMG).

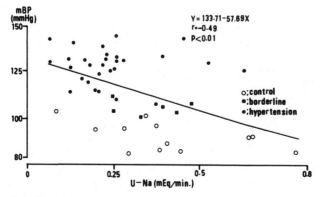

Figure 7. Relationship between mean blood pressure and urinary excretion rate of sodium.

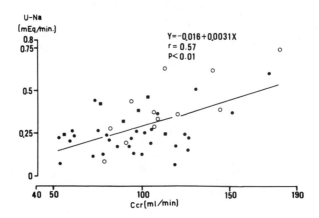

Figure 8. Relationship between creatinine clearance (Ccr) and urinary excretion rate of sodium.

state itself can cause various glomerular and tubulo-interstitial changes in
the kidney. To evaluate these we measured the urinary mAlb as a protein of
glomerular origin as well as NAG and BMG as proteins of tubular origin.

The urinary excretion of mAlb is thought to be regulated by hydraulic
pressure. It has been known that the urinary excretion of mAlb is increased
in hypertension, and that it has a significant correlation with blood pressure[4]. ACE inhibitors such as captopril, which decrease hydraulic pressure
have been reported to decrease urinary mAlb in hypertension[9]. This study
showed that the ratio of urinary mAlb to urinary Cr in hypertensive subject
was higher than in controls (p < 0.05) (Table 5). The urinary excretion of
mAlb tended to have a positive correlation with mean blood pressure (r=0.20
p < 0.05) as well as plasma Ang. II (r = 0.50, p < 0.01). However, the
urinary excretion of BMG and NAG have no relation to plasma Ang. II or mean
blood pressure.

On the other hand, the decrease in urinary GAA seen in hypertension
showed a significant negative correlation with mean blood pressure (r=0.44,
p < 0.01) (Fig. 9). Ccr had also a significant positive correlation with
urinary GAA excretion (r = 0.57, p < 0.01) (Fig. 1), but not with that of
NAG, BMG and mAlb. Urinary NAG and serum BMG had been reported to be elevate
in proximal tublar damage before the point of decreased GFR in some renal
diseases[10,11]. Since a high level of glycine amidinotransferase (GAT) is
present in the kidney, the kidney is suggested to be the primary site of GA
synthesis. Recently, the localization of GAT immmunoreactivity was confirmed
in the proximal tubules of the kidney in rats by an immunochemical study wit
monoclonal antibodies[12].

These findings suggest that the urinary excretion of GAA will be a more
useful marker of hypertensive renal damage than that of BMG, MAG or mAlb.

SUMMARY

This study was undertaken to evaluate the urinary excretion of GAA in
hypertensive patients compared to normal controls.

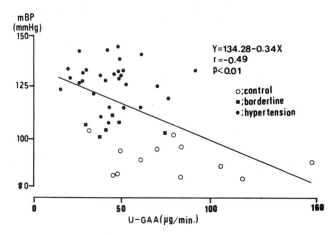

Figure 9. Relationship between urinary excretion rate of guanidinoacetic acid (GAA) and mean blood pressure.

   In hypertensive patients the urinary excretion rate of GAA was 0.042 ± 0.017 mg/min, which markedly lower than in controls (0.082 ± 0.038 mg/min) (p < 0.01). However, there was no significant difference in Ccr, serum Cr, serum BMG, urinary excretio of BMG, NAG, PG-E2 and mAlb between these groups. The urinary excretion rate of GAA had also a significant positive correlation with Ccr (r = 0.62, p < 0.01). The urinary excretion of GAA had a significant negative correlation with the urinary excretion of NAG (r=-0.24, p < 0.05) (Fig. 5) and serum BMG (r = -0.31, p < 0.05) (Fig. 6)

   These findings suggest that the urinary excretion rate of GAA may be a more useful marker of renal damage than that of BMG, NAG, and mAlb in hypertensive patients.

REFERENCE

1. Castleman, B.J. and Smithwick, R.H., The relation of vasculalar disease to the hypertensive state based on a study of renal biopsies from one hundred hypertensive patients, JAMA 121 (1943) 1256-1261.
2. Moriz, A.R. and Oldt, M.R., Arteriolar sclerosis in hypertensive and non-hypertensive individuals, Am. J. Pathol. 13 (1937) 679-728.
3. Alderman, M.H., Melcher, L., Drayer, D.E. and Reidenberg, M.M., Increased excretion of urinary N-acetyl-beta-D-glucosaminidase in essential hypertension and its decline with antihypertensive therapy, New Engl. J. Med. 309 (1983) 1213-1217.
4. Daving, H.H., Mogensen, C.E., Jensen, H.A. and Evrin, P.E., Increased urinary albumin excretion rate in benign essential hypertension, Lancet I (1974) 1190-1192.
5. Bonas, J.E., Cohen, B.D. and Natelson, S., Separation and estimation of certain guanidino compound. Application to human urine, Microchem. J. (1963) 63-77.
6. Kasahara, Y. and Ashihara, Y., Colorimetry of Angiotensin-I converting enzyme activity in serum, Clin, Chemi. 27 (1981) 1922-1925.
7. Leaback, D.H. and Walker, P.G., Studies on glucosaminidase, 4. The fluorimetric assay of N-acetly beta-glucosaminidase, Biochm. J. 78 (1961) 151-156.

8. Shirakane, Y., Utsushikawa, M. and Nakajima, M., A new enzymatic determination of guanidinoacetic acid in urine, Clin. Chem. 33 (1987) 349-397.
9. Venuto, G.D., Andreotti, C., Mattarei, M. and Pegoretti, G., Longterm
   captopril therapy at low doses reduces albumin excretion, J. Hypertension 3 (1985) 143-145.
10. Wellwood, J.M., Ellis, B.G., price, R.G., Hammond, K., Tompson, A.E. and
    Jones, N.G., Urinary N-acetyl-beta-D-glucosaminidase activites in pa-
    tients with renal disease, Br. Med. J. 3 (1975) 408-411.
11. Strober, W. and Waldmann, T.A., The role of the kidney in the metabolism
    of plasma protein, Nephron 13 (1974) 35-66.
12. McGuire, D.M., Gross, M.D., Elde, R.P. and Van Pilum, J.F., Localization
    of L-Arginine-Glycine amidinotransferase protein in rat tissue by immu-
    nofluorescence microscopy, J. Histchem. Cytochem. 34 (1986) 429-235.

GUANIDINO COMPOUNDS IN HEMORRHAGIC SHOCK

Seiji Sugimoto, Yoji Ochiai, Michio Kitaura, Koji Kabutan,
Chiaki Tsuji, Yoshinori Kosogabe and Futami Kosaka

Department of Anesthesiology and Resuscitology, Okayama
University Medical School, Okayama 700, Japan

INTRODUCTION

One of the most prominent features of hemorrhagic shock is the process
of differential vasoconstriction whereby, in order to maintain blood flow to
the brain and heart, circulation is shunted away from the skin muscles,
kidney and other splanchnic organs. Metabolic function is these organ will
be attenuated and metabolic balance between organs lost when the resulting
ischemia is prolonged. In 1981, Funahashi et al.[1] reported a relationship
between the kidney and liver regarding arginine metabolism. In this study,
we indeed hemorrhagic shock in rats and investigated its influence on the
metabolism of the liver and kidney by analyzing guanidino compounds.

MATERIALS AND METHODS

Male Wistar rats, weighing 300 g on the average, were used and given
free access to food and water. These rats were divided into two groups;
control group (C group) and hemorrhagic shock group (S group). Pentbarbital
(50 mg/kg) was intraperitoneally administered and surgical catheterization
was performed of the femoral and carotid arteries. The femoral artery catheters were used for arterial pressure monitoring and carotid artery catheters
were connected to disposable syringes and used for removing blood. Heparin
amounting to 3 mg/kg was intraarterially administered following the surgical
procedures. Rats in the C group were allowed to stand for 2 h without any
other procedures whereas the rats of the S group were gradually exsanguinated
through the carotid catheters. Their mean arterial blood presures were
maintained at 40 mmHg for 2 h by adjusting the height of the syringes. The
rats were sacrificed and the livers and kidneys perfused with cold saline
and removed immediately after their death.

ANALYSIS OF GUANIDINO COMPOUNDS

After the water in a volume 2 times the wet weight of tissues was added,
the liver and kidney were homogenized and centrifuged at 10,000 g at 5°C

for 30 min. The supernatant thus obtained (cytoplasmic and microsomal fractions) was used as a sample. Each sample was then deproteinized with trichloroacetic acid in a 10% final concentration and filtered through membrane filters. Guanidino compounds were determined by using high performance liquid chromatography, LC–3A (Shimadzu, Japan). The concentration of each guanidino compound was automatically calculated by Chromatopack C–RIA. Proteins contained in each sample were then quantitatively determined by the Biuret method, and the concentration of guanidino compounds at each protein level was obtained.

RESULTS

When guanidino compounds were analyzed, guanidinosuccinic acid (GSA), creatine (CTN), guanidinoacetic acid (GAA), guanidinobutyric acid (GBA) and arginine (Arg) showed definite peaks. For statistical analysis, mean values and standard deviations were obtained, and Student's t-tests were performed

Arg Levels

Arg levels in the liver were very low. Hepatic levels in the S group were significantly higher than in the C group. Arg levels in the kidney were higher in the S group as well.

CTN and GAA Levels

CTN and GAA levels in the liver showed no significant change. However, renal CTN levels were significantly lower in the S group and renal GAA level were appreciably higher in the S group.

GSA and GBA Levels

Hepatic GBA and GSA levels showed no significant change between the two groups. Renal GBA levels were significantly higher in the S group. Renal GSA did not show definite peaks.

Changes of CTN/GAA and Arg/GAA Ratios

Since CTN is produced in the liver by methylation of GAA and GAA is produced from Arg in the kidney, both CTN/GAA and Arg/GAA ratios were studied. However, no significant differences were recognized between the groups CTN/GAA ratios in the liver and Arg/GAA ratios in the kidney. The above findings are summarized in tables 1. and 2.

DISCUSSION

When circulating blood volume decreases by hemorrhage, peripheral vascular resistance is enhanced by activation of the sympathetic nervous system. As a result, the blood supply to the skin, muscle and kidney decreases and blood shifts to major organs such as the brain and heart. Although renal blood flow is maintained at a certain level when the blood pressure changes within a limited extent[2], it decreases rapidly and notably following massive hemorrhage. Blood flow to the liver also decreases along with a decline in cardiac output, but the ratio between hepatic blood flow and cardiac output remains unchanged, or slightly increased[3,4]. That is to say, the decrease in renal blood flow in hemorrhagic shock is larger than hepatic. It is therefore expected that changes in the distribution of blood to the liver and

Table 1. Guanidino compounds in liver tissue

	control group (n = 6)	hemorrhagic shock group (n = 6)
GSA	0.12 ± 0.04	0.15 ± 0.03
CTN	1.49 ± 0.61	1.12 ± 0.42
GAA	0.23 ± 0.02	0.03 ± 0.01
GBA	0.36 ± 0.12	0.36 ± 0.13
Arg	0.33 ± 0.09	0.68 ± 0.08**
CTN/GAA	39.96 ± 2.21	34.11 ± 11.26

**Significantly different (P<0.01) from the control. Values are expressed as nmol/mg protein (Mean ± SD). GSA; guanidinosuccinic acid, CTN; creatine, GAA; guanidinoacetic acid, GBA; guanidinobutyric acid and Arg; arginine.

Table 2. Guanidino compounds in kidney tissue

	control group (n = 6)	hemorrhagic shock group (n = 6)
CTN	13.30 ± 2.85	9.01 ± 2.67*
GAA	1.49 ± 0.60	5.19 ± 1.56**
GBA	0.18 ± 0.12	0.50 ± 0.06**
Arg	16.12 ± 4.80	38.09 ± 8.16**
Arg/GAA	8.44 ± 2.08	7.59 ± 1.30

Significantly different (*; $P < 0.05$, **; $P < 0.01$) from the control. Values are expressed as nmol/mg protein (Mean ± SD). The abbreviations of guanidino compounds are the same as shown in Table 1.

kidney will affect the metabolic relationship of these organs. Arg concentrations in the liver were very low, but hepatic Arg concentrations in the S group showed a significant increase. This was conceivably due to inhibition of the metabolism of liver by ischemia. It is said that mitochondrial functions are enhanced in the early stage of hemorrhagic shock, but inhibited when the shock become irreversible[5,6]. Since ornithine carbamoyl transferase in the urea cycle exists in hepatic mitochondria, this cycle is inhibited in hemorrhagic shock. It is conceivable that the increased Arg partially originated in the kidney, because a marked increase of GAA and Arg in the kidney was observed in the S group. In the kidney Arg is synthesized from citrulline[1]. Arg, thus synthesized, is released in the blood and again synthesized to GAA by glycine amidinotransferase. So enhanced renal metabolism of Arg and GAA in spite of ischemia might be explained as a compensation for a decline in the function of the hepatic urea cycle. It is said that disordered liver function is normalized easily soon after the removal of shock[7]. The renal function disorder, however, lasts longer and acute renal failure often develops. It is reported that in acute renal failure, GAA levels in

the liver and kidney are lower[8]. It is not known when GAA production decreases
during the transition from hemorrhagic shock to acute renal failure. Metabolic correlation between the liver and kidney in hepatic insufficiency has
been reported by several workers[9-10]. Although serum GAA levels are lower in
patients with acute renal failure without liver disorder, marked increases of
serum GAA are recognized in patients with acute hepatic failure[9]. This means
that GAA increases as citrulline metabolism in the kidney is activated in
compensation for declined liver function. However, serum CTN levels are
within normal limits as long as renal function is maintained[10], since GAA in
the liver is less converted to CTN. Nishitani et all.[11] reported their findings on the metabolic correlation between the liver and kidney in rats with
acute hepatic failure induced by D-galactosamine. They concluded that renal
metabolism is activated as a system of processing increased ammonia when the
urea cycle is inhibited in liver. They noted an increase of hepatic Arg and
serum GAA levels. Unlike the findings from our study, they reported no
appreciable increase of renal Arg and GAA levels. An increase of renal Arg
and GAA concentrations in hemorrhagic shock is possibly caused by exces-
sively enhanced metabolism rather than as compensation for a declined metabolic function in the liver. Excessive acceleration of kidney metabolism in
ischemia may be one of the factors in the development of acute renal failure
in a later stage.

SUMMARY

    We induced hemorrhagic shock in rats and the following findings were
obtained. An increase of Arg levels in liver and an increase of GAA levels
in kidney were observed in hemorrhagic shock group. These results suggest
that the metabolism of liver is inhibited and that of kidney accelerated in
hemorrhagic shock. It is therefore proposed that inhitition of the hepatic
urea cycle due to ischemia causes the acceleration of renal metabolism of
arginine and kidney compensated for liver functions.

REFERENCES

1. Funahashi, M., Kato, H., Shiosaka, S. and Nakagawa, H., Formation of
    arginine and GAA in the kidney in vivo. Their relations with the liver
    and their regulation, J. Biochem. 89 (1981) 1347-1356.
 2. Abe, y., Intrarenal blood flow and distribution and autoregulation of
    renal blood flow and glomelular filtration rate, Jpn. Circ. J. 35 (1971)
    1163-1173.
 3. Smith, L.L. and Veragut, U.P., The liver and shock, Prog. Surg. 4 (1964)
    55-107.
 4. Greenway, C.V. and Stark, R.D., Hepatic vascular bed, Physiol. Rev. 51
    (1971) 23-65.
 5. Baue, A.E. and Sayeed, M.M., Alterlations in the functional capacity of
    mitochondria in hemorrhagic shock, Surg. 68 (1970) 40.
 6. Hift, H. and Strawitz, J.G., Irreversible shock in dogs: structure and
    function of liver mitochondria, Am.J.Physiol. 200 (1961) 264.
 7. Donohoe, M.J., Rush, B.F. and Machiedo, G.W., Biochemical and morphologi
    change in hepatocytes from the shock injured liver, Surg. Gynecol.
    Obstet. 162 (1986) 323-333.
 8. Koide, H. and Azushima, C., Metabolic profiles of guanidino compounds in
    various tissues of uremic rats, in:"Guanidines," A. Mori, B.D. Cohen and
    A. Lowenthal Eds., Plenum Press, New York (1985) pp. 365-372.

9. Muramoto, H., Tofuku, Y., Kuroda, M. and Takeda, R., The role of the liver as a significant modulator of the guanidinoacetic acid level in man, in: "Guanidines," A. Mori, B.D.Cohen and A. Lowenthal Eds., Plenum Pres, New York (1985) pp. 105-112.

10. Ochiai, Y., Matsuda, R., Nishitani, K., Kosagabe, Y., Abe, S., Itano, Y., Yamada, T. and Kosaka, F., Metabolic change of guanidino compounds in acute renal failure complicated with hepatic disease, in: "Guanidines," A. Mori, B.D. Cohen and A. Lowenthal Eds., Plenum Press, New York (1985) pp. 317-325.

11. Nishitani, K., Metabolic role of the kidney in acute hepatic failure, Okayama Igakkai Zasshi 98 (1986) 135-143.

EFFECT OF GUANIDINO COMPOUNDS ON GABA-STIMULATED CHLORIDE INFLUX INTO
MEMBRANE VESICLES FROM RAT CEREBRAL CORTEX

Takaaki Obata, Akitane Mori* and Henry I. Yamamura

Department of Pharmacology, University of Arizona Health
Sciences Center, Tucson, Arizona, U.S.A. and *Institute for
Neurobiology, Okayama University Medical School, Okayama,
Japan

INTRODUCTION

More than one hundred guanidino compounds are known to occur naturally,
and the physiological role of some these compounds has been elucidated[1].
Since some of the guanidino compounds have the ability to induce seizures or
convulsions in animals, it has been suggested that guanidino compounds may
play a role in neurological disorders such as epilepsy and tetanus. Recently,
electrophysiological studies indicated that guanidino compounds such as γ-
guanidinobutyric acid, guanidinoethanesulfonic acid and δ-guanidinovaleric
acid induced spike discharges on the electroencephalogram[2-5]. Furthermore,
spike discharges induced by δ-guanidinovaleric acid were antagonized by γ-
aminobutyric acid (GABA) and muscimol, suggesting that δ-guanidinovaleric
acid may be an endogeneous GABA receptor antagonist[5]. GABA is known to be an
inhibitory neurotransmitter in the mammalian central nervous system and to
exert its effect by increasing chloride conductance in neuronal membranes. In
spite of cumulative evidence that convulsants such as bicuculline, picrotoxin, t-butylbicyclophosphorothionate and neuroinsecticides inhibit the
GABA-gated chloride channel[6,7], it is unknown whether guanidino compounds
have influence on the GABA-gated chloride channel. Thus, to elucidate the
mechanism of action of the guanidino compounds, by the use of a newly developed method[6], we examined the effect of ten guanidino compounds possessing
convulsant activity on GABA-stimulated $^{36}Cl^-$ uptake by membrane vesicles from
rat cerebral cortex.

METHODS

Preparation of Membrane Vesicles

Membrane vesicles from rat cerebral cortex were prepared by the procedure of Harris and Allan[6] with minor modifications. Male Sprague-Dawley rats
(200-250 g) were decapitated and their brains were removed. Cerebral cortices
were rapidly dissected and homogenized by hand (12 strokes) using a glass-
glass homogenizer in 10 vol (w/v) of ice-cold buffer (145 mM NaCl, 5 mM KCl,

1 mM $MgCl_2$, 10 mM D-glucose, 1 mM $CaCl_2$ and 10 mM HEPES adjusted to pH 7.5 with Tris base). The homogenate was centrifuged at 1,000 x g for 15 min at 4°C. The supernatant was decanted and the pellet resuspended in 10 vol (w/v) of buffer was centrifuged at 1,000 x g for 15 min at 4°C. The final pellet was resuspended in buffer to a final protein content of 8-9 mg/ml[8].

Measurement of $^{36}Cl^-$ Uptake

Aliquots of the membrane vesicle suspension (200 μl) were preincubated for 10 min at 30°C. After preincubation, $^{36}Cl^-$ uptake was initiated by the addition of 200 μl of a solution containing $^{36}Cl^-$ (0.2 μCi/ml) and GABA. After incubation for 3 s, uptake was terminated by the addition of ice-cold buffer (4 ml x 2) followed by rapid vacuum filtration through Whatman GF/C glass microfiber filter pretreated with 0.05% polyethylenimine. Filters were washed with ice-cold buffer (1 ml x 10) and $^{36}Cl^-$ content of the filters was determined by liquid scintillation counting. Guanidino compounds were preincubated with membrane vesicles for 10 min at 30°C and examined for their ability to inhibit $^{36}Cl^-$ uptake stimulated by 30 μM GABA.

RESULTS

GABA stimulated $^{36}Cl^-$ uptake by membrane vesicles from rat cerebral cortex in a concentration-dependent manner at 3-300 μM and the EC_{50} value for GABA was 37 μM (Fig. 1). The effect of ten guanidino compounds on GABA (30 μM)-stimulated $^{36}Cl^-$ uptake was examined. Their structures and inhibitor potencies are shown in Figure 2 and Table 1, respectively. The derivatives of guanidine and arginine (1-4) showed weak inhibition on GABA-stimulated $^{36}Cl^-$ uptake at 100 μM. Among guanidinocarboxylic acids, δ-guanidinovaleric acid (5) showed a 43.4% of significant inhibition on GABA-stimulated $^{36}Cl^-$ uptake at 100 μM. γ-Guanidinobutyric acid (6) and guanidinoacetic acid (7) were less potent than δ-guanidinovaleric acid. Guanidinosuccinic acid (8) and α-guanidinoglutaric acid (9) had no effect on this response at 100 μM. Guanidinoethanesulfonic acid (10) showed a 34.0% significant inhibition on GABA-stimulated $^{36}Cl^-$ uptake at 100 μM. In contrast to guanidino compounds, bicuculline, a GABA antagonist and picrotoxin, a GABA-gated chloride channel blocker potently inhibited GABA-stimulated $^{36}Cl^-$ uptake by 83.0% and 73.6% at 1 μM, respectively. δ-Guanidinovaleric acid was the most potent inhibitor of guanidino compounds tested in the present study, but more than 100 times less potent than bicuculline and picrotoxin.

DISCUSSION

Guanidino compounds tested in the present study have been found to induce seizures or convulsions in animals by intracisternal or intraventricular injections. However, the mechanism of the convulsant action by guanidino compounds remains unclear. A number of convulsants such as bicuculline, picrotoxin, t-butylbicyclophosphorothionate and neuroinsecticides are considered to exert their pharmacological effects by inhibiting the GABA-gated chloride channel[6,7]. Thus, the inhibitory potencies of ten guanidino compounds on the GABA-gated $^{36}Cl^-$ uptake by membrane vesicles from rat cerebral cortex were examined.

Among the guanidino compounds tested, δ-guanidinovaleric acid and guanidinoethenesulfonic acid significantly inhibited GABA-stimulated $^{36}Cl^-$ uptake at 100 μM. Other compounds showed weak inhibition or no effect on GABA-stimulated $^{36}Cl^-$ uptake. δ-Guanidinovaleric acid was the most potent inhi-

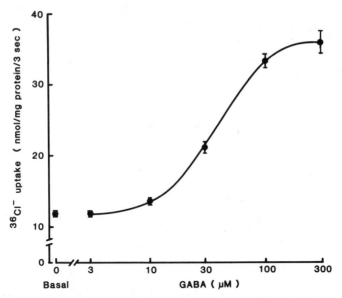

Figure 1. Concentration-response curve for GABA on $^{36}Cl^-$ uptake by membrane
vesicles from rat cerebral cortex. Each point represents the
mean ± SEM of five different preparations.

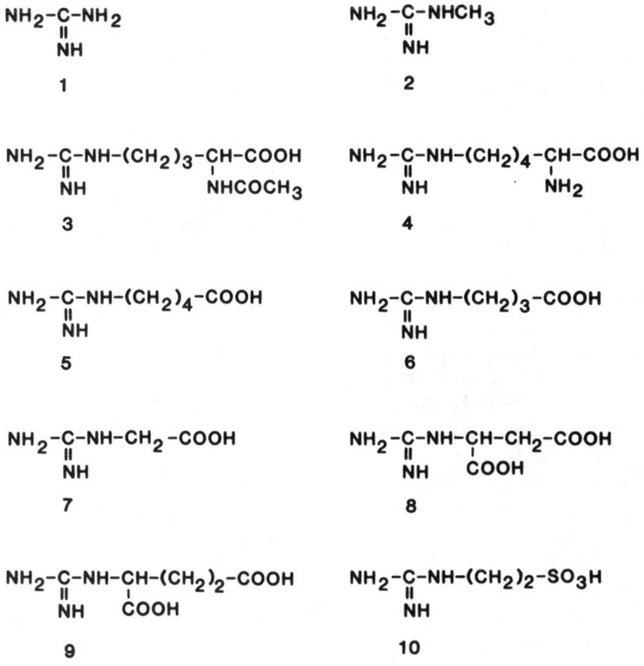

Figure 2. Structures of the guanidino compounds. Compounds numbers refer to
names given in Table 1.

Table 1. Effect of guanidino compounds, bicuculline and picrotoxin on GABA
(30 μM)-stimulated $^{36}Cl^-$ uptake by membrane vesicles from rat
cerebral cortex.

Compounds		Concentrations	% Inhibition of GABA-stimulated $^{36}Cl^-$ uptake
No.	Names	(μM)	
1	Guanidine	100	19.3
2	Methylguanidine	100	4.4
3	N-Acetylarginine	100	11.3
4	Homoarginine	100	12.6
5	δ-Guanidinovaleric acid	100	43.4*
6	γ-Guanidinobutyric acid	100	23.9
7	Guanidinoacetic acid	100	23.6
8	Guanidinosuccinic acid	100	1.0
9	α-Guanidinoglutaric acid	100	1.0
10	Guanidinoethanesulfonic acid	100	34.0*
	Bicuculline	1	83.0**
	Picrotoxin	1	76.3**

*,**) Significantly different from control at $p < 0.05$ and $P < 0.01$,
respectively.

bitor. Among the guanidinocarboxylic acids, guanidinoacetic acid and
γ-guanidinobutyric acid were expected to inhibit GABA-stimulated $^{36}Cl^-$ uptake
potently due to the structure similarity to GABA[9]. However, these compounds
were less potent than δ-guanidinovaleric acid and guanidinoethanesulfonic
acid. It is difficult to determine the structure-activity relationship of
guanidino compounds for GABA-stimulated $^{36}Cl^-$ uptake at this present time.

δ-Guanidinovaleric acid is thought to act on the GABAergic system
because of the structure similarity to GABA and the antagonism by GABA and
muscimol of the induction of spike discharges[4,5]. Although the inhibitory
potency of δ-guanidinovaleric acid for GABA-stimulated $^{36}Cl^-$ uptake was more
than 100 times less potent than those of bicuculline and picrotoxin which act
at different sites on the GABAergic system, our results support the above
speculation.

Guanidinoethanesulfonic acid is known to be a competitive inhibitor of
taurine uptake and pathophysiological increase was observed in the cerebrospinal fluid of epileptic patients[2]. Like GABA, taurine is an inhibitory
amino acid possessing antiepileptic activity. But taurine had little effect
on chloride influx into mouse brain vesicles[10]. Therefore, the inhibitory
potency of guanidinoethanesulfonic acid appears to be independent of the
antagonist activity of taurine. Considering the chemical eqivalence of sulfonic acid to carboxylic acid and the structure similarity to GABA, guanidinoethanesulfonic acid appears to act on a GABA receptor.

In conclusion, among the guanidino compounds tested, δ-guanidinovaleric
acid and guanidinoethanesulfonic acid showed significant inhibition on GABA-
stimulated $^{36}Cl^-$ uptake by membrane vesicles from rat cerebral cortex at 100
μM. The convulsant activity of δ-guanidinovaleric acid and guanidinoethane-

sulfonic acid may be mediated by inhibition of the GABA-gated chloride channel through the GABA receptors.

REFERENCES

1. Robin, Y. and Marescau, B., Guanidines: Historical, biological, bio-chemical and clinical aspects of the naturally occurring guanidino compounds, in: "Guanidines," A. Mori, B.D. Cohen and A. Lowenthal Eds., Plenum Press, New York (1985) pp. 383-438.
2. Mori, A., Watanabe, Y. and Akagi, M., Guanidino compound anomalies in epilepsy, in: "Advances in Epilepsy : XIIIth Epilepsy International Symposium," H. Akimoto, H. Kazamatsuri, M. Seino and A. Ward Eds., Raven Press, New York (1982) pp. 347-351.
3. Mizuno, A., Mukawa, J., Kobayashi, K. and Mori, A., Convulsive activity of taurocyamine in cats and rabbits, IRCS Med. Sci. 3 (1975) 385.
4. Shindo, S., Tsuruta, K., Yokoi, I. and Mori, A., Synthesis of δ-guanidinovaleric acid and its effect on EEG of rats, Neurosciences 10 (1984) 177-182.
5. Yokoi, I., Tsuruta, K., Siraga, H. and Mori, A., δ-Guanidinovaleric acid as an endogeneous and specific GABA-receptor antagonist: Electroenceph-alographic study, Epilepsy Res. 1 (1987) 114-120.
6. Harris, R.A. and Allan, A.M., Functional coupling of γ-aminobutyric acid receptors to chloride channels in brain membranes, Science 228 (1985) 1108-1110.
7. Bloomquist, J.R. and Soderlund, D.M., Neurotoxic insecticides inhibit GABA-dependent chloride uptake by mouse brain vesicles, Biochem. Biophys. Res. Commun. 133 (1985) 37-43.
8. Lowry, O.H., Rosebrough, N.J., Farr, A.L. and Randall, R.J., Protein measurement with the Folin phenol reagent, J. Biol. Chem. 193 (1951) 265-275.
9. Takeguchi, A. and Takeguchi, N., The structure-activity relationship for GABA and related compounds in the crayfish muscle, Neuropharmacol. 14 (1975) 627.
10. Allan, A.M. and Harris, R.A., γ-Aminobutyric acid agonists and antago-nists alter chloride flux across brain membranes, Mol. Pharmacol. 29 29 (1986) 497-505.

EFFECT OF IMMERSION STRESS ON GUANIDINO COMPOUNDS IN RAT ORGANS AND URINE

Hideki Sugi, Isao Yokoi, Yoko Watanabe and Akitane Mori

Department of Neurochemistry, Institute for Neurobiology
Okayama University Medical School, 2-5-1 Shikatacho
Okayama 700, Japan

INTRODUCTION

Some pathological conditions have been reported to cause variations of guanidino compound levels in human and experimental animals. Guanidino-succinic acid[1] and methylguanidine[2] increase in blood and urine of patients with renal failure. It is also observed that several guanidino compounds accumulate in patients with hyperargininemia which results from a deficency of arginase[3]. These observations suggest that guanidino compounds are closely related to aspects of nitrogen metabolism involving arginine (Arg) metabolism. Further, significant variations of both guanidinoethanesulfonic acid and creatinine (CRN) were observed in the brains of experimentally convulsive mice[4]. Chronic alcohol administration brought on variations of guanidino compound levels in rat organs, specifically a remarkable increase of homoarginine (HArg) in the kidneys[5].

On the other hand, various types of stressful stimuli have been shown to affect the functions of the central nervous system. It is reported that an acute stress enhances the turnover of noradrenaline[6] and serotonin[7] in the brain. These changes in the central nervous system result in various functional changes in peripheral organs through hormonal regulation and action of the autonomic nervous system. In this study, we examined the effect of immersion stress on guanidino compound levels in rat organs.

EXPERIMENTAL METHODS

Animals

Male Sprague-Dawley rats weighing 160-180g were used in all experiments. They were maintained on food and water ad libitum in a light (lights on 07:00-19:00) and temperature (25 ± 2°C) regulated room for one week before the experiment.

159

Immersion Stress

Immersion stress was given to rats according to a modified Porsolt's method[8]; i.e. rats were individually forced to swim inside a glass beaker (height: 27cm; diameter: 17.5cm) containing 15cm of water maintained at 25°C After 15min in the water they were removed, allowed to dry for 15min in a heated enclosure (32°C) and returned to their home cages. After 24h (2nd day), they were placed in the beaker again. After 5min or 60min immersion in the water, the rats were killed by microwave irradiation (5kW, 1.5s), and the brain, liver, kidney and pancreas were removed. The brains were dissected into the following 5 parts, cerebral cortex, hippocampus, striatum, midbrain + pons-medulla oblongata and cerebellum. Non-immersed rats were used as control animals. All animals were killed between 13:30-15:00 to avoi the change of contents in guanidino compounds produced by circadian rhythms.

To collect 24hr urines, rats were kept in metabolic cages. The urine collecting vessels contained a sodium piperacillin solution to control bacteria. A control 24hr urine was collected just before each rat was given immersion stress. The 24hr urine after the 60min immersion on the second day was used as the experimental urine.

Determination of Guanidino Compounds

Each weighed tissue sample was homogenized with 10 volumes of 1% picric acid. After centrifugation, the supernatant was passed through a column of Dowex 2X8 (Cl⁻ form) to remove excess picric acid. The colorless eluate was evaporated to dryness and the residue was dissolved in dilute HCl solution (pH 2.2) for guanidino compound analysis.

Urine were deproteinized by adding an equal volume of 20% trichloro-acetic acid. After centrifugation the supernatant was used for guanidino compound analysis.

Guanidino compounds were analyzed fluorometrically with an HPLC system (Jasco, Tokyo, Japan)[9], based on the reaction with 9,10-phenanthrenequinone.

RESULTS

Behavior of Rats

Rats which were placed in the beaker for the first time struggled violently, vigorously swimming around and scrabbling at the walls. After 2-3min their activity began to subside being interspersed with phases of immobility thrusting only their face, above the surface of the water and floating with their limbs. Ten min after, the duration of immobility reached a plateau where the rats remained immobile for approximately 75% of the time and this persisted for up to 15min thereafter on the first day. On the second day the rats rapidly become immobile only 2-3min after exposure to the water and remained so for 75% of the time. The immobility continued up to 60min thereafter.

Effect of Immersion Stress on Guanidino Compounds in Rat Organs

The effects of immersion stress on guanidino compounds in rat organs are shown in Fig. 1. After the 5min immersion, CRN levels increased in the cerebral cortex and HArg in the cerebellum. Other guanidino compounds did

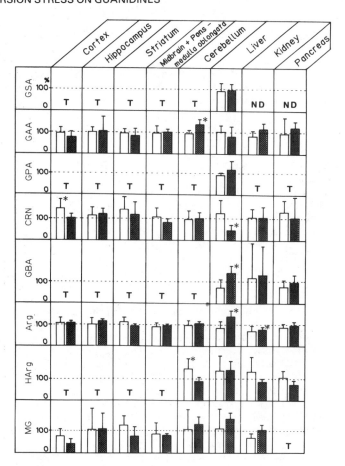

Figure 1.   Effect of immersion stress on guanidino compounds in the rat
            brain regions, liver, kidney and pancreas.
            The data are expressed as percentage of control values (mean ± SD).
            ☐ 5min immersion (n=9), ▨ 60min immersion (n=6).
            ND; not detected, T; trace. Statistical analysis was by ANOVA
            using original data. *p < 0.05, **p < 0.01 compared to control.
            The abbreviations of guanidino compounds are the same as shown
            in Table 1.

not change in the cerebral cortex and cerebellum. No significant changes
were observed in the hippocampus, striatum, midbrain + pons-medulla oblongata,
liver, kidney and pancreas after the 5min immersion. After the 60min immersion guanidinoacetic acid (GAA) increased in the cerebellum, but other guanidino compounds did not change in content. There was no significant change
of guanidino compound levels in the cerebral cortex, hippocampus, striatum
and midbrain+pons-medulla oblongata. In the liver, CRN levels decreased and
γ-guanidinobutyric acid (GBA) and Arg levels increased after the 60min immersion. Arg level significantly decreased in the kidney, but other guanidino
compounds did not change after the 60min immersion. No change was observed in
the pancreas after the 60min immersion.

Effect of Immersion Stress on the Urinary Guanidino Compounds

Table 1. shows the 24hr excretion of guanidino compounds in rat urines. Only GBA levels were found to be significantly increased after the 60min immersion stress. Other guanidino compounds were not affected by the immersion stress.

DISCUSSION

Porsolt et al.[8] reported that immobility in water reflects a stage of lowered mood in the rat which is selectively sensitive to antidepressant treatment. It was shown that the duration of immobility in a 5min exposure to water on the second day is reduced by various antidepressant agents[8-10] and other treatments which are generally thought to be effective in depression, such as electroconvulsive shock, REM sleep deprivation and "enrichment" of the environment. We observed that the rat in water remained immobile up to 60min on the second day. Therefore we examined both effects of mild (5min) and more intense (60min) immersion stress on guanidino compounds in rat organs.

CRN in the cerebral cortex and HArg in the cerebellum transiently increased after the 5min immersion stress but they fell to control levels after the 60min immersion. GAA in the cerebellum, which was not affected after the 5min immersion, markedly increased after the 60min immersion stress. These variations seem specific in the different brain regions, although guanidino compound levels are almost the same in these regions. Further it is noted that these changes in the brain did not relate to the changes observed in the kidney and liver which are major organs responsible for guanidino compound metabolism. It is conceivable that there is a selective transport system for guanidino compounds in both uptake from and release to blood or cerebrospinal fluid. CRN is the final metabolite of creatine phosphate which is ultimately excreted via the kidney. Our finding of an

Table 1. Effects of immersion stress on the excretion of guanidino compounds in urine

	Control	Stress
GSA	0.655 ± 0.332	0.667 ± 0.252
GAA	9.300 ± 2.450	10.542 ± 3.624
GPA	0.022 ± 0.018	0.017 ± 0.010
CRN	42.953 ± 6.271	48.562 ± 13.472
GBA	3.413 ± 0.598	4.798 ± 0.728*
Arg	0.838 ± 0.169	0.840 ± 0.106
HArg	0.071 ± 0.050	0.066 ± 0.020
GEt	0.004 ± 0.001	0.004 ± 0.001
Gua	0.792 ± 0.149	0.919 ± 0.205
MG	0.233 ± 0.056	0.271 ± 0.061

Values are expressed as μmol/24 hr urine (mean ± SD, n=6), statistical analysis was by paired t-test. *p < 0.05 compared to control. GSA; guanidinosuccinic acid, GAA; guanidinoacetic acid, GPA; ß-guanidinopropionic acid, CRN; creatinine, GBA; γ-guanidinobutyric acid, Arg; arginine, HArg; homoarginine, GEt; 2-guanidinoethanol, Gua; guanidine and MG; methylguanidine increase of CRN in the cerebral cortex indicates that it is not transported from brain tissue to circulation immediately after its formation. However, the possibility exists that the synthesis of these compounds in brain might be enhanced by stress. Some enzymes involving guanidino compound metabolism, such as arginase[11] and amidinotransferase[12], were reported to exist in brain although their activities are much lower than those in liver and kidney. The relation between the variation of guanidino compounds in the brain regions and central neuronal function is obscure.

It was observed that Arg increased in the liver, whereas it decreased in the kidney after the 60min immersion stress. Arg is synthesized from citrulline by argininosuccinate synthetase and argininosuccinase in the kidney[13,14], and utilized as a precursor of creatine biosynthesis or supplied to various organs for protein synthesis. Arg is also synthesized in the liver by the same enzymes, but it is rapidly cleaved to ornithine and urea by the potent activity of arginase. Therefore our results, variations of Arg in the liver and kidney, may suggest that breakdown of protein was enhanced in the liver by stress. However, the enhancement of protein breakdown seems transient since Arg in the 24hr urine was not detected.

It is well known that CRN is an end product of protein catabolism and that the daily out put of CRN is almost constant. From another point of view, CRN is synthesized non-enzymatically from creatine, which is formed from creatinephosphate after releasing its energy for muscle movement. During immersion the movement of the rat is depressed so the catabolism of creatinephosphate for muscle energy remains at a low level. In that case, CRN excretion should be reduced. However, since CRN in the 24hr urine was not affected after the 60min immersion stress, minor CRN changes due to a decrease of creatine synthesis in the liver, in which the CRN content was far lower than in the urine, should be obscured by the large amount of CRN in the urine. Further detailed examinations will be needed to elucidate the effect of stress on guanidino compound metabolisms and determine the effects of hormonal and autonomic nerve regulation.

We observed that GBA levels increased in the liver and 24hr urine after the 60min immersion. This result suggests that the immersion stress cause enhancement of GBA synthesis. Two synthetic pathways of GBA are proposed[15]; i.e. one by non-enzymatic oxidative decarboxylation of α-keto-δ-guanidino-valeric acid which is formed by the oxidative deamination of Arg and the other by synthesis from Arg and γ-aminobutyric acid (GABA) via a transamidination reaction. Which pathway is responsible for the synthesis of GBA in the rat is still obscure. However, an increase in Arg might result in an increase of GBA in liver through the former pathway. Further, it is reported that transaminase activities in serum elevate in stressfull circumstances[16]. It should be also considered that the transamidination reaction from Arg to GABA might be activated by immersion stress. GABA which is an inhibitory neurotransmitter and abundant in vertebrate central nervous systems is catalyzed by an L-glutamic acid decarboxylase. However the decarboxylation of L-glutamic acid has been shown to occur in several nonneuronal tissues such as the kidney and liver[17]. This observation suggests the possibility that GABA is synthesized and utilized for GBA formation in these organs.

SUMMARY

We examined the effect of immersion stress on the guanidino compounds in rat organs and urine.

1. CRN in the cerebral cortex and HArg in the cerebellum increased after
   the 5min immersion stress, and GAA increased in the cerebellum after
   the 60min immersion stress.
2. Five min immersion stress did not affect the guanidino compounds in the
   liver, kidney and pancreas. However, after the 60min immersion stress
   Arg and GBA increased in the liver, CRN decreased in the liver and Arg
   decreased in the kidney.
3. GBA excretion increased in the urine after the 60min immersion stress.

These findings suggest that guanidino compounds in the brain are affected in a manner independent of that in other organs by immersion stress, and
that long term immersion stress affects guanidino compound metabolism in the
liver and kidney implicating a break down of protein. Furthermore it is indicated that immersion stress enhances the biosynthesis of GBA in the liver.

REFERENCES

1. Cohen, B.D., Stein, I.M. and Bonas, J.E., Guanidinosuccinic aciduria in
   uremia. A possible alternate pathway for urea synthesis, Am. J. Med.
   45 (1968) 63-68.
2. Giovannetti, S., Balestri, P.L. and Barsotti, G., Methylguanidine in
   uremia, Arch. Intern. Med. 131 (1973) 709-713.
3. Marescau, B., Qureshi, I.A., De Deyn, P., Letarte, J., Ryba, R. and
   Lowenthal, A., Guanidino compounds in plasma, urine and cerebrospinal
   fluid of hyperargininemic patients during therapy, Clinica Chimica Acta
   146 (1985) 21-27.
4. Hiramatsu, C., Hiramatsu, M., Watanabe, Y., Katayama, Y. and Mori, A.,
   Variations of brain guanidino compounds by pentylenetetrazol-induced
   seizure and ECS, Neurosciences (Kobe) 6 (1980) 112-114 (in Japanese).
5. Yokoi, T., Toma, J. and Mori, A., Effect of chronic alcohol administration on the concentrations of guanidino compounds in rat organs, in:
   "Guanidines," A. Mori, B.D. Cohen and A. Lowenthal Eds., Plenum Press,
   New York (1985) pp. 249-261.
6. Tanaka, M., Kohno, Y., Nakagawa, R., Ida, Y., Takeda, S. and Nagasaki,
   N., Time-related differences in noradrenaline turnover in rat brain
   regions by stress, Biochem. Behav. Pharmac. 16 (1982) 315- 319.
7. Morgan, W.W., Rudeen, P.K. and Pfeil, K.A., Effect of immobilization
   stress on serotonin content and turnover in regions of the rat brain,
   Life Sci. 17 (1975) 143-150.
8. Porsolt, R.D., Anton, G., Blavet, N. and Jalfre, M., Behavioural
   despair in rats: A new model sensitive to antidepressant treatments,
   Eur. J. Pharmac. 47 (1978) 379-391.
9. Higashidate, S., Maekubo, T., Saito, M., Senda, M. and Hoshino, T., New
   high-speed fully automated guanidino compound analyzer, in:"Guanidines,"
   A. Mori, B.D. Cohen and A. Lowenthal Eds, Plenum Press, New York (1985)
   pp. 3-13.
10. Ogawa, N., Mizuno S., Mori, A., Nukina, I, Ota, Z. and Yamamoto, M.,
    Potential anti-depressive effects of thyrotropine releasing hormone
    (TRH) and its analogues, Peptides 5 (1984) 743-746.
11. Chang, J.S., Effects of sodium valproate on guanidino compounds
    unrelated hyperammonemia in mouse organs, Neurosciences 11 (1985)
    323-334.
12. Van Pilsum, J.F., Stephens, G.C. and Taylor, D., Distribution of crea-
    tine, guanidinoacetate and the enzymes for their biosynthesis in the
    animal kingdom, Biochem. J. 126 (1972) 325-345.
13. Ratner, S., Enzymes of arginine and urea synthesis, in: "Advances in
    Enzymology, "A. Meister Ed., vol 39, John Wiley and Sons, New York (1973).

14. Funahashi, M., Kato, H., Shiosaka, S. and Nakagawa, H., Formation of arginine and guanidinoacetic acid in the kidney in vivo, J. Biochem. 89 (1981) 1347-1356.
15. Robin, Y. and Marescau, B., Natural guanidino compounds, in:"Guanidines," A. Mori, B.D. Cohen and A. Lowenthal Eds., Plenum Press, New York (1985) pp. 383-438.
16. Namiki, M. and Sekiya, C., Experimental approach to stress-physiologically, Igaku-no-ayumi 125 (1983) 338-345 (in Japanese).
17. Lancaster, G., Mohyuddin, F., Scriver, C.R. and Whelan, D.T., A γ-aminobutyrate pathway in mammalian kidney cortex, Biochem. Biophys. Acta 297 (1973) 229-240.

IV. INVOLVEMENT OF SEIZURE MECHANISM

EFFECTS OF ANTICONVULSANTS ON CONVULSIVE ACTIVITY INDUCED BY 2-GUANIDINO-ETHANOL

Isao Yokoi, Akira Edaki, Yoko Watanabe, Yoshihisa Shimizu, Hiroko Toda and Akitane Mori

Department of Neurochemistry, Institute for Neurobiology
Okayama University Medical School, Okayama 700, Japan

INTRODUCTION

In the animal brain, the presence of more than 14 kinds of guanidino compounds has been reported[1]. The study of the role of guanidino compounds in epilepsy started from the report of guanidine-like substances in the blood in essential epilepsy by Murray and Hoffmann[2]. In the early reports of experimental epilepsy induced by guanidino compounds, we observed an increase in γ-guanidinobutanoic acid (GBA), i.e., γ-amidino-GABA, in the brains of cats during convulsions induced by pentylenetetrazole (PTZ), and we showed that GBA induces convulsions after intracysternal injection in cats[3]. The amidine group of arginine is transferred to GABA by a transamidination reaction catalyzed by glycine amidinotransferase to form GBA, so it is suspected that the GABAergic neurotransmission system in the brain participates in the induction of convulsions by GBA[3]. Thereafter, we reported that brain guanidinoethanesulfonic acid (GES) increases prior to convulsions induced by PTZ and decreases during convulsions induced by electrical stimulation[4], and that brain α-guanidinoglutaric acid (GGA) increases during convulsions induced by cobalt implantation[5~7]. We have also demonstrated a role for GABAergic neurotransmission in the development of spike discharges induced by δ-guanidinovaleric acid (DGVA) that is α-deamino-Arg or the analog of GBA having one more carbon in its chain[8´9].

2-Guanidinoethanol (GEt) was identified in urine from rats, mice, rabbits, cats and humans[10]. GEt, is formed from Arg and ethanolamine[11] and has a molecular configuration similar to GABA making it suspect as an epileptogenic.

In this report, we observed the epileptogenecity of GEt and the effects of GABA, GABA agonists and some anticonvulsants on the spike discharges induced by GEt in rats in order to investigate the mechanism involved in the development of spike discharges.

MATERIALS AND METHODS

Male Sprague-Dawley rats weighing 250-400 g were immobilized with suc-
cinyl choline chloride following tracheal intubation and artificial respir-
ation. Four electrodes were placed epidurally at the following sites: 2 mm
anterior to the coronal suture and 2 mm on either side of the sagital suture
and 6 mm behind the coronal suture and 3 mm on either side of the sagital
suture. EEGs were recorded with a model ME95D electroencephalograph (Nihon
Kōden, Tokyo) from three bipolar leads (from left frontal (LF) to left occipital (LO), from right frontal (RF) to right occipital (RO), and from LF to
RF) and four unipolar leads (LF- indifferent electrode (E), LO-E, RF-E, and
RO-E). Electrocardiograms were also recorded simultaneously. For the topical
application of substances, a trephine hole, 3 mm in diameter, was made over
the left sensorimotor cortex, and the dura mater was removed (Fig. 1, lowest
left corner). Through this hole, filter-paper (Toyo Filter Paper No. 2, 2x2
mm) soaked with 10μl of the different drug solutions was placed on the exposed cortex. All surgical preparations were completed under light ether
anesthesia. Each animal was allowed to recover for at least 2 h after preparation until EEGs showed no sign of anesthesia.

To evaluate the suppressing effect on seizure discharges by GABA, GABA
agonists or anticonvulsants, drugs were applied in two ways: 1) After-test;
in this test, each drug was applied after spike discharges induced by GEt had
stabilized at their plateau level (about 20-40 min after the GEt application)
and 2) Before-test; in this test, each drugs was applied 20 to 40 min prior
to the GEt application.

GEt was synthesized from 0-methylisourea sulfate (Calbiochem., USA) and
ethanolamine according to Weiss and Krommer[12]. GEt (300mM) was dissolved in
water and neutralized with HCl. GABA (100mM), (3R)-(-)-4-amino-3-hydroxy-
butanoic acid (L-GABOB) (100mM) (Ono Pharmaceutical Co., Ltd., Japan), and
muscimol (5mM)(Sigma Chemical Co., St. Louis, MO), were dissolved in physiological saline solution for topical application. Diazepam (DZP) (10mg/kg)
(Horizon Inj., Yamanouchi Pharmaceutical Co., Ltd., Japan), phenytoin (PHT)
(25mg/kg) (Aleviatin Inj., Dainippon Pharamaceutical Co., Ltd., Japan), and
sodium valproate (VPA) (200mg/kg) (dissolved in saline), were injected intraperitoneally. Phenobarbital (PB) (20mg/kg) (10% Phenobal Inj., Sankyo Co.
Ltd., Japan) was injected subcutaneously.

A group of 5 to 6 rats was used for each experiment.

RESULTS

Sporadic spike discharges began 2-10 min after GEt (3μmol) application
on the pia mater of the sensorimotor cortex on the same side as the application (Fig. 1, B), and thereafter polyspikes appeared and continued for 20
seconds. Polyspikes were repeated every 2-5 min, and then after the onset of
one group of polyspikes from the side where GEt was applied, polyspikes rose
from the opposite cerebral hemisphere (Fig. 1, C). The group of polyspikes
from the opposite hemisphere stopped when those from the same side did. When
the next group of polyspikes from applied side appeared, polyspikes from the
contralateral hemisphere started with latency. Polyspikes from both sides of
the hemisphere stopped at the same time. This manner of appearance of recurrent polyspikes was observed several times, and then grouped polyspikes on
the two sides became progressively time-locked. Finally the start and the en
of grouped polyspikes from both sides occurred simultaneously about 2-3 h
after GEt application, so the propagation of seizure activity from the ap-

plication side to the contralateral side ("propagation") was completed.
During the above mentioned "propagation" process, polyspikes progressed in a
recurrent ictal pattern (IP) alternating with postictal depression. The
timelength of the bursts also increased (Fig. 1,D). IP lasted until the end
of the recording at 3-4 h. The ED_{50} to induce spike discharges was 75mM (GEt:
750nmol).

In the After-test, the spike discharges were completely suppressed and
"propagation" was not observed for a couple of hours after supplementation of

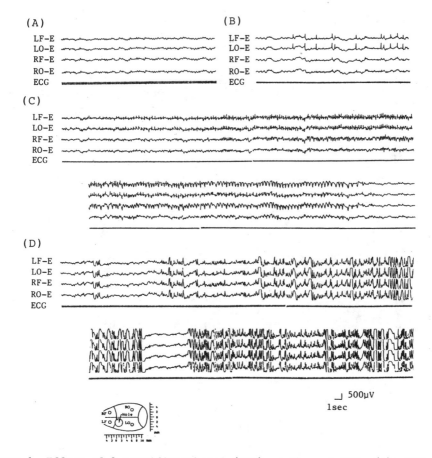

Figure 1. Effect of 2-guanidinoethanol (GEt) on the rat EEG. (A): EEG recorded before the GEt application. (B): Three min after the topical
application of 3μmol of GEt to the pia mater of the sensorimotor
cortex of the rat, sporadic spike discharges were observed. (C):
One hundred min after the GEt application, spike discharges induced
by GEt started to propagate to the contralateral cortex. (D): EEG
recorded 225 min after the GEt application. Spike bursts were
completely propagated to the contralateral cortex. The recordings
were from 4 epidural electrode; LF-E, unipolar recording from left
frontal electrode; LO-E, unipolar recording from left occipital
electrode; RF-E, unipolar recording from right frontal electrode,
and RO-E, unipolar recording from right occipital electrode. Posi-
tions of each electrode and the hole for the topical application
of drugs is shown in the lower left corner. ECG: electrocardiogram.

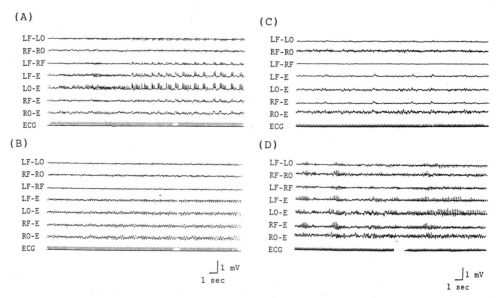

Figure 2. Effect of (3R)-(-)-4-amino-3-hydroxybutanoic acid (L-GABOB) on the
spike discharges induced by 2-guanidinoethanol (GEt). (A): EEG
obtained 44 min after the topical application of GEt to the pia
mater of the sensorimotor cortex of the rat. Poly-spike discharges
were observed. (B): EEG obtained 40 min after L-GABOB supplementation (105 min after first GEt application). Spike discharges induced by GEt were suppressed. (C): EEG obtained 28 min after the
topical application of L-GABOB. (D): EEG obtained 80 min after GEt
supplementation (113 min after first L-GABOB application), poly-
spike discharges induced by GEt had already started after a long
latent time. The recordings were from 4 epidural electrodes: LF-LO
bipolar recording from left frontal electrode (LF) to left occipital electrode (LO); RF-RO, bipolar recording from right frontal
electrode (RF) to right occipital electrode (RO); LF-RF, bipolar
recording from LF to RF; LF-E, unipolar recording from LF; LO-E,
unipolar recording from LO; RF-E, unipolar recording from RF; RO-E
unipolar recording from RO; and ECG, electrocadiogram.

the GABA or L-GABOB solution (Fig. 2, A,B). When GABA or L-GABOB was applied
prior to the GEt application, GEt induced spike discharges with a longer
latent time compared with that observed in controls, and "progagation" was
completed (Fig. 2, C,D).

In the After-test, spikes were completely suppressed a couple of min
after supplementation with muscimol (Fig. 3, A,B). In rats given muscimol 38
min prior to GEt application, spike activity was not induced by GEt until the
end of the recording at 4 h (Fig. 3, C,D).

When DZP was injected in the After-test, DZP showed a suppressing effect
on spike discharges within 30 min (Fig. 4, A,B). DZP, injected in the Before-
test, prolonged the latent time, and some of the spike discharges were induced, but they disappeared within 60 min (Fig. 4, C,D).

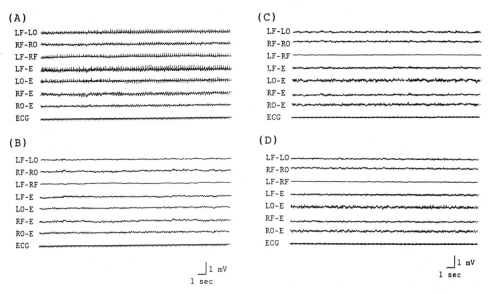

Figure 3. Effect of muscimol on the spike discharges induced by 2-guanidino-
ethanol (GEt). (A): EEG obtained 30 min after the topical application of GEt to the pia mater of the sensorimotor cortex of the rat,
poly-spike discharges were observed. (B): EEG obtained 40 min after
muscimol supplementation (79 min after first GEt application).
Spike discharges induced by GEt were suppressed. (C): EEG obtained
30 min after the topical application of muscimol. (D): EEG obtained 42 min after GEt supplementation (80 min after the first
muscimol application). GEt did not induce spike activity. The
recording method was the same as for Figure 2.

Though the slow wave component in the EEG increased after 20mg/kg of PB
applied in the After-test, PB showed a suppressing effect on spike discharges
within 40 min (Fig. 5, A,B). PB, injected in the Before-test, induced high
voltage slow waves within 15 min. A few grouped spike discharges lasting for
20 seconds were induced 3-15 min after GEt application, but these grouped
spike discharges were suppressed after 90 min (Fig. 5, C,D).

PHT, injected in the After-test, suppressed spike activity gradually and
"propagation" did not occurred (Fig. 6, A,B). Injected PHT prior to the GEt
application prolonged the latent time required induce spike activities and
these disappeared within 60 min (Fig. 6, C,D).

Though spike discharges were induced and "propagation" was completed in
the Before-test, VPA injected in the After-test suppressed spike activities
gradually and did not complete "propagation" (Fig. 7, A-E).

DISCUSSION

Though there are many reports about experimental convulsions and convulsive manifestations in EEGs induced by guanidino compounds (Table 1),
there are only a few guanidino compounds whose convulsive mechanisms have
been reported. One group of guanidino compounds influences GABAergic neuro-

Table 1. Guanidino Compounds Induced Seizures

Guanidino Compound	Experimental animal	Dosage	Latent time(min)	Type of seizure	Duration of seizures	Presence in brain	references
α-N-Acetylarginine	rabbit	3mg/kg i.c.	0-5	CC,TC	-60	o	28, 29
Creatine	rabbit	13mg/kg i.c.	1-10	TC,CC	-120	o	30
Creatine-phosphate	rabbit	12mg/kg i.c.	0	TC,CC	60-90	o	30
Creatinine	rabbit	13mg/kg i.c.	0-6	TC,CC	10-30 DWC	o	30
N,N'-dibenzoylguanidine	mouse	100mg/kg i.p.	3	TC,CC	-120	x	14
Guanidinoacetic acid	rabbit	5mg/kg i.c.	0-5	CC,TC	10-15	o	30
γ-Guanidinobutyric acid	rabbit	5-7mg/kg i.c.	10-17	TC,CC	60-90	o	3, 31
Guanidinoethanesulfonic acid	rabbit	6-7mg/kg i.c.	0-40	TC	60-180	o	4, 32
2-Guanidinoethanol	rat	0.3M, 10µl topical	10	SD	-240	o	10
α-Guanidinoglutaric acid	rabbit	0.2M, topical	10	SD	-180	o	5, 6, 7
δ-Guanidinovaleric acid	rat	57µM, 10µl topical	15	SD	-180	?	8, 9
Homoarginine	rat	0.1M, 10µl topical	15	SD	-120	o	33, 34
α-Keto-δ-guanidinovaleric acid	rabbit	0.2M, topical	8	SD	-120	?	35
Methylguanidine	rabbit	1.5mg/kg i.c.	0-2	CC	30 DWC	o	36, 37

TC: tonic convulsion, CC: clonic convulsion, SD: seizure discharges in EEG, DWC: died with tonic convulsion, i.c.: intra cisternal injection, i.p.: intra peritoneal injection, topical: topical application on brain surface.

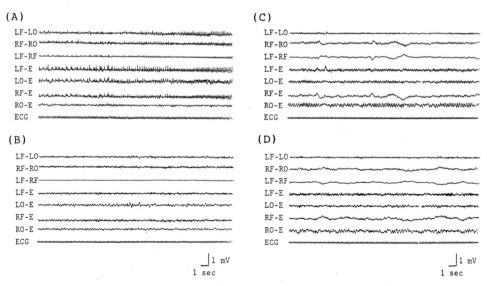

Figure 4. Effect of diazepam (DZP) on the spike discharges induced by 2-guanidinoethanol (GEt). (A): EEG obtained 27 min after the topical application of GEt to the pia mater of the sensorimotor cortex of the rat, poly-spike discharges were observed. (B): EEG obtained 142 min after DZP injection (170 min after first GEt application), spike discharges induced by GEt were suppressed. (C): EEG obtained 25 min after DZP injection. (D): EEG obtained 40 min after GEt application (70 min after first DZP injection). No spike activity was observed. The recording method was the same as for Figure 2.

transmission. GBA, that is γ-amidino-GABA, is thought to act on the GABAergic system to induce convulsion because of its structural similarity to GABA[3]. DGVA is thought to be an endogeneous and specific GABA-receptor antagonist from results of study using GABA, GABA agonists and anticonvulsants[8,9]. On the other hand, another group of guanidino compounds is thought to act on serotonergic neurotransmission. GGA injected in rat ventricles first increases serotonin transiently in the cerebral cortex, then decreases serotonin in the midbrain and finally in the whole brain[13]. Brain serotonin level increases during myoclonus and tonic convulsive stages induced by N'N-dibenzoyl-guanidine, but the catecholamine level is not affected[14].

As GEt has the almost same arrangement of atoms as that of GABA (Fig.8), it is possible to say that GEt is a structual analogue of GABA. From this point of view, GEt was expected to act upon the GABAergic system to induce seizure activity. First we observed that GEt induced spike activities in the rat EEG when it was applied on the left sensorimotor cortex, then the induced spike discharges were propagated from the left to the right cortex. The effect of GABA, GABA agonists and some anticonvulsants on spike discharges induced by GEt were examined in order to study the mechanism involved in the development of spike discharges.

GABA administered on the surface of the cortex was thought to be detoxified rapidly by GABA transaminase, as L-GABOB had a strong suppressive effect on spike discharges induced by penicillin injected into the cerebral

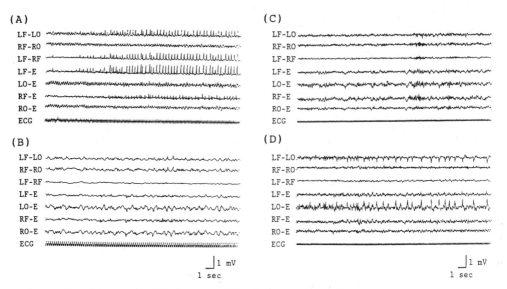

Figure 5. Effect of phenobarbital (PB) on the spike discharges induced by 2-guanidinoethanol (GEt). (A): EEG obtained 18 min after the GEt application, poly-spike discharges were observed. (B): EEG obtained 40 min after PB injection (59 min after first GEt application). Spike discharges induced by GEt were suppressed. (C): EEG obtained 26 min after PB injection. High voltage slow waves were observed. (D): EEG obtained 39 min after GEt application (83 min after first PB injection), sporadic apike discharges and high voltage slow waves were observed. There-after the spike discharges disappeared. The recording method was the same as for Figure 2.

cortex of cats while GABA had no effect[15]. Since suppressing effects of GABA or L-GABOB on spike discharges were not as strong compared to 3 anticonvulsants, the target neurons of GEt in the cortex must be in a deep layer.

Muscimol, which is known as a $GABA_A$ receptor agonist[16,17], suppressed the spike discharges induced by GEt. GEt appears to act on the physiological function of the GABA receptor system. Muscimol (0.25–1.5 mg/kg) injected intraperitoneally suppresses convulsions induced by pentylenetetrazol, bicuculline, picrotoxin or strychnine[18], but muscimol-induced EEG changes have been reported[19,20]. When 1.0 mg/kg of muscimol was injected i.p., the EEG of rats showed sporadic spike discharges with high voltage slow waves with a latency of about 20–30 min. As 50nmol of muscimol applied on the pia mater had no effects on the EEGs, we applied 50nmol of muscimol directly to the pia mater.

As it is well accepted that the GABA receptor forms a complex with the DZP receptor, PB receptor and Cl channel[21,22], the effect of DZP and PB on spike discharges induced by GEt was investigated. DZP showed a suppressive effect on the spike discharges. Though the slow wave component in the EEG increased after 20 mg/kg of PB was applied, PB also showed a suppressive effect on spike discharges.

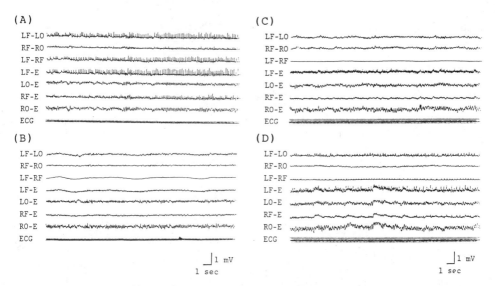

Figure 6. Effect of phenytoin (PHT) on the spike discharges induced by 2-
guanidinoethanol (GEt). (A): EEG obtained 22 min after the GEt
application, poly-spike discharges were observed. (B): EEG obtained
70 min after PHT injection (96 min after first GEt application).
Spike discharges induced by GEt were suppressed. (C): EEG obtained
17 min after PHT injection. (D): EEG obtained 41 min after GEt application (63 min after first PHT injection). A few low voltage
sporadic spike discharges were observed, and there-after the spike
discharges disappeared. The recording method was the same as for
Figure 2.

An anticonvulsive effect of VPA has been demonstrated during electroconvulsive and pentylenetetrazol-induced seizures[23], and VPA is thought to
increase the GABA content in the brain through inhibition of succinicsemi-
aldehyde dehydrogenase and glutamic acid decarboxylase activity[24,25]. Though
VPA failed to suppress entirely the spike discharges induced by GEt, it has a
weak suppressive effect. This suggests that the basic mechanisms for the induction and continuation of spike discharges induced by GEt are different.

Though the main action of PHT is thought to be the closing of the Na
channel in the synaptosomal membrane[26], and another action is reported to be
an increase in the rate of glutamate and GABA high-affinity uptake into the
synaptosome[27], PHT had a suppressing effect on the spike discharges induced
by GEt. Considering the results obtained using the 4 anticonvulsants, it
appears that GEt seems to be a GABA antagonist in a broad sense.

Recently, GEt in fresh human urine was identified by us using the Rf
values obtained by paper chromatography, the retention time obtained by high
pressure liquid chromatography utilizing 9',10'-phenanthrenequinone as the
fluorogenic reagent, and gas chromatography-mass spectrography after converting the sample to the trifluoroacetylated dimethylpyrimidyl butylester[10]. We
showed GEt is synthesized from Arg and ethanolamine by transamidination catalyzed by glycine amidinotransferase[11].

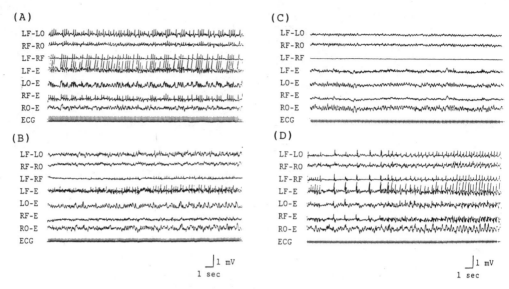

Figure 7. Effect of valproate (VPA) on the spike discharges induced by 2-guanidinoethanol (GEt). (A): EEG obtained 26 min after the GEt application, poly-spike discharges were observed. (B): EEG obtained 114 min after VPA injection (150 min after first GEt application). The amplitude of spike discharges induced by GEt decreased. There-after the spike discharges were suppressed. (C): EEG obtained 43 min after VPA injection. (D): EEG obtained 45 min after GEt application (85 min after first VPA injection). Spike discharges were induced and their propagation from the side of GEt application to the contralateral cortex had already started. The recording method was the same as for Figure 2.

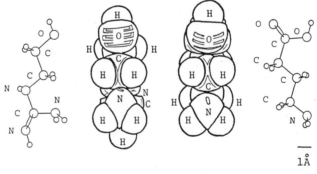

2-guanidinoethanol   γ-aminobutanoic acid

Figure 8. Spatial and skeletal molecular models of 2-guanidinoethanol and γ-aminobutanoic acid (GABA). In the model, N: nitrogen atom, 0: oxygen atom, C: carbon atom, and H: hydrogen atom.

As GEt occurs in the animal kingdom and might be an endogenous GABA antagonist as verified by the present electroencephalograhic study, bio-chemical studies concerning the effect of GEt on GABA binding are needed.

SUMMARY

The action of 2-guanidinoethanol (GEt), first identified in human urine, on the central nervous system and effects of some anticonvulsants or GABA relatives on the spike discharges (Sp-D) induced by GEt were investigated in the rat. Electroencephalograms with epidural electrodes revealed an initia-tion of Sp-D in 2-10 min on the side of GEt (3μmol) application to the pia mater of the sensorimotor cortex, and then a burst of activity of multiple Sp-D developed. During this burst of activity, Sp-D propagated to the oppo-site cerebral hemisphere, and continued for 2-3 hours after the application of GEt. In rats given muscimol (50nmol, on pia mater) together with GEt, Sp-D were not induced. After the completion of Sp-D induced by GEt, supplementary muscimol suppressed the Sp-D completely within 10 min and no propagation oc-cured. While GABA (1μmol, on pia mater), (3R)-(-)-4-amino-3-hydroxybutanoic acid (1μmol, on pia mater) and valproate (VPA) (200mg/kg, i.p.) showed a suppressive effect on Sp-D and the propagation induced by GEt, GEt induced Sp-D and the propagation after the pre-application of these three substances. About 20 min after injection of diazepam (DZP) (10mg/kg, i.p.), diphenyl-hydantoin (PHT) (20mg/kg, i.p.) or phenobarbital (PB) (20mg/kg, i.m.), GEt was applied to the pia mater, but Sp-D were not induced. PB, DZP, PHT or VPA, injected after the completion of Sp-D induced by GEt, suppressed Sp-D and the propagation. These findings suggest that GEt might act on the GABAergic system to induce convulsive activity.

REFERENCES

1. Robin, Y. and Marescau, B., Natural guanidino compounds, In: "Guan-idines," A. Mori, B.D. Cohen and A. Lowenthal Eds., Plenum Press, New York (1985) pp. 383-438.
2. Murray, M. and Hoffmann, A.B., The occurrence of guanidine-like sub-stances in the blood in essential epilepsy, J. Lab. Clin. Med. 25 (1940) 1072-1073.
3. Jinnai, D., Sawai, A. and Mori, A., γ-Guanidinobutyric acid as a convul-sive substance, Nature 212 (1966) 617.
4. Hiramatsu, T., Guanidino compounds in mouse brain II. Guanidino compounds levels in brain relation to convulsions, Okayama Igakkai Zasshi 92 (1980) 427-434.
5. Mori, A., Akagi, M., Katayama, Y. and Watanabe, Y., α-Guanidinoglutaric acid in cobalt-induced epileptogenic cerebral cortex of cats, J. Neurochem. 35 (1980) 603-605.
6. Mori, A., Watanabe, Y., Shindo, S., Akagi, M. and Hiramatus, M., α-Guanidinoglutaric acid and epilepsy. in: "Urea Cycle Diseases," A. Lowenthal, A. Mori and B. Marescau Eds., Plenum Press, New York (1982) pp. 419-426.
7. Shiraga, H. and Mori, A., Convulsive activity of α-guanidinoglutaric acid in rats, IRCS Med. Sci. 10 (1982) 855-856.
8. Shindo, S. Tsuruta, K, Yokoi, I and Mori, A., Synthesis of δ-guanidino-valeric acid and its effect on EEG of rats, Neurosciences (Kobe) 10 (1984) 177-182.
9. Yokoi, I., Tsuruta, K., Shiraga, H. and Mori, A., δ-Guanidinovaleric acid as an endogeneous and specific GABA-receptor antagonist, Electro-encephalographic study, Epilepsy Res. 1 (1987) 114-120.

10. Watanabe, Y., Shindo, S. and Mori, A., Identification of 2-guanidino-ethanol in human urine, Eur. J. Biochem. 147 (1985) 465-468.
11. Watanabe, Y., Yokoi, I. and Mori, A., The biosynthesis of 2-guanidino-ethanol in intact mice and isolated perfused rabbit kidney, life Sci. 40 (1987) 293-299.
12. Weiss, S. and Krommer, H., Zur Guanylierung von Aminen mit O-Methyl-isoharnstoff-sulfat, Chemiker-Zeitung 98 (1974) 617-618.
13. Shiraga, H., Hiramatus, M. and Mori, A., Convulsive activity of α-guanidinoglutaric acid and the possible involvement of 5-hydroxy-tryptamine in the α-guanidinoglutaric acid-induced seizure mechanism, J. Neurochem, 47 (1986) 1832-1836.
14. Nakae, I., Synthesis of N, N'-dibenzoylguanidine and its convulsive action, Neurosciences 7 (1981) 205-217.
15. Katayama, Y., Metabolism of 4-amino-3-hydroxybutanoic acids in the mouse organs and their inhibitory effects on penicillin-induced spike activity of the cat cerebral cortex and on electrical activity of an identified snail giant neurone, Okayama Igakkai Zasshi, 88 (1976) 209-221.
16. Snodgrass, S.R., Use of ^{3}H-muscimol for GABA receptor studies, Nature 273 (1978) 392-394.
17. Beamont, K., Chilton, W.S., Yamamura, H.I. and Enna, S.J., Muscimol binding in rat brain, Association with synaptic GABA receptors, Brain Res. 148 (1978) 153-162.
18. Lloyd, K.G., Munari, C., Worms, P., Bassi, L., Bancard, J., Talairach, J. and Morselli, P.L., The role of GABA mediated neurotransmission in convulsive states, in: "GABA and Benzodiazepine Receptors (Advances in Biochemical Psychopharmacology Vol. 26)," E. Costa, G. Di Chiara and G.L Gessa Eds., Raven Press, New York (1981) pp. 199-206.
19. Scotti de Carolis, A., Lipparini, F. and Longo, V.G., Neuropharmacolog-ical investigations on muscimol, a psychotropic drug extracted from Amanita muscaria. Psychopharmacol. 15 (1969) 186-195.
20. Shoulson, I., Goldblatt, D., Charlton, M. and Joynt, R.J., Huntington's disease, Treatment with muscimol, a GABA-mimetic drug, Ann. Neurol. 4 (1978) 279-284.
21. Olsen, R.W., GABA-benzodiazepine-barbiturate receptor interaction, J. Neurochem. 37 (1981) 1-13.
22. Olsen, R.W., The GABA postsynaptic membrane receptor-ionophore complex, Mol. Cell Biochem. 39 (1981) 261-279.
23. Kupferberg, H.J., Sodium valproate. In: "Antiepileptic Drugs, Mechanisms of Action, "G.H. Laser, J.K. Penny and D.M. Woodbury Eds., Raven Press, New York (1980) pp. 643-654.
24. Van der Laan, J.W., De Boer, T.H. and Bruinuels, J., Di-n-propylacetate and GABA degradation, Preferential inhibition of succinic semialdehyde dehydrogenase and indirect inhibition of GABA-transaminase, J. Neurochem 32 (1979) 1769-1780.
25. Loscher, W. and Frey, H.-H., On the mechanism of action of valproic acid, Arzneimittelforsch. 27 (1977) 1081-1082.
26. De Weer, P., Phenytoin, Blockage of resting sodium channels, in: "Antiepileptic Drugs, Mechanisms of Action," G.H. Laser, J.K. Penry and D.M. Woodbury Eds., Raven Press, New York (1980) pp. 353-361.
27. Weinberger, J., Nichlas, W.J. and Berl, S., Mechanism of action of anti-convulsants, Neurology (Minneap.) 26 (1976) 162-166.
28. Ohkus, H. and Mori, A., Isolation of α-N-acetyl-L-arginine from cattle brain, J. Neurochem. 16 (1969) 413-424.
29. Mori, A. and Ohkusu, H., Isolation and identification of alpha-N-acetyl-L-arginine and its effect on convulsive seizure, Adv. Neurol. Sci. (Tokyo) 15 (1971) 303-306.

30. Jinnai D, Mori A, Mukawa J. Ohkusu, H., Hosotani, M., Mizuno, A. and Tye, L.C., Biochemical and physiological studies on guanidino compounds induced convulsions, Jpn. J. Brain Physiol. 160 (1969) 3668-3673.
31. Mizuno, A., Mukawa, J., Kobayashi, K. and Mori, A., Convulsive activity of taurocyamine in cats and rabbits, IRCS Med. Sci. 3 (1975) 385.
32. Mori, A., Watanabe, Y. and Akagi, M., Guanidino compound anomalies in epilepsy, in: "Advances in Epileptology: XIIIth Epilepsy International Symposium, "H. Akimoto, H. Kazamatsuri, M. Seino and A. Ward Eds., Raven Press, New York (1982) pp. 347-351.
33. Mori, A., Ichimura, I. and Matsumoto, H., Gas chromatography-mass spectrometry of guanidino compounds in brain, Anal. Biochem. 89 (1978) 393-399.
34. Yokoi, I., Toma, J. and Mori, A., The effect of homoarginine on the EEG of rats, Neurochem. Pathol. 2 (1984) 295-300.
35. Marescau, B., Hiramatsu, M. and Mori, A., α-Keto-δ-guanidinovaleric acid-induced electroencephalographic epileptiform discharges in rabbits, Neurochem. Pathol. 1 (1983) 203-209.
36. Matsumoto, M., Kishikawa, H. and Mori, A., Guanidino compounds in the sera of uremic patients and in the sera and brain of experimental uremic rabbits, Biochem. Med. 16 (1976) 1-8.
37. Matsumoto, M., Kobayashi, K., Kishikawa, H. and Mori, A., Convulsive activity of methylguanidine in cats and rabbits, IRCS Med. Sci. 4 (1976) 65.

EFFECT OF GUANIDINOETHANE SULFONATE ON THE GENETICALLY SEIZURE SUSCEPTIBLE
RAT

Shoichiro Shindo, Douglas W. Bonhaus and Ryan J. Huxtable

Department of Pharmacology, College of Medicine
University of Arizona, Tucson, Arizona 85724

Taurine occurs in high concentration in many mammalian tissues. It has numerous pharmacological actions. Taurine is powerfully anticonvulsant in both animal and human epilepsies[1]. Taurine reduces the rate of lesion development in genetically cardiomyopathic hamsters[2]. Taurine has inotropic actions[3], and protects the heart in stress conditions such as calcium overload or increased workload[4].

The transport of taurine is decreased in various types of epilepsies[5], congenital myotonia, and Friedreich's ataxia[6]. These data are indicative of a physiological function for taurine, but its physiological importance is still uncertain. Reasons for this lack of knowledge include the difficulties of modifying organ concentrations of taurine and the absence of a good antagonist.

Huxtable et al.[7] reported that guanidinoethane sulfonate (GES), the amidino analog of taurine, rapidly depleted taurine from organs and was simultaneously accumulated in organs. This compound was found to be a good tool for lowering organ taurine concentrations and for studying the homeostasis of taurine.

The genetically seizure susceptible (SS) rat shows a lowered seizure threshold to a variety of chemical, electrical and auditory stimuli[8]. Taurine is a potent anticonvulsant in this strain[9]. This suggests that the SS rat may have a defect in taurine handling.

In this experiment, we have studied the taurine transport system and the homeostasis of taurine in the SS rat by using GES as a taurine antagonist.

TAURINE TRANSPORT AND BIOSYNTHESIS IN THE SS RAT

The in vivo uptake of [^{3}H] taurine into the brain of SS rat was less than that of the seizure resistant (SR) rat (Fig. 1). Taurine uptake into the P_2 fraction of the SS rat is also lower than in the SR rat (Table 1). Moreover, taurine uptake into platelets is lower in the SS rat than the SR rat (Fig. 2). The rate of taurine synthesis from cysteine was studied in

183

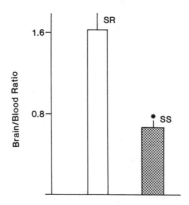

Figure 1. Taurine uptake into the brain in vivo. The brain to blood ratio of
          [^{3}H] taurine 4h after i.p. administration was determined. *P < 0.01
          SS; seizure susceptible and SR; seizure resistant rats.

brain and liver homogenates. No difference was found between SS and SR rats
(Table 2). The taurine level in the P_2 and P_2B subfraction of the SS rat was
lower than in the SR rat (Table 3). These results suggest that generalized
defects of taurine transport could be partially responsible for the seizure
susceptibility of the SS rat.

TAURINE DEPLETION BY GES IN THE SS RAT

        GES caused a remarkable increase in urinary excretion of taurine in
both the SS and SR rat (Fig.3). The taurine level in the urine of SS rat after
one day of GES treatment is higher than that found for the SR rat. On subsequent days, however, urinary excretion was the same in the two strains.

        We calculated the depletion of intrinsic taurine; that is, the amount
of excreted taurine minus the amount received from food (Fig. 4). For the
day prior to GES treatment, urinary excretion was almost the same as dietary
intake. During GES treatment, the intrinsic taurine in the SS rat was ini-
tially depleted more than in the SR rat. However, depletion of intrinsic
taurine fell during the course of GES treatmet in the SS rat, whereas the
daily loss of intrinsic taurine remained constant in the SR rat. This indicates that the GES-displaceable taurine pool is larger in the SS rat.

Table 1. Taurine transport in brain subfractions

	SR	SS
P_2	108 ± 28	36 ± 14*
P_2B	378 ± 100	339 ± 128

nmol/kg protein/s. *P < 0.02.
The P_2 and P_2B (synaptosomal) fractions were prepared by the method of Gray
and Whittaker[9]. Taurine uptake was determined by incubation with 4.0 μM [^{3}H]
taurine at 37°C for 5 min. SR; seizure resistant and SS; seizure susceptible
rats.

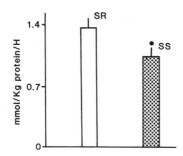

Figure 2. Taurine uptake into platelets. Taurine transport was determined by
    incubation of platelet rich plasma with 10 μM [^{3}H] taurine.
    *P < 0.05. SS; seizure susceptible and SR; seizure resistant rats.

Table 2. Taurine biosynthesis in brain and liver homogenates from seizure
    susceptible (SS) and seizure resistant (SR) rats.

	SR	SS
Brain	272 ± 53	217 ± 53
Liver	950 ± 306	1280 ± 314

nmol/kg protein/s.
Taurine biosynthesis was determined by incubation of homogenate with 2 mM
[^{35}S] cysteine, and [^{35}S] taurine was separated by anion-cation exchange
resin. This fraction was analyzed by scintillation counter.

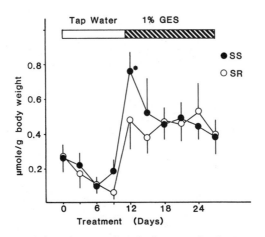

Figure 3. Taurine excretion into urine before and after exposure to 1% GES.
    Twenty four hour urines were collected by metabolic cages and
    deproteinized with equal volumes of 7% sulfosalicylic acid. After
    centrifugation the supernatants were analyzed by amino acid
    analyzer. *P < 0.02. GES; guanidinoethane sulfonic acid, SS; seizure
    susceptible and SR; seizure resistant rats.

Table 3. Subcellular distribution of taurine in the brain

	SR	SS
P_2	82.4 ± 8.6	57.9 ± 6.2*
P_2A	38.0 ± 7.2	35.1 ± 10.5
P_2B	62.1 ± 4.9	43.8 ± 6.5*
P_2C	29.5 ± 3.6	18.6 ± 5.5*

nmol/mg protein. *P < 0.05.
Fractions were prepared by the method of Gray and Whittaker. Samples were
prepared for amino acid analysis by homogenization with an equal volume of 1
sulfosalicylic acid, followed by centrifugation at 10,000 g for 10 min. SR;
seizure resistant and SS; seizure susceptible rats.

Tissue taurine level after GES treatment was analyzed (Table 4). Taurine
levels in the liver and muscle of the SS rat were higher than the SR rat, but
no difference was observed in the other organs examined.

Huxtable and Lippincott[10] report that the rate of exchange between
tissues is faster than the rate of elimination of taurine from the body. Our
results suggest that the exchange rate of taurine between tissues in the SS
rat may be slower than in the SR rat. In the absence of GES, this difference would not be reflected in the tissue content or urinary excretion of
taurine.

These results suggest that SS rat exhibits both transport and biosynthetic defects in the handling of taurine. These inbalances may be one
element contributing to the lower seizure threshold in the SS rat.

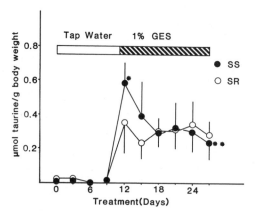

Figure 4. Intrinsic taurine excretion before and after exposure to 1% GES.
These values are calculated from the amount of taurine excreted
in the urine and the amount of taurine present in the food intake
for the day. *P < 0.05, **P < 0.01 compared to day 12 excretion of
SR rat. GES; guanidinoethane sulfonic acic, SS; seizure susceptible
and SR; seizure resistant rats.

Table 4. Tissue taurine level after GES treatment

	SR	SS
Liver	0.40 ± 0.08	1.74 ± 1.00*
Kidney	6.3 ± 0.7	5.6 ± 1.6
Heart	12.5 ± 1.8	12.0 ± 1.4
Cerebrum	3.0 ± 0.5	2.9 ± 0.4
Cerebellum	2.9 ± 0.4	3.1 ± 1.5
Muscle	8.0 ± 1.0	10.0 ± 1.3*

μmol/g tissue. * $P < 0.01$.
Tissues were homogenized with an equal volume of 7% sulfosalicylic acid and centrifuged at 10,000 g for 10 min. The supernatants were analyzed by amino acid analyzer. GES; guanidinoethane sulfonic acid, SR; seizure resistant and SS; seizure susceptible rats.

CONCLUSION

Urinary taurine excretion in the SS rat is not different from the SR rat under normal dietary conditions. However, we find some differences in taurine homeostasis when GES is used as a taurine antagonist.

REFERENCES

1. Huxtable, R.J., Insights on function: Metabolism and pharmacology of taurine in the brain, in "The Role of Peptides and Amino Acids as Neurotransmitters," J.B. Lombardini and A. Kenny Eds., Alan R. Liss, New York (1981) pp 53–98.
2. Azari, J., Brumbaugh P. and Huxtable, R.J., Prophylaxis by taurine in the hearts of cardiomyopathic hamsters, J. Molec. Cell Cardiol. 12 (1980) 1353–1366.
3. Dietrich,J. and Diacono, J., Comparison between ouabain and taurine effects on isolated rat and guinea pig hearts in low calcium medium, Life Sci. 10 (1971) 499–507.
4. Kramer, J.H., Chovan J.P. and Schaffer, S.W., The effect of taurine on calcium paradox and ischemic heart failure, Am. J. Physiol. 240 (1981) H238–H246.
5. Airaksinen, E.M., Uptake of taurine, GABA, 5-HT and dopamine by blood platelets in progressive myoclonus epilepsy, Epilepsia 20 (1979) 503–510.
6. Filla, A., Butterworth R.F. and Barbeau, A., Pilot studies on membranes and some transport mechanisms in Friedreich's Ataxia, Canad. J. Neurol. Sci. 6A (1979) 285–289.
7. Huxtable, R.J., Guanidinoethane sulfonate and the disposition of dietary taurine in the rat, J. Nutr. 112 (1982) 1193–2300.
8. Laird H.E. and Huxtable, R.J., Taurine and audiogenic epilepsy, in "Taurine and Neurological Disorders," A. Barbeau and R.J. Huxtable Eds., Raven Press, New York (1978) pp 339–357.
9. Gray E.G. and Whittaker, V.P., The isolation of nerve endings from brain: an electron-microscopic study of cell fragments derived by homogenization and centrifugation, J. Anat. 96 (1962) 79–88.
10. Huxtable R.J. and Lippincott, S.E., Relative contribution of diet and biosynthesis to the taurine content of the adult rat, Drug-Nutrient Interactions 1 (1982) 153–168.

GUANIDINOETHANE SULFONATE AND THE INVESTIGATION OF TAURINE AND OTHER

NEUROACTIVE AMINO ACIDS

Ryan J. Huxtable, Anders Lehmann*, Mats Sandberg* and
Shoichiro Shindo**

Department of Pharmacology, College of Medicine, University
of Arizona, Tucson, Arizona 85724
*Institute of Neurobiology, University of Goteborg
Goteborg, Sweden

The compound variously known as taurocyamine, amidinotaurine and guanidinoethane sulfonate (GES), and occasionally and inaccurately named guanidinotaurine, is found in low amounts in both vertebrates[1-3] and invertebrates[4]. It is, however, possible that certain literature values for GES
concentrations are artifactually high due to insufficient resolution of the
various guanidino compounds present[5].

In invertebrates, GES is biosynthesized from hypotaurine via guanidinoethane sulfinate. In mammals, GES is formed by transamidation of taurine,
which, along with inorganic sulfate, is one of the excreted products of
mammalian sulfur amino acid metabolism[6]. The N-phosphoryl derivative of GES
functions as a phosphagen in various marine worms such as Annelida[7]. It is
an unsolved biochemical puzzle as to why the eight guanidino compounds used
as phosphagens in invertebrates are replaced by the methylguanidino, creatine, in vertebrates. Creatine is an energetically expensive compound, whose
biosynthesis consumes 80% of methyl group neogenesis in mammals. However,
the unmethylated analog, ß-guanidinopropionate, can substitute for creatine
in rat muscle, indicating that methylation is not essential for function-
ality[8]. We have discussed elsewhere a possible reason for the switch from
simple to methylated phosphagens in the course of vertebrate evolution,
suggesting that the use of creatine provides a means of releasing excess
methyl groups from methionine to provide sufficient homocysteine as a substrate for transsulfuration[6].

No function has been demonstrated for GES in mammals. It is present in
low concentrations in various tissues, and small amounts are excreted in the
urine. For humans, approximately 1 to 3 mg/day are so excreted. There is an
unfortunate misprint in the paper of Guidotti and Costagli[9] implying that GES
concentrations are a thousand times higher in mammalian tissues than in fact
they are.

** Current address: Tsumura Research Institute for Pharmacology, 3586
    Ami-machi, Inashiki, Ibaraki 300-11, Japan

GUANIDINOETHANE SULFONATE AND TAURINE DEPLETION

GES is proving a valuable tool in the investigation of excitatory and inhibitory amino acids in the central nervous system due to its direct and indirect action on excitability[10-19]. GES, like other guanidino compounds, is an excitotoxin, producing convulsions when injected into the brain[12-19]. In addition, GES is an effective competitive antagonist for transport of the inhibitory amino acid, taurine, in the isolated perfused heart[19], brain synaptosomes[18,20], and isolated retina[21,22]. This suggested to us that GES could be used in vivo to produce lowering of taurine levels in as much as the greatest biosynthesis of taurine occurs in liver, followed by transport to other tissues. Indeed, maintaining rats and mice on drinking water containing 1% GES led to a smooth decline in taurine concentrations of both central and peripheral tissues over a two week period, thus providing taurine researchers with a convenient and simple tool for producing taurine depletion[1] GES produces a greater decrease of taurine content in peripheral tissues then in brain (Fig. 1). This may indicate a greater capacity for taurine biosynthesis in brain as compared to tissues such as the heart. Such a conclusion is supported by the observation that the tissue to serum distribution ratio of [^{3}H] taurine given i.p. is lower for brain then for other organs (Table 1). Thus, the average ratio of 19.6 found in peripheral tissues drops to 12.1 following GES treatment. Average ratios for brain areas,

Table 1. Effect of guanidinoethane sulfonate (GES) on tissue-to-serum ratio of [^{3}H] taurine

	Control	GES-treated	%	P
Thymus	17.3 ± 2.6	13.3 ± 1.0	77	
Lung	34.8 ± 4.6	19.0 ± 0.8	55	< 0.005
Kidney	19.0 ± 4.4	11.9 ± 0.7	63	< 0.05
Stomach	18.8 ± 1.4	11.1 ± 1.2	59	< 0.001
Heart	21.5 ± 1.7	13.6 ± 1.4	63	< 0.005
Spleen	23.7 ± 3.7	13.3 ± 0.2	56	< 0.005
Muscle	8.6 ± 1.8	7.8 ± 0.9	91	
Pancreas	8.7 ± 1.9	5.7 ± 2.3	66	
Liver	14.5 ± 2.4	5.2 ± 2.0	36	< 0.005
Small intestine	29.3 ± 1.8	19.7 ± 1.4	67	< 0.001
Frontal cortex	5.3 ± 1.9	4.6 ± 0.4	87	
Cerebellum	4.3 ± 0.8	3.9 ± 0.2	91	
Cerebral hemispheres	4.7 ± 0.4	4.5 ± 0.2	96	
Pons medulla	2.8 ± 0.8	3.2 ± 0.5	114	
Spinal cord	2.7	3.9 ± 0.7	114	
Hypothalamus	3.2 ± 0.1	3.1 ± 0.4	97	
Inferior colliculus	3.3 ± 0.6	3.9 ± 0.6	118	
Midbrain	3.7 ± 1.7	3.0 ± 0.7	81	

Male Sprague-Dawley rats were maintained on a taurine-free diet. GES-treated animals were supplied for 6 days with drinking water containing 1% GES. On day 5, each animal was given [^{3}H] taurine (11.9 μCi/100g body weight i.p., 10 mCi/mmol.). Animals were killed 24 hours later and tissues analyzed for [^{3}H] taurine content. Data are means ± SD for 3 animals per group. From Huxtable, 1982[41].

however, remain constant at 3.75. The decrease in taurine content of control brain tissue shown on Figure 1 illustrates the well-established fall in CNS taurine concentration with development. Most workers[18,19,23] report that in rats and mice GES decreases taurine concentrations uniquely, other amino acids being unaffected (see e.g. Table 2). Others report that in mice[24] and rats[18], GES alters concentrations of a number of amino acids in the brain. However, this phenomenon was not dose-related, a higher level of GES having an opposite effect.

GES is ineffective as a taurine depleter in certain species. The cat, for example, is able to use GES as a transamidination substrate, thereby metabolizing it to taurine[25].

Amoung the drawbacks to the use of GES as a taurine depleter are that it is transported into tissues by the beta amino acid transport system (and possibly also by the creatine transport system), and is accumulated in place of taurine (Fig. 2)[12].

GUANIDINOETHANE SULFONATE AND THE NEUROCHEMISTRY OF MAINO ACIDS

Despite the drawbacks listed above, numerous investigations on the physiology and pharmacology of taurine have employed GES. In the rat, for example, depletion of retinal taurine contents with GES leads to retinal degeneration identical to that seen in taurine-deficient cats[13,26]. In the latter species, depletion can be produced by dietary deprivation, as the cat is unable to synthesize taurine[27]. However, this approach does not work with small laboratory mammals such as the rat and mouse which are capable of taurine synthesis.

The genetically seizure-susceptible (SS) rat exhibits a defect in the transporting ability of the brain for taurine[28,29] and a decrease in synaptosomal taurine concentration[29-31]. These defects in the neurochemistry of a neuroinhibitory amino acid may be related to the increased seizure suscep-

Table 2. Effect of guanidinoethane sulfonate (GES) treatment on total free amino acid content of rat brain

	nmol/brain	
	Control	GES
Glu	6413 ± 757	6877 ± 426
Tau	3796 ± 403	2164 ± 226*
Asp	3583 ± 594	3650 ± 391
GABA	3528 ± 564	3626 ± 189
Gln	2285 ± 355	2486 ± 632
Gly	1528 ± 290	1390 ± 89
Ser	1112 ± 272	1040 ± 76
Ala	809 ± 148	758 ± 48
Thr	463 ± 97	445 ± 45

Male Sprague-Dawley rats were used, eight per group. Treated animals were maintained on drinking water containing 1% GES for 14 days. At that time, animals were killed, and total free amino acid content of brain homogenate determined by HPLC. *P < 0.05 compared to control.

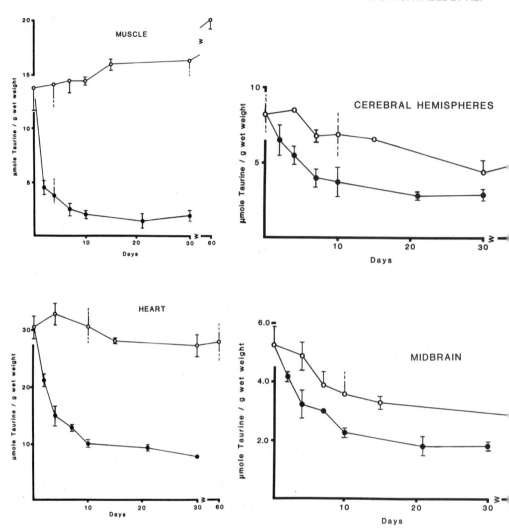

Figure 1. Taurine concentrations in tissues of control and guanidinoethane
        sulfonate (GES)-treated rats. GES produced less depletion in the CNS
        then in peripheral tissues. Modified from Huxtable (1982)[41].

tibility of this strain of rat. GES (1% in drinking water for 2 weeks) low-
ers synaptosomal (P_2B) taurine concentration by 50% in the SS rat and by 63%
in the seizure-resistant (SR) rat; an indication of the greater lability of
taurine in the SS rat brain. In addition, GES has a differential action on
synaptosomal glutamate levels in the two strains, being without significant
effect in the SR rat, but lowering levels in the SS rat to 68% of control.

    This latter finding may be related to three other phenomena: the greater
potassium-stimulated release of excitatory amino acids from the hippocampus
observed in the SS rat[32]; the interrelationship of taurine and glutamate
concentrations[33]; and the decrease in intracellular:extracellular ratio of
glutamate concentration in the hippocampus of GES-pretreated animals[23].

    The extracellular levels of substances in the brain can be examined un-

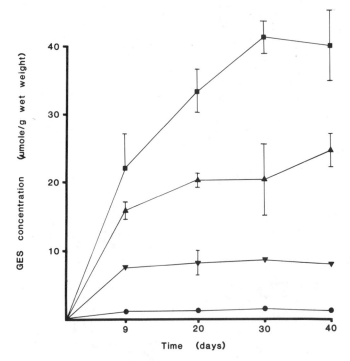

Figure 2. Accumulation by organs of guanidinoethane sulfonate (GES) in GES-
exposed animals. Concentrations for brain represents averages of
separately determined brain regions. Data are shown as means ± SD
for 4 animals per point. ▲ Heart; ■ muscle; ▼ spleen; ● brain. Data
adapted from Huxtable et al., (1979)[12].

der a variety of conditions using the brain microdialysis technique[23,32,34].
The basal release of most amino acids is unaffected by GES pretreatment, as
is the release following neuronal destruction with kainic acid[23]. Basal and
kainate-induced release of taurine is, however, depressed in animals pre-
exposed to GES. Conversely, the release of glutamate is increased. In these
experiments, the dialysis probe was placed in the hippocompus. If the ratio
of total tissue content to the perfusate concentration for each amino acid
is calculated, GES pretreatment does not alter the ratio for taurine (i.e.
both tissue content and perfusate concentration are reduced to the same extent) (Fig. 3). For glutamate, however, there is a marked fall in the ratio.
As the gross tissue content of glutamate is unaltered, this decrease in ratio
must be caused by an increase in net glutamate efflux. Is this increase in
extracellular glutamate a direct effect of GES, or secondary to the depression of taurine concentrations? This experiment does not directly address
this point. However, Van Gelder has suggested a modulatory action of taurine
on brain glutamate content and release[35-37]. Relationships between taurine
content in the brain have been established experimentally by several laboratories[33,36,38,39]. Although a small amount of the total taurine in the brain
may serve a neurotransmitter function, it appears that much of the taurine
serves a neuromodulator function[32]. In other words, taurine modifies the
release of other neuroactive substances. If taurine, a membrane stabilizer,
inhibits the release of glutamate, the lowering of taurine content should

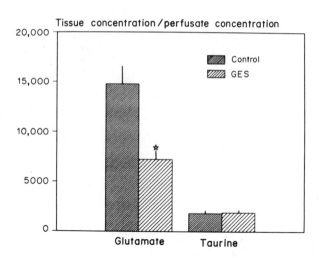

Figure 3. Effect of guanidinoethane sulfonate (GES) pretreatment (1% in drinking water for 9 days) on the hippocampal: perfusate concentration ratio for glutamate and taurine. Tissue concentration refers to content of free amino acid in the hippocampus. Date are means ± SEM for 7 animals per group. From Lehmann et al., (1987)[23].

reduce the degree of inhibition, leading to increased glutamate release. This is what is observed in this experiment. Increased release of glutamate is, of course, excitatory.

A further area of neurochemistry in which GES has proved useful is the tasting of taurine analogs as anticonvulsant agents[40].

CONCLUSION

A number of examples of the value of GES as a tool for investigating the biochemistry and physiology of neuroactive amino acids in the brain has been given. GES is not without drawbacks - it has a marked pharmacology of its own, and it may produce changes in brain biochemistry beyond modification of amino acid content. Until more specific and more powerful means of modifying taurine concentrations in the brains of laboratory mammals are available, however, GES will continue as part of the armamentarium of the neurochemist.

ACKNOWLEDGMENT

Supported in part by the Flinn Foundation (Arizona Heart Association).

REFERENCES

1. Wiechert, P., De Deyn, P.P., Marescau, B. and Lowenthal, A., Taurocyamine in cerebrospinal fluid of neurological and psychiatric patients, Neurol. and Psychiatry 49 (1986) 716-717.
2. Koide, H. and Azushima, C., Metabolic profiles of guanidino compounds in various tissues of uremic rats, In: "Guanidines," A. Mori, B.D. Cohen

and A. Lowenthal Eds., Plenum Press, New York (1985) pp 365-372.

3. Plum, C.M., Studies on the occurrence of guanidino compounds in serum and urine of patients with epilepsy, Neuronal ceroid-lipofuscinosis and uremic patients, In "Guanidines," A. Mori, B.D. Cohen and A. Lowenthal Eds., Plenum Press, New York (1985) pp 159-169.

4. Robin, Y. and Marescau, B., Natural guanidino compounds, In:"Guanidines," A. Mori, B.D. Cohen and A. Lowenthal Eds., Plenum Press, New York (1985) pp 383-438.

5. Marescau, B., De Deyn, P.P., Wiechert, P., Van Gorp, L.E. and Lowenthal, A., Quantitative overestimation of free taurocyamine using short liquid cation-exchange chromatography columns, J. Chromatogr. 345 (1985)215-217.

6. Huxtable, R.J., Biochemistry of Sulfur, Plenum Press, New York, (1986).

7. Thoai, N.V. and Robin, Y., Guanidine compounds and phosphagens, In: "Chemical Zoology, IV," M. Florkin and B.T. Scheer Eds., Academic Press, New York. (1969).

8. Fitch, C.D., Jellinek, M., Fitts, R.H., Baldwin K.M. and Holloszy, J.O., Phosphorylated γ-guanidino propionate as a substitute for phosphocreatine in rat muscle, Am. J. Physiol. 228 (1975) 1123-1125.

9. Guidotti, A. and Costagli, P.F., Occurrence of guanidotaurine in mammals: Variation of urinary and tissue concentration after guanidotaurine administration, Pharm. Res. Commun. 2 (1970) 341-354.

10. Lake, N., Taurine depletion of lactating rats: Effects on developing pups, Neurochem. Res. 8 (1983) 881-889.

11. Huxtable, R.J., Bonhaus, D., Nakagawa, K., Laird, H.E. and Pasantes-Morales, H., Taurine and the actions of guanidinoethane sulfonate, In: "Guanidines," A. Mori, B.D. Cohen and A. Lowenthal Eds., Plenum Press, New York (1985) pp 213-226.

12. Huxtable, R.J., Laird, H.E. and Lippincott, S.E., The transport of taurine in the heart and the rapid depletion of tissue taurine content by guanidinoethyl sulfonate, J. Pharmacol. Exptl. Therap. 211 (1979) 465-471.

13. Bonhaus, D.W., Pasantes-Morales, H. and Huxtable, R.J., Actions of guanidinoethane sulfonate on taurine concentration, retinal morphology and seizure threshold in the neonatal rat, Neurochem. Int. 7 (1985) 263-270.

14. Baba, A., Yamamoto T., Morimoto, T., Matsuda, T. and Iwata, H., Release of [³H] γ-aminobutyric acid by taurocyamine in rat brain slices, Jpn. J. Pharmacol. 35 (1984) 465-467.

15. Okamoto, K. and Sakai, Y., Inhibitory actions of taurocyamine, hypotaurine, homotaurine, taurine and GABA on spike discharges of purkinje cells, and localization of sensitive sites, Brain Res. 206 (1981) 371-386

16. Izumi, K., Kishita, C., Nakagawa, K, Huxtable, R.J., Shimizu, T., koja, T. and Fukuda, T., Modification of the antiepileptic actions of phenobarbital and phenytoin by the taurine transport inhibitor, guanidinoethane sulfonate, Eur. J. Pharm. 110 (1985) 219-224.

17. Lake, N., Electroretinographic deficits in rats treated with guanidinoethyl sulfonate, a depletor of taurine, Exp. Eye Res. 42 (1986) 87-92.

18. Marnela, K.M. and Kontro, P., Free amino acids and the uptake and binding of taurine in the central nervous system of rats treated with guanidinoethane sulfonate, Neuroscience 12 (1984) 323-328.

19. Mori A., Katayama, Y., Yokoi, I. and Matsumoto, M., Inhibition of taurocyamine (guanidinotaurine) induced seizures by taurine, In: "The Effects of Taurine on Excitable Tissues," S.W. Schaffer, S.I. Baskin and J.J. Kocsis Eds., Spectrum Press, Philadelphia (1981) pp 41-48.

20. Hruska, R.E., Padjen, A., Bressler, R. and Yamamura, H.I., Taurine: Sodium-dependent, high-affinity transport into rat brain synaptosomes, Mol. Pharmacol. 14 (1978) 77-85.

21. Quesada, O., Huxtable R.J. and Pasantes-Morales, H., Effect of guanidinoethane sulfonate on taurine uptake by rat retina, J. Neurosci. Res. 11 (1984) 179-186.
22. Lake, N., Depletion of retinal taurine by treatment with guanidinoethyl sulfonate, Life Sci. 29 (1981) 445-448.
23. Lehmann, A., Hagberg, H., Huxtable, R.J. and Sandberg, M., Reduction of brain taurine: effects on neurotoxic and metabolic actions of kainate, Neurochem. Int. 10 (1987) 265-274.
24. Hiramatsu, M., Niiya-Nishihara, H. and Mori, A., Effect of taurocyamine on taurine and other amino acids in the brain, liver and muscle of mice, Neurosciences 8 (1982) 289-294.
25. Huxtable, R.J. and Lippincott, S.E., Comparative metabolism and taurine depleting effects of guanidinoethane sulfonate in cats, mice, and guinea pigs, Arch. Biochem. Biophys. 210 (1981) 698-709.
26. Pasantes-Morales, H., Quesada, O., Carabez A. and Huxtable, R.J., Effect of the taurine transport antagonists, guanidinoethane sulfonate and beta-alanine, on the morphology of the rat retina, J. Neurosci. Res. 9 (1983) 135-143.
27. Schmidt, S.Y., Berson, E.L. and Hayes, K.C., Retinal degeneration in cats fed casein. I: Taurine deficiency, Invest. Opthalmol. 15 (1976) 47-52.
28. Bonhaus, D.W. and Huxtable, R.J., Seizure susceptibility and decreased taurine transport in the genetically epileptic rat, Neurochem. Int. 6 (1984) 365-368.
29. Bonhaus, D.W., Laird, H., Mimaki, T., Yamamura, H.I. and Huxtable, R.J., Possible bases for the anticonvulsant action of taurine, In: "Sulfur Amino Acids: Biochemical and Clinical Aspects," K. Kuriyama, R.J. Huxtable and H. Iwata Eds., A.R. Liss, New York (1983) pp 195-209.
30. Bonhaus, D.W., Lippincott, S.E. and Huxtable, R.J., Subcellular distribution of neuractive amino acids in brains of genetically epileptic rats, Epilepsia 25 (1984) 564-568.
31. Huxtable, R.J., Bonhaus, D. and Lehmann, A., Excitatory and inhibitory amino acids in the genetically seizure-susceptible rat, In: "Neurotransmitters in Epilepsy," G. Nistico, P.L. Morselli, K.G. Lloyd, R.G. Fariello and J. Engel Eds., Raven Press, New York (1986) pp 369-385.
32. Lehmann, A., Sandberg, M. and Huxtable, R.J., In vivo release of neuroactive amines and amino acids from the hippocampus of seizure-resistant and seizure-susceptible rats, Neurochem. Int. 8 (1986) 513-520.
33. Huxtable, R.J., Laird, H., Bonhaus D. and Thies, A.C., Correlations between amino acid concentrations in brains of seizure-susceptible and seizure-resistant rats, Neurochem. Int. 4 (1982) 73-78.
34. Hamberger, A., Berthold, C.-H., Karlsson, B., Lehmann A. and Nystrom, B., Extracellular GABA, glutamate and glutamine in vivo-perfusion dialysis of the rabbit hippocampus, Neurol. Neurobiol. 7 (1983) 473-492.
35. Van Gelder, N.M., The role of taurine and glutamic acid in the epileptic process: a genetic predisposition, Rev. Pure Appl. Pharmacol. Sci. 2 (1981) 293-316.
36. Van Gelder, N.M., Siatitsas, I., Manini, C. and Gloor, P., Feline generalized penicillin epilepsy: changes of glutamic acid and taurine parallel the progressive increase in excitability of the cortex, Epilepsia 24 (1983) 200-213.
37. Van Gelder, N.M., Changed taurine-glutamic acid content and altered nervous tissue cytoarchitecture, In: "Taurine in Nutrition and Neurology," R.J. Huxtable and H. Pasantes-Morales Eds., Plenum Press, New York (1982) pp 239-256.
38. Rassin, D.K., Sturman, J.A. and Gaull, G.E., Sulfur amino acid metabolism in the developing rhesus monkey brain: interrelationship of taurine and glutamate, Neurochem. Res. 7 (1982) 1107-1118.

39. Rassin, D.K., Taurine, systeine sulfinic acid decarboxylase and glutamic acid in brain, In: "Taurine, in Nutrition and Neurology," R.J. Huxtable and H. Pasantes-Morales Eds., Plenum Press, New York (1982) pp 257-268.

40. Nakagawa, K. and Huxtable, R.J., The anticonvulsant actions of two lipophilic taurine derivatives, Neurochem. Int. 7 (1985) 819-824.

41. Huxtable, R.J., Guanidinoethane sulfonate and the disposition of dietary taurine in the rat, J. Nutrition 12 (1982) 2293-2300.

GUANIDINO COMPOUNDS IN AUDIOGENIC SENSITIVE RATS

Peter Wiechert, Bart Marescau*, Peter P. De Deyn*, Luth
Van Gorp*, Werner P. De Potter* and Armand Lowenthal*

Clinic of Psychiatry and Neurology, Dept. of Laboratory
Diagnostics and Clinical Neurochemistry, W.-Pieck-University
2500 Rostock, G.D.R. and *Laboratory of Neurochemistry, Born-
Bunge Foundation and Laboratory of Neuropharmacology, U.I.A.
2610 Antwerp, Belgium

The epileptogenic action of several guanidino compounds, e.g. taurocyamine[1], guanidinoacetic acid (GAA)[2], γ-guanidinobutyric acid (GBA)[3], N-α-
acetylarginine[4], α-guanidinoglutaric acid[5], methylguanidine (MG)[6] and
α-keto-δ-guanidinovaleric acid[7] is experimentally demonstrated. In previous
works we used audiogenic sensitive rats as experimental models for epilepsy.
We found changes in several guanidino compounds occurring in serum as well
as in brain[8,9]. Interpreting the results was difficult. The observed change
could be caused by activation of muscular metabolism or by cell destruction
during seizures. In this study we try to control the influence of muscular
contraction on the changes in guanidino compound levels. Therefore we used
audiogenic sensitive rats treated with curare.

MATERIAL AND METHODS

Audiogenic sensitive Wistar rats from both sexes, weighing 200 to 250 g
and bred in our clinic were used for these experiments. The animals' muscu-
lature was relaxed with succinylcholinechloride i.p. and the audiogenic seizures were controlled by E.E.G. Brain dissection, homogenisation, deprotein-
isation and qualitative and quantitative analyses of the guanidino compounds
in serum and brain were carried out as described earlier.[8-10].

RESULTS AND DISCUSSION

Ten guanidino compounds were qualitatively and quantitatively determined
in serum and brain extracts of audiogenic sensitive rats. Table 1 gives the
levels (mean ± S.D.) of the guanidino compounds in serum of two series of
audiogenic sensitive rats: 1. serum from rats during tonic seizures and 2.
serum from curarized rats during electroencephalographically controlled
seizures. Both groups were compared to a control group without seizures. A
significant decrease of creatine (CTN) in serum was seen in non-curarized as

Table 1. Levels of guanidino compounds in serum of audiogenic sensitive rats
with seizures.

Guanidino compounds	Control rats	Rats with tonic seizures	Rats with seizures and curare
Guanidinosuccinic acid	0.18 ± 0.22	0.18 ± 0.10	0.23 ± 0.14
Creatine	458 ± 102	284 ± 68*	270 ± 87*
Guanidinoacetic acid	6.60 ± 2.30	9.25 ± 1.30*	11.5 ± 3.3*
Argininic acid	< 0.03 - 0.04	< 0.03 - 0.07	< 0.03 - 0.07
Creatinine	42.8 ± 11.3	147 ± 42*	52.7 ± 21.8
γ-Guanidinobutyric acid	0.80 ± 0.33	1.57 ± 0.37*	1.65 ± 0.79
Arginine	263 ± 59	314 ± 29*	264 ± 43
Homoarginine	2.64 ± 1.28	2.45 ± 0.66	2.94 ± 1.12
Guanidine	0.40 ± 0.16	1.62 ± 0.81*	< 0.20 - 1.70
Methylguanidine	0.50 ± 0.15	1.04 ± 0.41	< 0.10 - 0.30

(μM/1) *:P < 0,001 to controls

well as in curarized rats. GAA and GBA, on the contrary, were increased in
both experimental groups. However, increased levels of creatinine (CRN),
arginine, guanidine and MG found in non-curarized rats, were not found in
curarized rats. No significant differences were found between serum guanidinosuccinic acid, argininic acid and homoarginine levels of experimental
and control rats.

Table 2 shows the levels of the guanidino compounds in brain during
seizures of audiogenic sensitive rats with and without curare compared to
control rats. Creatine was significantly decreased in the rat brains treated

Table 2. Levels of guanidino compounds in brain of audiogenic sensitive rats
with seizures.

Guanidino compounds	Control rats	Rats with tonic seizures	Rats with seizures and curare
Guanidinosuccinic acid	< 0.06 - 0.5	< 0.06 - 2.1	< 0.06 - 0.1
Creatine	10123 ± 1433	9366 ± 2433	9020 ± 492*
Guanidinoacetic acid	10.7 ± 5.3	13.0 ± 3.4	9.5 ± 1.8
Argininic acid	1.0 ± 0.4	1.0 ± 0.4	0.9 ± 0.3
Creatinine	347 ± 204	1292 ± 555**	741 ± 239*
γ-Guanidinobutyric acid	5.1 ± 1.1	4.6 ± 1.2	5.3 ± 0.6
Arginine	188 ± 37	154 ± 49	197 ± 46
Homoarginine	1.7 ± 0.8	1.7 ± 0.7	1.6 ± 0.6
Guanidine	< 0.6 - 1.5	2.2 ± 1.0	2.8 ± 1.6
Methylguanidine	2.9 ± 2.9	2.9 ± 1.1	1.6 ± 0.4

(nmol/g tissue) *:p < 0,01 to controls,**:P < 0,001 to controls

with curare. In rats without curare there was a not significant decrease of
CTN. CRN was about 3.5 times higher in the brain of non-curarized rats than
in controls. When rats were treated with curare a similar increase of CRN was
seen. However the increase was only two-fold. The other guanidino compounds
were not significantly changed.

The changes in the guanidino compound levels observed in serum are much
clearer than in brain. Independent of the use of curare a clear decrease of
CTN levels in serum and brain is seen during audiogenic convulsions. The
large increase in CRN in serum and brain of non-curarized rats is less significant in curarized rats.

In conclusion we can say that during seizures the decrease in CTN cannot
be explained by muscular contraction or cell destruction. Neither can an
increase of muscular activity account for the higher levels of GAA and GBA in
serum since the increases were seen in both experimental groups. However the
increases in CRN, arginine, guanidine and MG in serum were only found in
non-curarized rats. Consequently these increases could be related to muscular
activity. We furthermore call attention to the increase in CRN in brain of
rats treated with curare. We believe that this increase could be related to
the intensified cerebral metabolism during seizures.

ACKNOWLEDGEMENTS

This work was supported by the Ministerium fuer Gesundheitswesen der
DDR, "Forschungsrichtung 04 Hirngeschädigte Kinder der HFR M 30 Schwanger-
schaft und frühkindliche Entwicklung", the "Fonds voor Geneeskundig Weten-
schappelijk Onderzoek" (Grant No. 3.00019.86), The "Ministerie van Nationale
Opvoeding en Nederlandse Cultuur", the "Nationale Loterij" Grant no.
9.0017.83, the "Concerted Action U.I.A., 1984-1989", the "Ministerie van de
Vlaamse Gemeenschap", the University of Antwerp; and the Born-Bunge Foundation.

REFERENCES

1. Mizuno, A., Mukawa, J., Kobayashi, K. and Mori, A., Convulsive activity
   of taurocyamine in cats and rabbits, IRCS Med. 3 (1975) 385.
2. Jinnai, D., Mori, A., Mukawa, J., Ohkusu, H., Hosotani, M., Mizuno, A.
   and Tye, L.L., Biochemical and physiological studies on guanidino compounds induced convulsions, Jpn. J. Brain Physiol. 106 (1969) 3668-3673.
3. Jinnai, D., Sawai, A. and Mori, A., γ-Guanidinobutyric acid as a convulsive substance, Nature 212 (1966) 617.
4. Mori, A. and Ohkusu, H., Isolation and identification of alpha-N-acetyl-
   L-arginine and its effect on convulsive seizure, Adv. Neurol. Sci.
   (Tokyo) 15 (1971) 303-306.
5. Mori, A., Watanabe, Y., Shindo, S., Akagi, M. and Hiramatsu, M.,
   α-Guanidinoglutaric acid and epilepsy, in: "Urea cycle diseases," A.
   Lowenthal, A. Mori and B. Marescau Eds., Plenum Press, New York (1982)
   pp. 419-426.
6. Matsumoto, M., Kobayashi, K., Kishikawa, H. and Mori, A., Convulsive
   activity of methylguanidine in cats and rabbits, IRCS Med. Sci. 4
   (1976) 65.
7. Marescau, B., Hiramatsu, M. and Mori, A., α-Keto-δ-guanidinovaleric
   acid induced electroencephalographic, epileptiform discharges in
   rabbits, Neurochem. Pathol. 1 (1983) 203-209.

8. Wiechert, P., Marescau, B., De Deyn, P. and Lowenthal, A., Guanidino compounds in serum and brain of audiogenically sensitive rats, Biomed. Biochim. Acta 45 (1986) 1339-1342.
9. Wiechert, P., Marescau, B., De Deyn, P. and Lowenthal, A., Guanidino compounds in serum and brain of audiogenic sensitive rats during the preconvulsive running and tonic phase of cerebral seizures, Neurosciences 13 (1987) 35-39.
10. Marescau, B., Qureshi, J.A., De Deyn, P., Letarte, J., Ryba, R. and Lowenthal, A., Guanidino compounds in plasma, urine and cerebrospinal fluid of hyperargininemic patients during therapy, Clin. Chim. Acta. 146 (1985) 21-27.

GUANIDINO COMPOUNDS IN SERUM AND CEREBROSPINAL FLUID OF EPILEPTIC AND SOME

OTHER NEUROLOGICAL PATIENTS

Bart Marescau, Peter P. De Deyn, Peter Wiechert*, Midori
Hiramatsu**, Lut Van Gorp, Werner P. De Potter and Armand
Lowenthal

Laboratory of Neurochemistry, Born-Bunge Foundation, and
Laboratory of Neuropharmacology, University of Antwerp (UIA)
Universiteitsplein 1, B-2610 Antwerp, Belgium
*Clinic of Psychiatry and Neurology, Department of Laboratory
Diagnostics and Clinical Neurochemistry, W.-Pieck-University
2500 Rostock, D.D.R.
**Department of Neurochemistry, Institute for Neurobiology
Okayama University Medical School, Okayama 700, Japan

INTRODUCTION

The study of the guanidino compounds in our laboratory started a few
years after the description[1] of the first patients with hyperargininemia[2].
These patients are characterized by neurological symptomatology with
epilepsy and by an accumulation of guanidino compounds in the biological
fluids. Most of the guanidino compounds such as taurocyamine[3], guanidino-
acetic acid (GAA)[4], γ-guanidinobutyric acid[5], N-α-acetylarginine[6], α-guan-
idinoglutaric acid[7], methylguanidine[8] and α-keto-δ-guanidinovaleric acid[9],
have been shown to be experimentally epileptogenic. Given these facts and
with the possibility of a sensitive fluorescence detection method, we
investigated the guanidino compounds in serum and cerebrospinal fluid (CSF)
of patients with some neurological disorders including epilepsy.

METHODS

The concentration of the guanidino compounds was determined using a
Biotronic LC 6001 amino acid analyser adapted for guanidino compound determination. The guanidino compounds were separated over a cation-exchange
column using sodium citrate buffers and were detected with the fluorescence
ninhydrin method as has been reported in detail earlier[10]. The procedure of
collection and preparation of samples has also been described[11].

RESULTS AND DISCUSSION

1. Controls

The control group consisted of 38 patients (16 men and 22 women, mean age 36) presenting with neurological complaints in whom however, after performing a series of clinical and neurochemical diagnostic tests, no neuro logical, nephrological, hepatological or metabolic disease was diagnosed.

Differences, according to sex, in the serum levels of the guanidino compounds were found for some guanidino compounds. Table 1 gives the levels of the different guanidino compounds. We found creatine (CTN) to be significantly lower and GAA to be significantly higher in men compared to women. The other guanidino compounds were not significantly changed. The same phenomenon was observed in the epileptic and other neurological patient groups (Table 2). The levels of the different guanidino compounds in CSF of men and women are represented in Table 3. For all the determined guanidino compounds we found no significant differences between sexes.

2. Neurological, Non-epileptic Patients

The non-epileptic neurological patient group consisted of 46 individual suffering from a variety of central nervous system diseases (26 men and 21 women, mean age 53).

Comparing the serum and CSF guanidino compound levels from neurological patients with controls, we did not find significant differences except for guanidinosuccinic acid (GSA) in serum and CSF and for CTN in CSF (Table 4,5) Both compounds were higher than in controls.

Table 1. Levels of guanidino compounds in serum of both sexes of controls.

Guanidino compounds	Men n=16	Women n=22	p	total control group (n=38)
α-K-δ-GVA	< 0.05	< 0.05		< 0.05
GSA	0.26 ± 0.09	0.28 ± 0.12	N.S.	0.27 ± 0.11
CTN	37.2 ± 12.5	58.7 ± 19.8	< 0.01	
GAA	2.60 ± 0.57	1.85 ± 0.67	< 0.01	
NAA	< 0.025 − 0.64	< 0.025 − 0.33		< 0.025 − 0.64
ArgA	< 0.025 − 0.19	< 0.025 − 0.17		< 0.025 − 0.19
CRN	85.6 ± 21.3	81.5 ± 27.9	N.S.	83.1 ± 25.3
GBA	< 0.025 − 0.06	< 0.025		< 0.025 − 0.06
Arg	113 ± 12.7	114 ± 31	N.S.	114 ± 26
HArg	2.22 ± 0.69	1.70 ± 0.63	N.S.	1.97 ± 0.72
Gua	< 0.20	< 0.20		< 0.20
MG	< 0.10	< 0.10		< 0.10

(μM), α-K-δ-GVA; α-Keto-δ-guanidinovaleric acid, GSA; guanidinosuccinic acid, CTN; creatine, GAA; guanidinoacetic acid, NAA; N-α-acetylarginine, ArgA; argininic acid, CRN; creatinine, GBA; γ-guanidinobutyric acid, Arg; arginine, HArg; homoarginine, Gua; guanidine, MG; methylguanidine and N.S.; not significant.

Table 2. Levels of creatine (CTN) and guanidinoacetic acid (GAA) in serum
          of both sexes of controls, epileptic and some other neurological
          patients (μM).

		CTN	GAA
controls, men	(n=16)	37.2 ± 12.5	2.60 ± 0.57
controls, women	(n=22)	58.7 ± 19.8*	1.85 ± 0.67*
neurological, men	(n=26)	42.4 ± 22.9	2.43 ± 0.52
neurological, women	(n=21)	67.3 ± 31.9*	1.91 ± 0.60*
epileptic, men	(n=55)	33.2 ± 18.9	2.47 ± 0.71
epileptic, women	(n=47)	50.4 ± 27.9*	1.94 ± 0.63*

*$p < 0.01$ to men.

Table 3. Levels of guanidino compounds in cerebrospinal fluid of both sexes
          of controls.

Guanidino compounds	Men n=13	Women n=27	p	total control group n=40
α-K-δ-GVA	< 0.025	< 0.025		< 0.025
GSA	0.09 ± 0.04	0.07 ± 0.03	N.S.	0.07 ± 0.03
CTN	51.6 ± 7.7	56.3 ± 10.8	N.S.	54.1 ± 9.1
GAA	0.05 ± 0.02	0.04 ± 0.03	N.S.	0.05 ± 0.02
NAA	0.10 ± 0.04	0.11 ± 0.03	N.S.	0.11 ± 0.04
ArgA	< 0.013 - 0.04	< 0.013 - 0.07		< 0.013 - 0.07
CRN	64.0 ± 18.9	69.8 ± 29.6	N.S.	68.3 ± 27.0
GBA	< 0.013 - 0.10	< 0.013 - 0.10		< 0.013 - 0.10
Arg	22.2 ± 3.9	20.3 ± 5.0	N.S.	20.9 ± 4.7
HArg	0.28 ± 0.13	0.29 ± 0.12	N.S.	0.28 ± 0.11
Gua	< 0.10	< 0.10		< 0.10
MG	< 0.05	< 0.05		< 0.05

(μM) The abbreviations of guanidino compounds are the same as shown in
Table 1.

3. Epileptic Patients

     Serum was obtained from 95 epileptic patients (55 men and 41 women, mean
age 40) with different clinical seizure types: Grand Mal (essential)
(n=35); Grand Mal (symptomatic) (n=16); Complex Focal (essential) (n=22);
Complex Focal (symptomatic) (n=7); Simple Focal (essential) (n=11) and Petit
Mal (n=4). CSF was obtained from 56 epileptic patients (32 men and 24 women,
mean age 40) with different clinical seizure types: Grand Mal (essential) (n=28); Grand Mal (symptomatic) (n=9); Complex Focal (essential) (n=9);
Complex Focal (symptomatic) (n=2); Simple Focal (essential + symptomatic)
(n=4) and Petit Mal (n=4). With exception of a few patients, time span between last seizure and moment of sampling was 12 h or more. Samples of
fifteen patients of the epileptic group were obtained before anticonvulsive
therapy.

Table 4. Levels of guanidino compounds in serum of non-epileptic neurolog-
          ical patients.

Guanidino compounds	Controls n=38	Neurological patients n=46	p
α-K-δ-GVA	< 0.05	< 0.05	
GSA	0.27 ± 0.11	0.34 ± 0.17	< 0.05
CTN, men	37.2 ± 12.5	42.4 ± 22.9	N.S.
CTN, women	58.7 ± 19.8	67.3 ± 31.9	N.S.
GAA, men	2.60 ± 0.57	2.43 ± 0.52	N.S.
GAA, women	1.85 ± 0.67	1.91 ± 0.60	N.S.
NAA	< 0.025 - 0.64	< 0.025 - 0.87	
ArgA	< 0.025 - 0.19	< 0.025 - 0.20	
CRN	83.1 ± 25.3	77.4 ± 29.5	N.S.
GBA	< 0.025 - 0.06	< 0.025 - 0.15	
Arg	114 ± 26	124 ± 27	N.S.
HArg	1.97 ± 0.72	1.83 ± 0.77	N.S.
Gua	< 0.20	< 0.20	
MG	< 0.10	< 0.10	

(μM) The abbreviations of guanidino compounds are the same as shown in
Table 1.

Table 5. Levels of guanidino compounds in cerebrospinal fluid of non-
          epileptic neurological patients.

Guanidino compounds	Controls n=40	Neurological patients n=46	p
α-K-δ-GVA	< 0.025	< 0.025	
GSA	0.07 ± 0.03	0.11 ± 0.06	< 0.01
CTN	54.1 ± 9.1	60.3 ± 12.3	< 0.05
GAA	0.05 ± 0.02	0.05 ± 0.03	N.S.
NAA	0.11 ± 0.04	< 0.013 - 0.36	
ArgA	< 0.013 - 0.07	< 0.013 - 0.03	
CRN	68.3 ± 27.0	68.6 ± 24.1	N.S.
GBA	< 0.013 - 0.10	< 0.013 - 0.05	
Arg	20.9 ± 4.7	21.0 ± 3.9	N.S.
HArg	0.28 ± 0.11	0.26 ± 0.10	N.S.
Gua	< 0.10	< 0.10	
MG	< 0.05	< 0.05	

(μM) The abbreviations of guanidino compounds are the same as shown in
Table 1.

     Since significant decreases were found for GSA, creatinine (CRN) and
homoarginine (HArg) in serum of the total epileptic group (Table 6) we subdivided the epileptic patients according to the clinical seizure types. Table
and 8 however give only the levels from three seizure types. For the other

Table 6. Levels of guanidino compounds in serum of epileptic patients.

Guanidino compounds	Controls n=38	Epileptic patients n=95	p
α-K-δ-GVA	< 0.05	< 0.05	
GSA	0.27 ± 0.11	0.19 ± 0.09	< 0.01
CTN, men	37.2 ± 12.5	33.2 ± 18.9	N.S.
CTN, women	58.7 ± 19.8	50.4 ± 27.9	N.S.
GAA, men	2.60 ± 0.57	2.47 ± 0.71	N.S.
GAA, women	1.85 ± 0.67	1.94 ± 0.63	N.S.
NAA	< 0.025 - 0.64	< 0.025 - 0.72	
ArgA	< 0.025 - 0.19	< 0.025 - 0.28	
CRN	83.1 ± 25.3	68.6 ± 25.4	< 0.01
GBA	< 0.025 - 0.06	< 0.025 - 0.26	
Arg	114 ± 26	104 ± 28	N.S.
HArg	1.97 ± 0.72	1.46 ± 0.59	< 0.01
Gua	< 0.20	< 0.20	
MG	< 0.10	< 0.10	

(μM) The abbreviations of guanidino compounds are the same as shown in Table 1.

Table 7. Levels of guanidino compounds in serum of patients with different clinical types of epilepsy.

Guanidino compounds	Controls n=38	Grand Mal Essential n=35	Grand Mal Symptomatic n=16	Complex Focal Essential n=22
α-K-δ-GVA	< 0.05	< 0.05	< 0.05	< 0.05
GSA	0.27 ± 0.11	0.21 ± 0.10*	0.16 ± 0.09**	0.17 ± 0.09**
CTN, men	37.2 ± 12.5	34.4 ± 16.8	44.2 ± 26.4	22.9 ± 7.86**
CTN, women	58.7 ± 19.8	47.2 ± 17.0	60.4 ± 20.6	33.2 ± 11.0**
GAA, men	2.60 ± 0.57	2.61 ± 0.71	2.59 ± 0.96	2.42 ± 0.43
GAA, women	1.85 ± 0.67	1.77 ± 0.67	2.16 ± 0.71	1.92 ± 0.46
NAA	< 0.025 - 0.64	< 0.025 - 0.50	< 0.025 - 0.53	< 0.025 - 0.72
ArgA	< 0.025 - 0.19	< 0.025 - 0.30	< 0.025 - 0.22	< 0.025 - 0.23
CRN	83.1 ± 25.3	74.1 ± 24.7	66.4 ± 16.7*	71.4 ± 33.1
GBA	< 0.025 - 0.06	< 0.025 - 0.20	< 0.025 - 0.14	< 0.025 - 0.26
Arg	114 ± 26.4	103 ± 21.8*	124 ± 57.0	109 ± 20.1
HArg	1.97 ± 0.72	1.48 ± 0.59**	1.36 ± 0.60**	1.55 ± 0.58*
Gua	< 0.20	< 0.20	< 0.20	< 0.20
MG	< 0.10	< 0.10	< 0.10	< 0.10

(μM) *$P < 0.05$ to controls, **$P < 0.01$ to controls. The abbreviations of guanidino compounds are the same as shown in Table 1.

Table 8. Levels of guanidino compounds in cerebrospinal fluid of patients
          with different clinical types of epilepsy.

Guanidino compounds	Controls n=38	Grand Mal Essential n=27	Grand Mal Symptomatic n=8	Complex Focal Essential n=9
α-K-δ-GVA	< 0.025	< 0.025	< 0.025	< 0.025
GSA	0.07 ± 0.03	0.07 ± 0.03	0.06 ± 0.03	0.07 ± 0.04
CTN	54.1 ± 9.1	52.2 ± 8.2	57.0 ± 14.7	60.3 ± 8.5
GAA	0.05 ± 0.02	0.08 ± 0.07	0.06 ± 0.05	0.07 ± 0.05
NAA	0.11 ± 0.04	0.10 ± 0.03	0.09 ± 0.04	0.14 ± 0.08
ArgA	< 0.013 – 0.07	< 0.013 – 0.05	< 0.013 – 0.04	< 0.013 – 0.03
CRN	68.3 ± 27.0	82.2 ± 32.2	55.4 ± 25.6	79.7 ± 30.5
GBA	< 0.013 – 0.10	< 0.013 – 0.04	< 0.013 – 0.04	< 0.013 – 0.03
Arg	20.9 ± 4.7	19.6 ± 5.1	21.6 ± 9.6	21.4 ± 5.4
HArg	0.28 ± 0.11	0.27 ± 0.16	0.25 ± 0.11	0.32 ± 0.11
Gua	< 0.10	< 0.10	< 0.10	< 0.10
MG	< 0.05	< 0.05	< 0.05	< 0.05

(μM) The abbreviations of guanidino compounds are the same as shown in
Table 1.

Table 9. Levels of guanidino compounds in cerebrospinal fluid of epileptic
          patients.

Guanidino compounds	Controls n=35	Epileptic patients n=56	P
α-K-δ-GVA	< 0.025	< 0.025	
GSA	0.07 ± 0.03	0.07 ± 0.03	N.S.
CTN	54.1 ± 9.1	55.0 ± 10.7	N.S.
GAA	0.05 ± 0.02	0.08 ± 0.07	N.S.
NAA	0.11 ± 0.04	0.11 ± 0.05	N.S.
ArgA	< 0.013 – 0.07	< 0.013 – 0.15	
CRN	68.3 ± 27.0	75.2 ± 30.6	N.S.
GBA	< 0.013 – 0.10	< 0.013 – 0.04	
Arg	20.9 ± 4.7	19.4 ± 4.5	N.S.
HArg	0.28 ± 0.11	0.26 ± 0.11	N.S.
Gua	< 0.10	< 0.10	
MG	< 0.05	< 0.05	

(μM) The abbreviations of guanidino compounds are the same as shown in
Table 1.

classes, the number of patients was too small for statistical analysis. Also
in the three studied groups, GSA and HArg is significantly decreased to controls. The levels of all the guanidino compounds determined in CSF of the to
tal epileptic group were not significantly different from those of controls
(Table 9).

Since GSA, CRN and HArg were significantly decreased in serum of the total epileptic group, we investigated the possible influence of seizure frequency, medication and time between the last seizure and taking of the sample on the levels of the guanidino compounds. As well for serum as for CSF we found no significant differences between the groups of epileptic patients classified according to seizure frequency. Also, in serum and CSF, we found no significant differences between the groups of epileptic patients classified with respect to time between the last seizure and the sampling moment, except for HArg in serum that significantly decreased with shorter times (Table 10). However, it might be possible that shorter time spans than those used in this study, could be characterized by other differences in guanidino compound patterns. When the epileptic patients were subdivided according to their medication, no significant differences in the CSF levels of all the guanidino compounds were found as compared to controls. However in serum a significant decrease of GSA was observed in all the groups, except in the epileptic group taking phenobarbital (PB) (Table 11). Interesting as well is that the serum levels of GSA in epileptic patients taking no anti-epileptic drugs were comparable to the normal values. The epileptic patients taking no anti-epileptic drugs had also serum CRN levels comparable to control values (Table 12). For al the different groups of patients taking anti-convulsive drugs, the serum levels of CRN were decreased. The decreases

Table 10.   Levels of homoarginine in serum of epileptic patients classified according to time span between last seizure and sampling moment.

Time	Serum	n	p
Control	1.97 ± 0.72	38	
> 30 days	1.77 ± 1.05	19	N.S.
20–30 days	1.03 ± 0.04	3	< 0.05
15–20 days	1.40 ± 0.05	9	< 0.05
7–15 days	1.39 ± 0.78	12	< 0.05
1–7 days	1.46 ± 0.62	30	< 0.01
< 1 days	1.41 ± 0.55	5	N.S.

(µM)   N.S., not significant.

Table 11.   Levels of guanidinosuccinic acid in serum of epileptic patients classified according to anti-epileptic treatment.

Anti-epileptic drugs	Serum	n	p
Control	0.27 ± 0.11	38	
No anti E	0.32 ± 0.15	15	N.S.
PB	0.26 ± 0.11	8	N.S.
Val	0.19 ± 0.06	11	< 0.05
Carb.	0.17 ± 0.06	7	< 0.05
PHT + PB	0.16 ± 0.10	13	< 0.01
PHT + PB + Carb.	0.15 ± 0.09	8	< 0.01

(µM)   PHT; Phenytoin, PB; phenobarbital; Carb.; carbamazepine, Val; sodium valproate, No anti E; no anti-epileptic drugs and N.S.; not significant.

of CRN in serum were significant for those taking valproate and the combination of phenytoin (PHT) and PB. In contrast to GSA and CRN, which levels were comparable to controls, the concentration of HArg was significantly decreased in epileptic patients taking no anti-epileptic drugs (Table 13). Further the levels of HArg were significantly decreased during treatment with carbamazepine and with the combination of PHT, PB and carbamazepine. It is worth noting that patients only treated with PB have HArg levels comparable to controls. It seems that the serum guanidino compound levels of epileptic patients taking only PB are similar to the levels of controls. However, other anti-epileptic drugs seem to influence the levels of guanidino compounds. Table 14 shows the tendency for the diminishing effect of valproate on the serum levels of the guanidino compounds in epileptic patients.

To conclude, we found that in serum of controls, epileptic patients and other neurological patients, the creatine levels are significantly lower and the GAA levels are significantly higher in men than in women. Further it seems that seizure frequency and time span between the last seizure and the sampling moment (at least 12 hours) do not influence the serum and cerebrospinal fluid levels of guanidino compounds. On the other hand, most of the anti-epileptic drugs seem to be associated with lower levels of GSA, CRN and HArg in serum.

Table 12.   Levels of creatinine in serum of epileptic patients classified according to anti-epileptic treatment.

Anti-epileptic drugs	Serum	n	p
Control	83.1 ± 25.3	38	
No anti E	79.3 ± 32.9	15	N.S.
PB	70.7 ± 15.1	8	N.S.
Val	60.7 ± 27.5	10	< 0.05
Carb.	69.8 ± 37.3	7	N.S.
PHT + PB	64.8 ± 23.0	13	< 0.05
PHT + PB + Carb.	76.3 ± 39.3	8	N.S.

(μM) The abbreviations of drugs are the same as shown in Table 11.

Table 13.   Levels of homoarginine in serum of epileptic patients classified according to anti-epileptic treatment.

Anti-epileptic drugs	Serum	n	p
Control	1.97 ± 0.72	38	
No anti E	1.33 ± 0.71	14	< 0.01
PB	1.97 ± 0.97	8	N.S.
Val	1.62 ± 0.52	11	N.S.
Carb.	1.21 ± 0.64	7	< 0.05
PHT + PB	1.87 ± 1.07	12	N.S.
PHT + PB + Carb.	1.37 ± 0.36	8	< 0.05

(μM) The abbreviations of drugs are the same as shown in Table 11.

Table 14.   Levels of guanidino compounds in epileptic patients under treatment only with valproate.

Guanidino compounds	Controls n=38	Valproate n=11	P
α-K-δ-GVA	< 0.05	< 0.05	
GSA	0.27 ± 0.11	0.19 ± 0.06	< 0.05
CTN, men	37.2 ± 12.5	34.3 ± 13.1	N.S.
CTN, women	58.7 ± 19.8	55.5 ± 15.3	N.S.
GAA, men	2.60 ± 0.57	2.94 ± 0.53	N.S.
GAA, women	1.85 ± 0.67	1.56 ± 0.38	N.S.
NAA	< 0.025 - 0.64	< 0.025 - 0.40	
ArgA	< 0.025 - 0.19	< 0.025 - 0.25	
CRN	83.1 ± 25.3	60.7 ± 27.5	< 0.05
GBA	< 0.025 - 0.06	< 0.025 - 0.20	
Arg	114 ± 26	96.5 ± 20.5	< 0.05
HArg	1.97 ± 0.72	1.62 ± 0.52	N.S.
Gua	< 0.20	< 0.20	
MG	< 0.10	< 0.10	

(μM) The abbreviations of guanidino compounds are the same as shown in Table 1.

ACKNOWLEDGEMENTS

     This work was supported by the "Fonds voor Geneeskundig Wetenschappelijk Onderzoek" (grant nr. 3.00019.86), the "Ministerie voor Nationale Opvoeding en Nederlandse Cultuur", the "Nationale Loterij" (grant nr. 9.0017.83), the NATO Research Grant nr. 83/0913, the "Concerted Action U. I. A." nr. 84/89-68, the "Ministerie van de Vlaamse Gemeenschap", the "Foundation Justine Lacoste-Beaubien", the Quebec-Belgium exchange programme, the University of Antwerp and the Born-Bunge Foundation.

REFERENCES

1. Terheggen, H.G., Lavinha, F., Colombo, J.P., Van Sande, M. and Lowenthal, A., Familial hyperargininemia, J. Génét. Hum. 20 (1972) 69-84.
2. Terheggen, H.G., Schwenk, A., Lowenthal, A., Van Sande, M. and Colombo, J.P., Argininemia with arginase deficiency, Lancet II (1969) 748-749.
3. Mizuno, A., Mukawa, J., Kobayashi, K. and Mori, A., Convulsive activity of taurocyamine in cats and rabbits, IRCS Med. Sci. 3 (1975) 385.
4. Jinnai, D., Mori, A., Mukawa, J., Ohkusu, H., Hosotani, M., Mizuno, A. and Tye, L.L., Biochemical and physiological studies on guanidino compounds induced convulsions, Jpn. J. Brain Physiol. 106 (1969) 3668-3673.
5. Jinnai, D., Sawai, A. and Mori, A., γ-Guanidinobutyric acid as a convulsive substance, Nature 212 (1966) 617.
6. Mori, A. and Ohkusu, H., Isolation and identification of α-N-acetyl-L-arginine and its effect on convulsive seizure, Neurol. Sci. 15 (1971) 303-305.
7. Mori, A., Watanabe, Y., Shindo, S., Akagi, M. and Hiramatsu, M., α-Guanidinoglutaric acid and epilepsy, In: "Urea Cycle Diseases," A.

Lowenthal, A. Mori and B. Marescau Eds., Plenum Press, New York (1982)
pp. 419-426.

8. Matsumoto, M., Kobayashi, K., Kishikiwa, H. and Mori, A., Convulsive
   activity of methylguanidine in cats and rabbits, IRCS Med. Sci, 4
   (1976) 65.

9. Marescau, B., Hiramatsu, M. and Mori, A., α-Keto-δ-guanidinovaleric acid
   induced electroencephalographic, epileptiform discharges in rabbits,
   Neurochem. Pathol. 1 (1983) 203-209.

10. Marescau, B., Qureshi, I.A., De Deyn, P., Letarte, J., Ryba, R. and
    Lowenthal, A., Guanidino compounds in plasma, urine and cerebrospinal
    fluid of hyperargininemic patients during therapy, Clin. Chim. Acta 146
    (1985) 21-27.

11. De Deyn, P.P., Marescau, B., Cuyckens, J.J., Van Gorp, L., Lowenthal,
    A. and De Potter, W.P., Guanidino compounds in serum and cerebrospinal
    fluid of non-dialyzed patients with renal insufficiency, Clin. Chim.
    Acta, 167 (1987) 81-88.

THE INVOLVEMENT OF CATECHOLAMINES IN THE SEIZURE MECHANISM INDUCED BY α-GUANIDINOGLUTARIC ACID IN RATS

Hiroshi Shiraga, Midori Hiramatsu and Akitane Mori

Department of Neurochemistry, Institute for Neurobiology
Okayama University Medical School, 2-5-1 Shikata-cho
Okayama 700, Japan

INTRODUCTION

Guanidino compounds, e.g., γ-guanidinobutyric acid[1], guanidinoacetic acid[2], N-acetylarginine[3], taurocyamine[4], methylguanidine[5], and homoarginine[6], are known to induce epileptic seizures when administered intracisternally into rabbits. α-Guanidinoglutaric acid (α-GGA) is a guanidino compound first found in the cobalt-induced epileptogenic focus tissue in the cerebral cortex of cats[7]. α-GGA was also found to induce epileptic seizures when administered into the sensory motor cortex of rabbits[8] and when administered intraventricularly into rats[9].

There are many studies concerning seizure mechanisms and brain monoamines are thought to be closely related to seizure susceptibility[10-17]. We have already reported the possible involvement of 5-hydroxytryptamine (5-HT) in the α-GGA-induced seizure mechanism[9]. In this experiment, we examined the effect of α-GGA on brain catecholamine levels after intraventricular administration in rats.

MATERIALS AND METHODS

Animals

Male Sprague-Dawley rats weighing 260-350 g were used. Animals were maintained on a 12-hr light/dark cycle at a constant temperature with food and water available ad libitum. Rats were anesthetized with diethylether and 10 μl of 1 M α-GGA or saline as a control was administered into the left lateral ventricle with the use of a stereotaxic apparatus. The rats were killed by microwave irradiation at 4.5 kW of power for 1.3 s (Model NJE 2601, New Japan Radio, Japan)[18] 3 min, 5 min, 10 min, 30 min, 100 min, and 200 min after the administration of α-GGA or saline. These times were selected with reference to the findings in the electroencephalograms[9], i.e., 3 min corresponds to the time before the appearance of spikes; 5 min corresponds to the time when sporadic spikes just appear; 10 min corresponds to the time when multiple spikes appear; 30 min corresponds to the time when multiple spikes appear at their maximal frequency; 100 min corresponds to the time when spikes disappear, and 200 min corresponds to the time 100 min after spikes disappear. The whole brain was rapidly removed and the cerebellum, pons-medulla oblongata, hypothalamus, striatum, midbrain, hippocampus, right cortex, and left cortex were dissected on an ice plate[19]; these samples were then kept at −80°C until catecholamine analysis. Catecholamines were analysed only in the pons–medulla oblongate 200 min after administration; all other regions were followed to a maximum of 100 min.

Catecholamine Analysis

Brain tissues were homogenized with 0.1 ml of 0.1 M EDTA disodium salt, 0.1 ml of 0.1 M $NaHSO_3$, 2.7 ml of 0.05 M $HClO_4$, and 0.1 ml of 7 µM 3,4–dihydroxybenzylamine in a Polytron homogenizer. 3,4–Dihydroxybenzylamine was used as the internal standard. After centrifugation at 8,000 rpm for 15 min, supernatants were mixed with 30 mg of activated aluminum oxide and 5 ml of 1 M Tris–HCl (pH 8.6). After shaking for 15 min, supernatants were removed through aspiration and activated aluminum oxide was washed twice in 3 ml of distilled water. Then catecholamine, absorbed to activated aluminum oxide, was eluted with 0.5 ml of 0.1 N HCl. After centrifugation at 5,000 rpm for 5 min, 10 µl of the supernatant was injected into an HPLC apparatus with electrochemical detection. The HPLC (Yanaco L–4000W, Kyoto) conditions were as follows: the column was ODS–A (4.6 x 250 mm), the solvent was 10 µM EDTA disodium salt/0.1 M phosphate buffer (pH 3.0), and the detector was VMD–501 and its voltage was 750 mV.

RESULTS

Fig. 1–8 show the effect of α–GGA on norepinephrine (NE) levels in different regions of the rat brain. The NE level decreased significantly in the pons–medulla oblongata 10 min after administration. No change in the NE level was found 3 min, 5 min, 30 min, 100 min, and 200 min after administration.

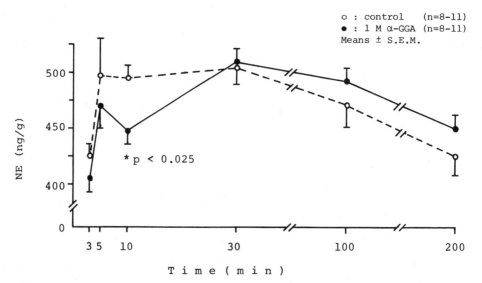

Figure 1.   Effect of α–guanidinoglutaric acid (α–GGA) on the norepinephrine (NE) level in the pons–medulla oblongata.

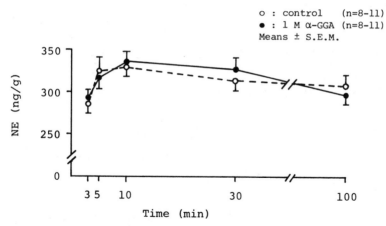

Figure 2.   Effect of α-guanidinoglutaric acid (α-GGA) on the norepinephrine
            (NE) level in the right cortex.

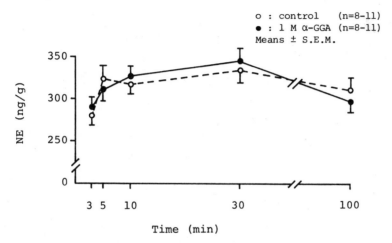

Figure 3.   Effect of α-guanidinoglutaric acid (α-GGA) on the norepinephrine
            (NE) level in the left cortex.

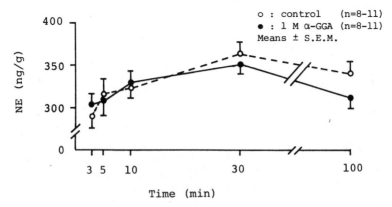

Figure 4.   Effect of α-guanidinoglutaric acid (α-GGA) on the norepinephrine
            (NE) level in the hippocampus.

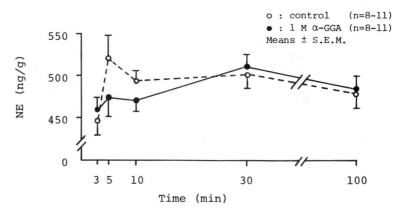

Figure 5.   Effect of α-guanidinoglutaric acid (α-GGA) on the norepinephrine
(NE) level in the midbrain.

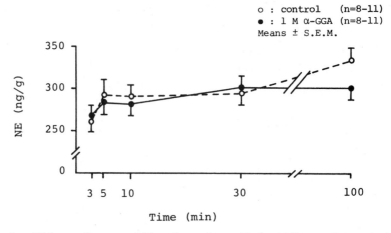

Figure 6.   Effect of α-guanidinoglutaric acid (α-GGA) on the norepinephrine
(NE) level in the striatum.

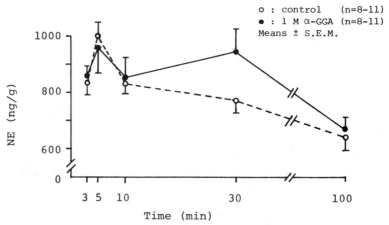

Figure 7.   Effect of α-guanidinoglutaric acid (α-GGA) on the norepinephrine
(NE) level in the hypothalamus.

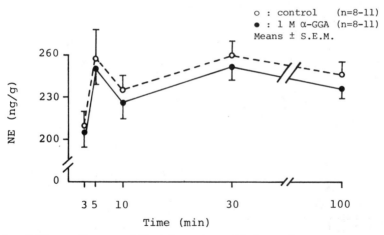

Figure 8.   Effect of α-guanidinoglutaric acid (α-GGA) on the norepinephrine
             (NE) level in the cerebellum.

        Fig. 9-16 show the effect of α-GGA on dopamine (DA) levels in different
regions of the rat brain. The DA level decreased significantly in the pons-
medulla oblongata 10 min after administration and increased significantly in
the same region 100 min after administration. No change in the DA level was
found 3 min, 5 min, 30 min, and 200 min.

DISCUSSION

        There are many reports concerning the relationship between seizures and
catecholamines. It is said that the seizure threshold increases when the
brain catecholamine level increases and the seizure threshold decreases when
the brain catecholamine level decreases[10-17].

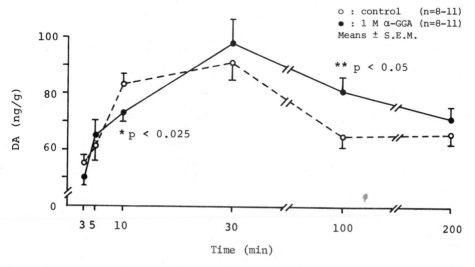

Figure 9.   Effect of α-guanidinoglutaric acid (α-GGA) on the dopamine (DA)
             level in the pons-medulla oblongata.

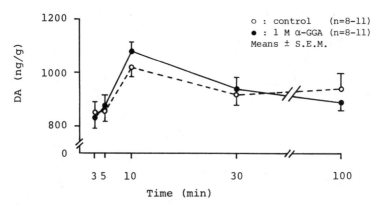

Figure 10. Effect of α-guanidinoglutaric acid (α-GGA) on the dopamine (DA) level in the right cortex.

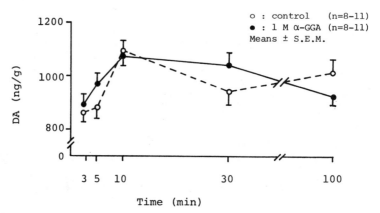

Figure 11. Effect of α-guanidinoglutaric acid (α-GGA) on the dopamine (DA) level in the left cortex.

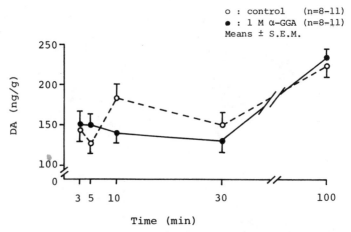

Figure 12. Effect of α-guanidinoglutaric acid (α-GGA) on the dopamine (DA) level in the hippocampus.

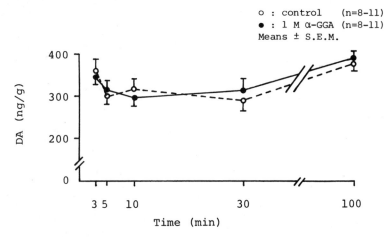

Figure 13. Effect of α-guanidinoglutaric acid (α-GGA) on the dopamine (DA) level in the midbrain.

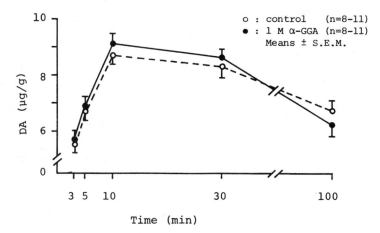

Figure 14. Effect of α-guanidinoglutaric acid (α-GGA) on the dopamine (DA) level in the striatum.

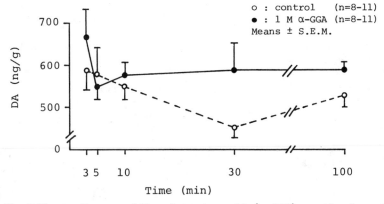

Figure 15. Effect of α-guanidinoglutaric acid (α-GGA) on the dopamine (DA) level in the hypothalamus.

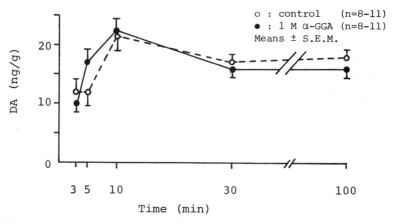

Figure 16. Effect of α-guanidinoglutaric acid (α-GGA) on the dopamine (DA) level in the cerebellum.

In this experiment, we found a decrease of DA and NE in the pons-medulla oblongata 10 min after administration.

There are many noradrenergic neurons and some dopaminergic terminals in the pons-medulla oblongata. This area is thought to be one of those where epileptic foci exist and seizures generalize[20,21]. Ten min after administration is the time for spikes to increase. As for the relationship between seizure progression and catecholamines, Michelson et al. reported[22] that 6-hydroxydopamine treatment accelerated the early stages of kindling but it had no effect on the later stages of kindling. Corcoran et al. also reported[15] that the facilitative effects of NE depletion were related to disinhibition of the spread of seizure discharges from the stimulated site in the amygdaloid kindling. These results suggest that the decrease of DA and NE is associated with α-GGA-induced seizures.

Previously we have already shown a close relationship between α-GAA-induced seizures and 5-HT[9]. That report showed that 5-HT level decreased in the midbrain 5 min after administration and also decreased in all brain regions 10 min after administration. These results show that 5-HT seems to play an important role in α-GGA-induced seizures and NE and DA seem to be partially associated with α-GGA-induced seizures.

One hundred min after administration, DA levels increased in the pons-medulla oblongata. It is well known that subsequent seizures are inhibited just after a previous seizure, which is known as postictal suppression[23]. It is also well known that seizure susceptibility decreases when brain mono-amines increase. These results suggest that the increase in DA is associated with postictal suppression in α-GGA-induced seizures.

CONCLUSION

We examined the effect of α-GGA on brain catecholamine levels after intraventricular administration in rats.

The NE level decreased significantly in the pons-medulla oblongata 10 min after administration. The DA level decreased significantly in the pons-medulla oblongata 10 min after administration and increased significantly in the pons-medulla oblongata 100 min after administration.

These results suggest that the decrease of NE and DA is partially associated with α-GGA-induced seizures and also suggest that the increase of DA is associated with postictal suppression in α-GGA-induced seizures.

REFERENCES

1. Jinnai, D., Sawai, A. and Mori, A., γ-Guanidinobutyric acid as a convulsive substance, Nature 212 (1966) 617.
2. Jinnai, D., Mori, A., Mukawa, J., Ohkusu, H., Hosotani, M., Mizuno, A. and Tye, L.C., Biological and physiological studies on guanidino compounds induced convulsion, Jpn. J. Brain Physiol. 106 (1969) 3368-3673.
3. Ohkusu, H. and Mori, A., Isolation of α-N-acetyl-L-arginine from cattle brain, J. Neurochem. 16 (1969) 1485-1486.
4. Mizuno, A., Mukawa, J., Kobayashi, K. and Mori, A., Convulsive activity of taurocyamine in cats and rabbits, IRCS Med. Sci. 3 (1975) 385.
5. Matsumoto, M., Kobayashi, K., Kishikawa, H. and Mori, A., Convulsive activity of methylguanidine in cats and rabbits, IRCS Med. Sci. 4 (1976) 65.
6. Yokoi, I., Toma, J. and Mori, A., The effect of homoarginine on the EEG of rats, Neurochem. Pathol. 2 (1984) 295-300.
7. Mori, A., Akagi, M., Katayama, Y. and Watanabe, Y., α-Guanidinoglutaric acid in cobalt-induced epileptogenic cerebral cortex of cats, J. Neurochem. 35 (1980) 603-605.
8. Mori, A., Watanabe Y., Shindo, S., Akagi, M. and Hiramatsu, M., α-Guanidinoglutaric acid and epilepsy, in: "Advances in Experimental Medicine and Biology, Vol. 153, Urea Cycle Diseae," A. Lowenthal, A. Mori and B. Marescau Eds., Plenum Press, New York (1982) pp. 419-426.
9. Shiraga, H., Hiramatsu, M. and Mori, A., Convulsive activity of α-guanidinoglutaric acid and the possible involvement of 5-hydroxy-tryptamine in the α-guanidinoglutaric acid-induced seizure mechanism, J. Neurochem. 47 (1986) 1832-1836.
10. Chen, G., Ensor, C.R. and Bohner, B., A facilitation of reserpine on the central nervous system, Proc. Soc. Exp. Biol. Med. 86 (1954) 507-510.
11. Scudder, C.L., Karczmar, A.G., Everett, G.M., Gibson, J.E. and Rifkin, M., Brain catecholamines and serotonin levels in various strains and genera of mice and a possible interaction for the correlations of amine levels with electrschock latency and behavior, Int. J. Neuropharmacol. 5 (1966) 343-351.
12. Schlesinger, K., Boggan, W. and Freedman, D.X., Genetics of audiogenic seizures: II. Effects of pharmacological manipulation of brain serotonin, norepinephrine and gammaaminobutyric acid, Life Sci. 7 (1968) 437-447.
13. Azzaro, A.J., Wenger, G.R., Craig, C.R. and Stitzel, R.E., Reserpine-induced alterations in brain amines and their relationship to changes in the incidence of minimal electroshock seizures in mice, J. Pharmacol. Exp. Ther. 180 (1972) 558-568.
14. McIntyre, D.C., Saari, M. and Pappas, B.A., Potentiation of amygdala kindling in adult or infant rats by injections of 6-hydroxydopamine, Exp. Neurol. 63 (1979) 527-544.
15. Corcoran, M. and Mason, S.T., Role of forebrain catecholamines in amygdaloid kindling, Brain Res. 190 (1980) 473-484.

16. Sato, M., Tomoda, T., Hikasa, N. and Otsuki, S., Inhibition of amygda-
    loid kindling by chronic pretreatment with cocaine or methamphetamine,
    Epilepsia 21 (1980) 497-507.
17. Jobe, P.C., Laird, H.E., Ko, K.H., Ray, T. and Daily, J.W.,
    Abnormalities in monoamine levels in the central nervous system of
    the genetically epilepsy-prone rat, Epilepsia 23 (1982) 359-366.
18. Weintraub, S.T., Stavinoha, W.B., Pike, R.L., Morgan, W.W., Modak, A. T.
    Koslow, S.H. and Blank, L., Evaluation of the necessity for rapid
    inactivation of brain enzymes prior to analysis of norepinephrine,
    dopamine and serotonin in the mouse, Life Sci. 17 (1975) 1423-1428.
19. Glowinski, J. and Iversen, L.L., Regional studies of catecholamines in
    the rat brain-I, J. Neurochem. 13 (1966) 655-669.
20. Browning, R.A., Role of the brain-stem reticular formation in tonic-
    clonic seizures: lesion and pharmacological studies, Federation Proc.
    44 (1985) 2425-2431.
21. Browning, R.A., Nelson, D.K., Mogharreban, N., Jobe, P.C. and Laird II,
    H.E., Effect of midbrain and pontine tegmental lesions on audiogenic
    seizures in genetically epilepsy-prone rats, Epilepsia 26 (1985) 175-183
22. Michelson, H.B. and Buterbaugh, G.G., Amygdala kindling in juvenile rats
    following neonatal administration of 6-hydroxydopamine, Exp. Neurol. 90
    (1985) 588-593.
23. Stock, G., Klimpel, L., Sturm, V. and Schlor, K.-H., Resistance to tonic
    clonic seizures after amygdaloid kindling in cats, Exp. Neurol. 69 (1980
    239-246.

V. HYPERARGININEMIA

HYPERARGININEMIA

Armand Lowenthal and Bart Marescau

Laboratory of Neurochemistry, Born-Bunge Foundation
University of Antwerp (UIA), Universiteitsplein 1, B-2610
Antwerp

INTRODUCTION

In 1969 we published a paper concerning the first family with hyper-argininemia[1]. Describe once again the discovery of this disease seemed interesting. The first patient with hyperargininemia, a girl, was treated by Dr. Terheggen for epilepsy. Later, when her sister was brought to him for treatment also of epilepsy, he considered the possibility of a metabolic disease. An examination of the urine was carried out in one of the best biochemical laboratories in Germany. The answer was that the found cystinuria was irrelevant in a neurological metabolic disease with epilepsy and thus of no further interest.

At a meeting one of our collaborators, when speaking about metabolic diseases, was asked if he was interested in cystinuria, as we had already published a family with cystinuria and neurological deficiencies. According to the custom of our laboratory we proposed to examine not only urine but also blood. We received some samples: in urine we could confirm the existence of the cystine-lysine-ornithine-argininuria and in blood we found an hyperargininemia. From those results, for the first time, the diagnosis of hyperargininemia was established. Cystinuria is in fact well known now as a secondary symptom of hyperargininemia.

Since, we were able to demonstrate in hyperargininemia not only a cystinuria but also the following biochemical changes: a deficiency of arginase in the erythrocytes[1] and in the leukocytes[2], an increase of ammonemia[3-4] and at least a partial metabolism of arginine through secondary pathways, associated with the formation of guanidino compounds[5]. One of these guanidino compounds, identified in our laboratory, is practically never observed in mammals or in man and was identified as α-keto-δ-guanidinovaleric acid[6].

Until 1977, during 8 years, no new patients with hyperargininemia were reported[7-8]. Since then probably 27 cases have become known, belonging to 23 different families. Several questions concerning hyperargininemia have been discussed[9-22]: which is the pathologial role of arginine? of the guanidino compounds? what is the importance of orotic acid excretion observed in such patients? is it possible to compensate for the arginase which is lacking in such patients? is it possible to identify an arginase anomaly in any other way? can biopterins play a role[23]?

Some of these questions will be discussed during the meeting.

CLINICAL DATA

As we already mentioned, 27 cases are probably known. Before 1969, Peralta Serrano, in Spain, described a case where, after paper chromatography, an intense spot of arginine among the blood amino acids was observed, however no quantitative determinations were made[24]. Confirmation of the diagnosis by determination of the arginase activity was not carried out. The author never answered our letters. So we cannot accept this as a case of hyperargininemia.

Twenty-one cases from 17 families were published between 1969 and now. Six cases have not yet been published: one in Canada, studied by Dr. Qureshi, one in Argentina, studied by Dr. Chamoles, two probably by Dr. Bachmann and two in Japan. This brings the number of families presenting hyperargininemia to 23.

Hyperargininemia was described in western and southern Europe, Canada, United States, South America, Central America, Japan and Autralia. The disease does not seem to be tied to a geographical distribution or a particular race.

Girls are as frequently affected as boys and in the 21 cases published, 10 were girls and 10 were boys, one is unknown.

The clinical image described by us is an oligophrenia associated with epilepsy with a bipyramidal syndrome and periodic coma with vomiting, anorexia due to ammonemia. In many publications there is mention of ataxia. This ataxia is not very well documented. As far as we know, the Canadian cases have a similar clinical picture as ours, but two Japanese cases[13-15] are more severely affected. They show not only frequent epileptic fits, but are also practically decerebrated.

Several genealogical trees were published. In this respect the two cases of Snyderman, published in 1977 should be mentioned[8]. Two couples, formed by two sisters who married two brothers, had each one child. The two children, a boy and a girl were both affected with hyperargininemia. In general it cannot be said with certainty, with the exception of our family and the family published by Jorda[21], that other members of the family are found to be affected with hyperargininemia. However, the biochemistry of members of the families often shows that some are carriers, but not actively affected themselves. It was thus concluded that hyperargininemia is an autosomal, recessive disease. Occasionally there is a mention of consanguinity.

Strangely enough for urea cycle diseases, the survival time of the patients is rather long. This might explain why there are very few anatomopathological documents. It is possible that the case published by Jorda[21] was autopsied but the results have not yet been published. Very few biopsies were carried out, although it is possible, by a liver biopsy, to confirm the affection, ballooning hepatocytes and increased glycogene inclusions[11] being observed.

The disease evolves very slowly. The oligophrenia is not very apparent.

BIOCHEMISTRY

As reported, in the first cases, a <u>cystine-lysine-ornithine-argininuria</u> was demonstrated[1]. This cannot qualitatively be distinguished from a classical cystine-lysinuria. Sofar, there has been no mention of cystine-lithiase.

In the blood 5 to 10 times increase of <u>arginine</u> is observed. An exceptional case was the third child of the family we published[4]. Fourthy-one days after birth, when not yet on low protein diet, serum arginine increases up to 17 times the normal values were observed[4].

Referring to the literature, arginine increases of 4 to 8 times normal values were found in cerebrospinal fluid.

In our cases, an increase of <u>guanidino compounds</u> was demonstrated associated with the presence of α-keto-δ-guanidinovaleric acid[5,6]. This can be compared to what is seen in phenylketonuria where phenylalanine that cannot normally be metabolised and subsequently leads to the formation of phenylpyruvic acid is accumulated. Next to the increase in the body fluids of many guanidino compounds, decreased values of guanidinosuccinic acid have to be mentioned. The creatinine values are normal. As in controls, the values for guanidine and methylguanidine are lower than the detection limit[5,25] and therefore are not increased. Similar results for guanidines were found in 8 cases by us. It is interesting to note that important increases of guanidinosuccinic acid, creatinine, guanidine and methylguanidine are found in patients presenting a renal insufficiency.

In addition, <u>orotic acid</u> also was found in urines of hyperargininemic patients.

The <u>hyperammonemia</u>, usually found in all urea cycle diseases, is only periodically observed here.

PHYSIOPATHOLOGY

Hyperargininemia presents a series of physiopathological problems which cannot always be solved. The main question is to know which metabolite or metabolites play a role in triggering off the disease.

It is certain that <u>ammonia</u> is toxic and can explain the periodes of coma. It is also certain that children which present hyperammonemia, regardless of its origin, can be affected with oligophrenia associated with growth problems and developmental difficulties. The clinical picture is then completely different, more so because hyperammonemia only occurs either rarely. Nevertheless the toxic role of ammonia should not be overlooked in hyperargininemia.

Is <u>arginine</u> toxic in these patients? Tests have been often performed with arginine as it is often used for endocrinological diagnosis. In these tests regularly a cystine-lysinuria is found, but nothing proves clearly that arginine can be toxic for the nervous system. As mentioned before, cystinuria often goes on par with neurological symptoms and we have published two families with neurological symptoms and cystinuria[26]. Very often however cystinuria is not associated with a neurological disease. The disappearance of the cystinuria due to treatment does not seem to play a role in the clinical condition of patients with hyperargininemia.

The guanidino compounds found in patients with hyperargininemia are rather toxic and most of them are even epileptogenic[27]. This could certainly explain the epilepsy. It is more difficult to prove that the pyramidal syndrome and the oligophrenia could be caused by these guanidino compounds.

Although the patients demonstrate a block between on the one side arginine and on the other ornithine and urea, they synthesize still urea but their uremia is subnormal. How this urea synthesis takes place, is not known. There are several possible explanations. The most probable seems to be the formation of urea due to residual enzyme activity[28], or perhaps a different genetic control exists for the arginases produced in the different organs, or an alternative pathway such as an homologous urea cycle could perhaps induce the formation of urea[29].

Arginase activity decreases in erythrocytes. The leukocyte arginase is low. The arginase activity was shown to be low in one case in stratum corneum and in at least 3 cases where a liver biopsy was done. In the case described by Jorda et al.[21], low arginase activity was also found in liver autopsy material. Up to now it is not possible to state with certainty that no arginase activity is to be found in patients tissues, or whether an eventual decrease of arginase activity is due to a molecular abnormality or to an inhibiting factor for arginase. Studies to caracterize or eventually isolate arginase DNA are in progress. At the present moment it is known that the remaining arginase in the patients with hyperargininemia displays normal kinetics.

THERAPY

Different therapies for such patients were suggested. A really effective one has not yet been found.

A low protein diet and thus poor in arginine was prescribed. The serum arginine decreases with this diet, but the excretion of guanidino compounds remains high. Hyperammonemia becomes less frequent.

Kang[16] proposed a diet with increased intake of ornithine and lysine. This lead to a decrease of orotic acid but not of arginine.

To counter hyperammonemia a treatment using sodium benzoate or phenylacetic acid was suggested.

Sodium benzoate 250 mg/kg/day removes glycine under the form of hippuric acid. For the biosynthesis of glycine, ammonia is used. Phenylacetic acid probably forms phenylacetylglutamine, but this has not yet been proven, but could allow the removal of ammonia. The results of these therapies are relatively good when considering the general condition. Sodium benzoate therapy in hyperargininemia patients allows either the bypassing of the endogeneous arginine biosynthesis via the urea cycle, or its control.

Together with the control of exogeneous arginine via a diet, theoretically it should be possible either to control the serum arginine or to normalize it. This was demonstrated very clearly[25]. One patient studied by the group of Dr. Qureshi is up to now treated with sodium benzoate associated with a low protein diet with low arginine level. Since 3 years the serum arginine levels vary between 150 and 200 µM. The values for the other guanidino compounds are difficult to normalize and thus remain high. Bernar's patient[19] was also treated with sodium benzoate. This patient has arginine values around 300 µM. Mizutani's patient shows serum arginine values around 350 µM after sodium benzoate therapy.

It was proposed to substitute[4] arginase by blood transfusion[9]. This does not seem to be the solution. A treatment with Shope papilloma virus was considered: this way arginase is incorporated into the patient's cells. It works in fibroblast cultures as demontrated[30]. When applied under bad conditions it did not give satisfactory results[31]. Perhaps the treatment assay should be repeated.

It means that we have not yet found an effective therapy for hyperargininemic patients and therefore can only suggest a symptomatic treatment.

CONCLUDING REMARKS

Finally we would like to make some remarks concerning the cases which were published.

We already mentioned the two cousins, children of the two sisters married to two brothers, which were published by Snyderman in 1977[8].

In many cases, in ours[2-5], in those published by Cederbaum[7], by Michels[9] etc. determinations of arginase and arginine were carried out on several members of the family. Thus it could be demonstrated that some members, although they did not present any neurological symptoms, had relatively high arginine values and a low arginase activity in the blood. They are undoubtedly carriers of the disease.

The case published by Jorda in 1986[21] is difficult to classify. Here the biochemical anomalies are less severe than in others. In this family several members have died from an unknown disease. The proband died at an early age. We presume that an autopsy was performed but the results are still unknown. We are not convinced that it concerns a classical case of hyperargininemia.

Finally we would like to emphasize that, except for the case of Jorda[21], that of Enders[18], and that of Sakiyama[14], as far as we know, all published patients are still alive. This creates a clear difference from all other cases of urea cycle diseases.

In conclusion, the study of hyperargininemia has not only been important to acquire a better knowledge of the diseases of the urea cycle, but has brought up several questions with regard to genetics, enzymes, biochemistry, physiopathology, many of which are still unsolved. Perhaps the work of M. Mori[32] and S.D. Cederbaum[33] and their collaborators, still in progress, will bring a new light to some of these questions.

ACKNOWLEDGEMENTS

   This work was supported by grants awarded by the University of Antwerp, the National Fund for Scientific Research (Nr. 3.0019.86), the National Lottery (nr. 9.0017.83), the Ministry for National Education and the Programma for Scientific Activities (nr. 84/89-68), the NATO Research Grant N° 83/0913, the "Ministerie van de Vlaamse Gemeenschap", and the Born-Bunge Foundation.

REFERENCES

1. Terheggen, H.G., Schwenk, A., Lowenthal, A., van Sande, M. and Colombo, J.P., Argininaemia with arginase deficiency, Lancet II (1969) 748-749.
2. Marescau, B., Pintens, J., Lowenthal, A., Terheggen, H.G. and Adriaenssens, K., Arginase and free amino acids in hyperargininemia: Leukocytes arginine as a diagnostic parameter for heterozygotes, J. Clin. Chem. Clin. Biochem. 17 (1979) 211-217.
3. Terheggen, H.G., Schwenk, A., Lowenthal, A., van Sande, M. and Colombo, J.P., Hyperargininämie mit Arginasedefekt. Eine neue familiäre Stoffwechselstörung. I. Klinische Befunde + II. Biochemische Untersuchungen, Z. Kinderheilk. 107 (1970) 298-323.
4. Terheggen, H.G., Lowenthal, A., Lavinha, F. and Colombo, J.P., Familial hyperargininemia, Arch. Dis. Childh. 50 (1975) 57-62.
5. Terheggen, H.G., Lavinha, F., Colombo, J.P., van Sande, M. and Lowenthal A., Familial hyperargininemia, J. Génét. Hum. 20 (1972) 69-84.
6. Marescau, B., Pintens, J., Lowenthal, A., Esmans, E., Luyten, Y., Lemière, G., Dommisse, R., Alderweireldt, F. and Terheggen, H.G., Isolation and identification of 2-oxo-5-guanidinovaleric acid in urine of patients with hyperargininemia by chromatography and gas chromatography-mass spectrometry, J. Clin. Chem. Clin. Biochem. 19 (1981) 61-65.
7. Cederbaum, S.D., Shaw, K.N.F. and Valente, M., Hyperargininemia, J. Pediatr. 90 (1977) 569-573.
8. Snyderman, S.E., Sansaricq, C., Chen, W.J., Norton, P.M. and Phansalkar, S.V., Argininemia, J. Pediatr. 90 (1977) 563-568.
9. Michels, V.V. and Beaudet, A.L., Arginase deficiency in multiple tissues in argininemia, Clin, Genet. 13 (1978) 61-67.
10. Snyderman, S.E., Sansaricq, C., Norton , P.M. and Goldstein, F., Argininemia treated from birth, J. Pediatr. 95 (1979) 61-63.
11. Cederbaum, S.D., Shaw, K.N.F., Spector, E.B., Verity, M.A., Snod-grass P.J. and Sugarman, G.I., Hyperargininemia with arginase deficiency, Pediatr. Res. 13 (1979) 827-833.
12. Qureshi, I.A., Letarte, J., Ouellet, R., Lelièvre, M. and Laberge, C., Ammonemia metabolism in a family affected by hyperargininemia, Diabete Metab. 7 (1981) 5-11.
13. Yoshino, M., Kubota, K., Yoshida, I., Murakami, T. and Yamashita, F., Argininemia: Report of a new case and mechanisms of orotic aciduria and hyperammonemia, in: "Urea Cycle Diseases," A. Lowenthal, A. Mori and B. Marescau Eds., Plenum Press, New York (1982) pp. 121-125.
14. Sakiyama, T., Nakabayashi, H., Kondo, Y., Shimizu, H., Kodama, S. and Kitagawa, T., Argininemia: Clinical course and trial of enzyme replace-ment therapy, Biomed. Therapeut. (Tokyo) 8 (1982) 907-910.
15. Mizutani, N., Maehara, M., Hayakawa, C., Kato, T., Watanabe, K. and Suzuki, S., Hyperargininemia: Clinical course and treatment with sodium benzoate and phenylacetic acid, Brain Dev. 5 (1983) 553-563.
16. Kang, S.S., Wong, P.W.K. and Melyn, M.A., Hyperargininemia: Effect of ornithine and lysine supplementation, J. Pediatr. 103 (1983) 763-765.

17. Qureshi, I.A., Letarte, J., Ouellet, R., Larochelle, J. and Lemieux, B., A new French-Canadian family affected by hyperargininemia, J. Inher. Metab. Dis. 6 (1983) 179-182.

18. Endres, W., Schaller, R. and Shin, Y.S., Diagnosis and treatment of argininemia. Characteristics of arginase in human erythrocytes and tissues, J. Inher. Metab. Dis. 7 (1984) 8.

19. Bernar, J., Hanson, R.A., Kern, R., Phoenix, B., Shaw, K.N.F. and Cederbaum, S.D., Arginase deficiency in a 12-year-old boy with mild impairment of intellectual function, J. Pediatr. 108 (1986) 432-435.

20. Hyland, K., Smith, I., Clayton, P.T. and Leonard, J.V., Impaired neurotransmitter amine metabolism in arginase deficiency, J. Neurol. Neurosurg. Psychiatr. 49 (1986) 1188-1189.

21. Jorda, A., Rubio, V., Portoles, M., Vilas, J. and Garcia-Pino, J., A new case of arginase deficiency in a Spanish male, J. Inher. Metab. Dis. 9 (1986) 393-397.

22. Romano, C., Pescelto, T., Caruso, U., Cerone, R. and Caffarena, G., Clinical and biochemical studies on a new patient affected by argini-nemia, Persp. Met. Dis. 7, Ermes Pub., Milano (1986).

23. Hyland, K., Smith, I. and Leonard, J.V., Biopterins in arginase, dihydropteridine reductase and phenylalanine hydroxylase deficiency, J. Neurol. Neurosurg. Psychiatr. 50 (1987) 242.

24. Peralta-Serrano, A., Argininuria, convulsionesy oligofrenia: Un nuevo error innato del metabolismo, Rev. Clin. Esp. 97 (1965) 176-184.

25. Marescau, B., Qureshi, I.A., De Deyn, P., Letarte, J., Ryba, R. and Lowenthal, A., Guanidino compounds in plasma, urine, and cerebrospinal fluid of hyperargininemic patients during therapy, Clin. Chim. Acta 146 (1985) 21-27.

26. van Sande, M., Terheggen, H.G., Clara, R., Leroy, J.G. and Lowenthal, A., Lysine-cystine pattern associated with neurological disorders, in: "Inherited disorders of sulphur metabolism," N.A.J. Carson and D.N. Raine Eds., Churchill Livingstone, Edinburgh (1971) pp. 85-112.

27. Mori, A., Watanabe, Y. and Akagi, M., Guanidino Compound Anomalies in Epilepsy. In: "Advances in Epileptology." H. Akimoto, H. Kazamatsuri, M. Seino and A. Ward Eds., Raven Press, New York (1982) pp. 347-351.

28. Kennan, A.L. and Cohen, P.P., Ammonia detoxication in liver from humans, Proc. Soc. Exp. Biol. Med. 106 (1961) 170-177.

29. Scott-Emuakpor, A.B., On a late developing urea cycle, Pediatr. Res. 8 (1974) 858-859.

30. Rogers, S., Lowenthal, A., Terheggen, H.G. and Colombo, J.P., Induction of arginase activity with the Shope papilloma virus in tissue culture cells from an argininemic patient, J. Exp. Med. 137 (1973) 1091-1096.

31. Terheggen, H.G., Lowenthal, A., Lavinha, F., Colombo, J.P. and Rogers, S., Unsuccessful trial of gene replacement in arginase deficiency, Z. Kinderheilk. 119 (1975) 1-3.

32. Haraguchi, Y., Takiguchi, M., Amaya, Y., Kawamoto, S., Matsuda, I. and Mori, M., Molecular cloning and nucleotide sequence of cDNA for human liver arginase, Proc. Natl. Acad. Sci. 84 (1987) 412-415.

33. Sparkes, R.S., Dizikes, G.J., Klisak, I., Grady, W.W., Mohandas, T., Heinzmann, C., Zollman, S., Lusis, A.J. and Cederbaum, S.D., The gene for human liver arginase (ARG1) is assigned to chromosome band Gq23, Am. J. Hum. Genet. 39 (1986) 186-193.

NEUROTOXICITY OF GUANIDINO COMPOUNDS IN HYPERARGININEMIA

Naoki Mizutani, Chiemi Hayakawa, Yukihiro Oya,
Kazuyoshi Watanabe, Yoko Watanabe* and Akitane Mori*

Department of Pediatrics, Nagoya University Medical School
Nagoya, Japan
*Department of Neurochemistry, Institute for Neurobiology
Okayama University Medical School, Okayama, Japan

INTRODUCTION

Hyperargininemia is a rare hereditary disorder of the urea cycle due to a deficiency of arginase, which catalyzes arginine to ornithine and urea. This disorder is the most uncommon among the five inborn errors resulting from enzyme deficiencies of the urea cycle. Patients with this disorder all have similar clinical symptoms, which include vomiting, lethergy, irritability, convulsions and coma. Severe mental deterioration and a marked degree of spasticity of all four extremities are also characteristic.

Elevated plasma ammonia levels are considered toxic to the central nervous system. But the neurological symptoms are not adequately explained by hyperammonemia or hyperargininemia. It is suspected that other factors may contribute to these neurological symptoms.

In this study, long term changes in plasma and CSF concentrations and urinary excretion of guanidino compounds were investigated in a patient with hyperargininemia. Furthermore, in order to investigate the effect of some of the guanidino compounds on the centrl nervous system, we administered arginine (Arg) and homoarginine (HArg) orally to ddY mice, and measured plasma concentrations of the guanidino compounds before and after while observing their clinical symptoms.

CASE REPORT

The patient is a 9-year-old boy of consanguineous parents, presenting with ataxic gait and abnormal liver function tests. He was found to have hyperammonemia at the age of 3. The details of the laboratory findings and clinical course of the patient were reported elsewhere[1].

He has been treated with a low-protein diet, the oral administration of an essential amino acid mixture, sodium benzoate (250 mg/kg/day), and phenylacetic acid (100 mg/kg/day). Furthermore, enzyme replacement therapy such as exchange transfusions or erythrocyte transfusions were performed. His plasma ammonia levels were controlled by this treatment but his plasma Arg concentration ranged from 100 to 400 nmole/ml while the CSF concentration showed n change. At present, he is in a vegetative state.

METHODS

Plasma and CSF concentration and urinary excretion of guanidino compounds were determined in the patient with hyperargininemia and his family members using an automatic guanidino compound analyzer system[2].

Furthermore, the effects of Arg and HArg on the central nervous system of ddY mice were investigated. Three groups of 26-28 week old ddY mice weigh ing 20-30 g were used. One group was given a 5% solution of Arg, another a 5 solution of HArg, and the third or control group was given distilled water orally for three weeks. The plasma concentration of guanidino compounds before and after oral administration of these solution was determined.

RESULTS

Plasma and CSF Concentration and Urinary Excretion of Guanidino Compunds in Hyperargininemia

Plasma and CSF concentration and urinary excretion of guanidino compounds in a patient with hyperargininemia are shown in Table 1. In the patient, the plasma concentration of α-keto-δ-guanidinovaleric acid (GVA), N-α-acetylarginine (NAA), Arg, and HArg were elevated about 5 to 50 times the normal range. Guanidinoacetic acid (GAA) and γ-guanidinobutyric acid (GBA) showed no remarkable change. Guanidinosuccinic acid (GSA) concentratio

Table 1. Plasma and CSF concentrations and urinary excretion of guanidino compounds in a patient with hyperargininemia during treatment with low-protein diet and oral administration of an essential amino aci mixture and sodium benzoate (250 mg/kg/day).

	Plasma (nmole/ml)		CSF (nmole/ml)		Urine (μmole/g creatinine)	
	patient	controls	patient	controls*	patient	controls
GVA	2.7	<0.2	ND	<0.025	608.3	24.40± 13.40
GSA	ND	0.18± 0.14	ND	0.09±0.04	ND	59.55± 16.77
GAA	2.43	1.22± 0.40	0.03	0.05±0.03	991.55	784.82±397.54
NAA	5.34	1.25± 0.28	0.50	0.15±0.08	293.01	25.40± 16.10
GBA	ND	<0.025	ND	<0.025	118.35	<12.39
Arg	456.18	56.87±14.50	58.78	22.53±6.04	104.46	49.71± 23.40
HArg	20.09	1.12± 0.60	1.16	0.31±0.15	3.83	16.67± 14.36
G	ND	<1.81	ND	<0.10	10.93	41.03± 32.19
MG	ND	<0.05	ND	<0.05	3.06	3.08± 1.86

ND: not detected,   controls: mean ± SD,   * Marescau et al,

GVA: α-keto-δ-guanidinovaleric acid,   GSA: guanidinosuccinic acid,

GAA: guanidinoacetic acid,   NAA: N-α-acetylarginine,   GBA: γ-guanidinobutyric acid,

Arg: arginine,   HArg: homoarginine,   G: guanidine,   MG: methylguanidine

Table 2. Levels of guanidino compounds in plasma and urine of the family of the patient with hyperargininemia.

sample guanidino compounds	plasma (nmole/ml)				urine (μmole/g creatinine)			
	patient (homo)	father (hetero)	mother (hetero)	brother (hetero)	patient (homo)	father (hetero)	mother (hetero)	brother (hetero)
GVA	4.6	0.5	0.3	0.3	579.5	29.6	83.0	81.1
GSA	ND	ND	ND	ND	ND	12.25	12.98	100.79
GAA	1.77	3.93	1.77	0.85	1468.5	1293.0	1306.5	700.7
NAA	1.72	0.81	0.57	0.46	172.12	22.55	27.03	23.74
GBA	ND	ND	ND	ND	45.77	ND	ND	ND
Arg	260.50	135.65	94.05	88.63	69.81	49.97	38.32	38.56
HArg	27.35	6.75	5.40	2.36	23.78	5.16	17.95	5.96
G	ND	ND	ND	ND	4.97	9.30	15.40	-
MG	ND	ND	ND	ND	0.68	1.38	1.55	3.51

ND: not detected

GVA: α-keto-δ-guanidinovaleric acid, GSA: guanidinosuccinic acid,

GAA: guanidinoacetic acid, NAA: N-α-acetylarginine,

GBA: γ-guanidinobutyric acid, Arg: arginine, HArg: homoarginine,

G: guanidine, MG: methylguanidine

Table 3. Levels of guanidino compounds in plasma (nmole/ml) in a patient with hyperargininemia.

Guanidino Compounds	1983.					1984.				Controls	
	6.29.	7.23.	8.3.	8.25.	10.5.	3.6.	4.12.	5.21.	6.15.	3-4y (n=4)	3-11y (n=17)
GVA	3.3	4.2	2.1	2.6	4.2	2.3	2.3	2.0	4.6	< 0.2	< 0.2
GSA	ND	ND	ND	ND	ND	ND	ND	ND	ND	0.15 ± 0.06	0.18 ± 0.14
GAA	0.78	2.11	1.47	2.63	2.82	1.19	1.71	1.25	1.77	1.10 ± 0.70	1.22 ± 0.40
NAA	1.72	4.26	3.14	10.96	7.48	2.24	4.36	1.28	1.72	1.38 ± 0.44	1.25 ± 0.28
CRN	30.20	34.80	24.06	30.04	36.77	32.24	33.57	33.57	31.74	39.84 ± 7.83	43.54 ± 6.20
Arg	181.21	416.0	312.0	428.42	471.88	297.05	233.22	251.41	269.50	52.03 ± 26.06	56.87 ± 14.50
HArg	13.70	13.98	12.02	21.48	36.77	40.70	43.50	25.66	27.35	1.10 ± 0.62	1.12 ± 0.60
G	ND	0.34	ND	ND	ND	ND	ND	ND	ND	< 1.11	< 1.81
MG	0.02	0.03	ND	0.06	0.01	ND	ND	ND	ND	< 0.07	< 0.05
GE	0.04	0.06	0.15	0.02	0.08	0.04	-	-	-	ND	< 0.01

ND: not detected,   Controls: mean ± SD

Treatments:

exchange transfusion →	
Low-protein diet (0.8-1.0 g/kg/day) →	
Essential amino acid mixture (10 g/day) →	
Sodium benzoate (250 mg/kg/day) →	
erythrocyte transfusion →	

GVA: α-keto-δ-guanidinovaleric acid,   GSA: guanidinosuccinic acid,
GAA: guanidinoacetic acid,   NAA: N-α-acetylarginine,   CRN: creatinine,
Arg: arginine,   HArg: homoarginine,   G: guanidine,   MG: methylguanidine,
GE: guanidinoethanol

Table 4. Levels of guanidino compounds in urine (μmole/g creatinine) in a patient with hyperargininemia.

Guanidino Compounds	1983. 6.29.	7.27.	8.3.	8.25.	10.2.	1984 3.6.	4.12.	5.12.	6.15.	Controls 1-8y (n=8)
GVA	502.5	1072.7	766.3	1224.8	427.4	588.5	312.5	387.6	625.3	24.4 ± 13.4
GSA	ND	ND	ND	ND	ND	ND	ND	ND	ND	59.55 ± 16.77
GAA	69.71	358.76	661.05	667.3	956.83	318.18	1011.4	368.5	1468.5	784.82 ± 397.54
NAA	-	465.18	303.04	386.12	661.00	229.22	101.06	44.07	172.12	25.40 ± 16.10
GBA	72.14	30.98	92.95	461.02	276.72	684.24	51.79	12.16	45.77	<12.39
Arg	70.50	6.31	92.36	99.42	28.21	176.69	53.85	53.85	135.02	49.71 ± 23.40
HArg	8.03	4.59	9.94	22.09	24.36	87.00	25.31	42.40	23.78	16.67 ± 14.36
G	3.81	43.44	42.49	17.62	6.52	16.57	3.36	9.06	4.87	41.03 ± 32.19
MG	2.76	2.44	0.88	1.89	0.86	0.95	3.86	0.74	0.68	3.08 ± 1.86
GE	2.72	4.45	7.34	27.14	10.47	41.07	36.05	12.09	21.44	7.93 ± 3.61

ND: not detected,　Controls: mean ± SD

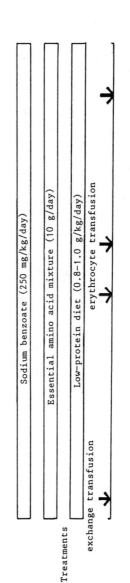

Treatments

Sodium benzoate (250 mg/kg/day)

Essential amino acid mixture (10 g/day)

Low-protein diet (0.8-1.0 g/kg/day)

erythrocyte transfusion

exchange transfusion

GVA: α-keto-δ-guanidinovaleric acid,　GSA: guanidinosuccinic acid,
GAA: guanidinoacetic acid,　NAA: N-α-acetylarginine,
GBA: γ-guanidinobutyric acid,　Arg: arginine,　HArg: homoarginine,
G: guanidine,　MG: methylguanidine,　GE: guanidinoethanol

Table 5. Levels of guanidino compounds in CSF (nmole/ml) of a patient with hyperargininemia.

Guanidino Compounds	1983					1984		Controls*
	6.29.	7.27.	8. 3.	8.25.	10. 5.	2.21.	4.12.	
GVA	ND	0.4	0.1	Tr	Tr	0.1	0.16	< 0.025
GSA	Tr	ND	ND	ND	ND	ND	ND	0.09± 0.04
GAA	0.11	0.18	0.13	0.08	0.09	0.07	0.63	0.05± 0.03
NAA	0.30	0.44	0.35	0.41	0.48	0.55	0.28	0.15± 0.08
CRN	64.57	25.51	37.20	38.03	32.42	39.49	39.75	54.86±17.00
Arg	43.62	64.85	54.87	53.00	55.65	37.10	25.03	22.53± 6.04
HArg	2.43	1.48	1.60	1.91	2.80	3.87	3.08	0.31± 0.15
G	ND	0.88	1.62	ND	4.54	Tr	ND	<0.10
MG	0.03	0.02	0.05	ND	0.04	Tr	ND	<0.05
GE	0.02	0.01	Tr	0.01	0.01	0.14	-	-

ND: not detected,   Tr: trace,   * Marescau et al., mean ± SD

Treatments:

Sodium benzoate (250 mg/kg/day)

Essential amino acid mixture (10 g/day)

Low-protein diet (0.8-1.0 g/kg/day)

exchange transfusion   erythrocyte transfusion

GVA: α-keto-δ-guanidinovaleric acid,   GSA: guanidinosuccinic acid,
GAA: guanidinoacetic acid,   NAA: N-α-acetylarginine,   CRN: creatinine,
Arg: arginine,   HArg: homoarginine,   G: guanidine,   MG: methylguanidine,
GE: guanidinoethanol

Table 6. Plasma concentrations of guanidino compounds in ddY mice
          before and after the oral administration of a 5% solution
          of arginine, homoarginine, and distilled water (controls).

		GVA		GSA		GAA		GBA		Arg		HArg		MG	
		before	after	before	after	before	after	before	after	before	after	before	after	before	after
Arg	1	0.50	0.20	ND	ND	3.87	2.31	ND	0.82	183.50	226.77	5.38	0.26	0.06	0.06
	2	ND	0.20	ND	ND	2.92	0.20	ND	0.99	135.41	142.67	1.29	0.50	0.06	0.06
	3	ND	0.02	ND	ND	3.03	2.94	ND	3.11	139.39	103.23	0.64	0.58	0.03	0.06
HArg	1	0.25	ND	ND	ND	4.04	4.82	ND	0.44	180.33	89.03	ND	640.00	0.09	0.10
	2	1.59	ND	ND	ND	4.52	7.59	ND	1.02	234.78	225.85	0.18	802.48	0.09	0.07
	3	0.59	ND	ND	ND	4.02	3.41	ND	0.44	220.70	178.07	0.58	632.89	0.06	0.04
Controls	1	1.60	ND	ND	ND	3.72	3.23	0.23	ND	264.39	186.43	0.65	ND	0.11	0.09
	2	ND	ND	ND	ND	4.61	2.37	0.60	ND	232.70	124.38	0.64	0.42	0.12	0.02
	3	ND	0.02	ND	ND	3.72	2.23	0.52	ND	169.67	86.58	0.67	ND	0.06	0.03

ND: not detected,

GVA: α-keto-δ-guanidinovaleric acid,   GSA: guanidinosuccinic acid,
GAA: guanidinoacetic acid,   GBA: γ-guanidinobutyric acid,   Arg: arginine,
HArg: homoarginine,   MG: methylguanidine

in plasma decreased to undetectable levels. CSF concentration of Arg and HArg also increased, but the other guanidino compounds such as GVA, GAA, NAA, GBA guanidine (Gua) and methylguanidine (MG) showed no remarkable change. Urinary excretion of GVA remarkably increased.

Arginase activity of erythrocytes of family members were less than half that of normal controls[1] suggesting that they are heterozygotes for hyperargininemia. Their plasma conçentration and urinary excretion of guanidino compounds were almost within the normal range (Table 2).

Long term change in plasma concentration of guanidino compounds are shown in Table 3. GVA, Arg, and HArg remain at high levels. On the other hand, NAA sometimes decreased into the normal range. There was no correlation between Arg and HArg level, and the latter increased gradually irrespective of increases in the former. Other guanidino compounds such as GSA, GAA, Gua, MG, and 2-guanidinoethanol (GEt) are almost within the normal range

Long term change in urinary excretion of guanidino compounds are shown in Table 4. Urinary excretion of GVA and NAA usually 10 to 20 times the normal range. Arg, HArg, and the other guanidino compounds were excreted within the normal range.

Long term change in CSF concentration of guanidino compounds are shown in Table 5. Arg and HArg levels were high, and GAA, Gua, MG, and GEt were almost within the normal range. GSA was not detected at all. GVA levels sometimes increased markedly.

In summary, in the patient with hyperargininemia, CSF and plasma concentration of Arg, HArg, and GVA were markedly elevated. Increased urinary excretion of GVA and NAA were also observed.

Animal Experiments

The group given the 5% solution of Arg showed no change in plasma concentration of Arg and HArg (Table 6). On the other hand, in the group given the 5% solution of HArg, plasma levels of HArg rose markedly to between 1,000 and 10,000 times the level before administration, but plasma Arg concentration showed no change (Fig. 1). The other guanidino compounds also showed no remarkable change in all mice (Table 6). Furthermore, in all animals there were no changes in body weight or activity and no neurological symptoms such as convulsions or spasticity of four extremities.

DISCUSSION

There are only a few reports concerning plasma and CSF concentration and urinary excretion of guanidino compounds in patients with hyperargininemia[3-5]. Arg is considered to be a representative guanidino compounds. Several reports indicate that plasma and CSF concentration and urinary excretion of guanidino compounds, especially Arg, HArg and GVA increase markedly in hyperargininemia[3-5]. Changes occurring among guanidino compounds as catabolites of Arg are very interesting in hyperargininemia. The role of most guanidino compounds in hyperargininemia as well as in normal metabolism is still unclear in spite of several hypotheses concerning the metabolic pathways of guanidino compounds[6-7]. However, some guanidino compounds such as HArg[8], GBA[9], GVA[10], and MG[11] are shown to cause convulsions in animals when injected intraventricularly.

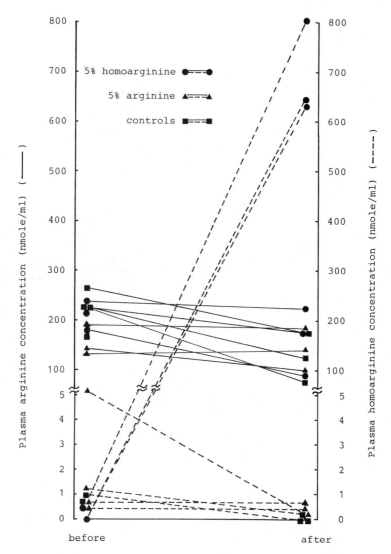

Figure 1. Plasma concentrations of arginine (———) and homoarginine (----)
before and after the oral administration of 5% arginine solution
(▲---▲), 5% homoarginine solution (●---●) and distilled water
(■---■) in mice.

In our case, GVA, which is not usually detected in normal subjects, was
detected in plasma and urine of the patient. Furthermore, a large amount of
GBA, which is also normally undetectable, was excreted in the urine of the
patient. On the other hand, GSA, which is usually present in normal urine,
was lacking in all samples. These findings seem to be characteristic in
hyperargininemia. But the most important change is the marked increase in
concentration of Arg and HArg in plasma and CSF.

We propose that some guanidino compounds are catabolites of Arg, and
accumulation of HArg as well as Arg produce the central nervous system damage
seen in hyperargininemia.

On the other hand, nitrogen is excreted normally as urea. The two nitrogen atoms of urea are derived from Arg. Instead of Arg hydrolysis, in hyperargininemia, Arg may be catabolized via other pathways and nitrogen excreted partially in the form of certain guanidino compounds.

Generally, hyperammonemia seems to be a cause of central nervous system damage in patients with enzyme defects of the urea cycle. However, in many patients with hyperargininemia, hyperammonemia is observed only intermittently whereas it is always observed in the other enzymopathies of the urea cycle. The pathogenesis of hyperammonemia is not fully understood. The accumulation of ammonia in body fluids may not be the sole etiology of the brain damage. Littel is also known about the neurotoxicity of guanidino compounds in humans.

We, therefore, performed experiments looking for the neurotoxicity of Arg and HArg in mice. First, ddY mice given a 5% solution of HArg became markedly hyperhomoargininemic but plasma Arg concentration unchanged. However, in ddY mice given a 5% solution of Arg, there was no change in plasma Arg or HArg concentration. Furthermore, no neurological symptoms were observed in any mice. Therefore, we could not confirm the neurotoxicity of Arg and HArg. The reason for the absence of neurological symptoms in spite of markedly elevated concentrations of plasma HArg is unclear. It may be related to the permeability of guanidino compounds and amino acids at the blood brain barrier since we could not demonstrate increases in CSF concentrations of the guanidino compounds. Furthermore, in order to produce central nervous system damage, elevated levels of HArg in tissue or cells may be required. Also, it may be necessary to induce elevated levels of all of these agents: Arg, HArg, and ammonia.

In conclusion, our results suggest that accumulation of Arg and its metabolites, the guanidino compounds, such as HArg, may produce the central nervous system damage in hyperargininemia. Furthermore, it is suspected that in hyperargininemia, Arg may be catabolized via other pathways and nitrogen may be excreted partially in the urine in the form of guanidino compounds.

REFERENCES

1. Mizutani, N., Maehara, M., Hayakawa, C., Kato, T., Watanabe, K. and Suzuki, S., Hyperargininemia: Clinical course and treatment with sodium benzoate and phenylacetic acid, Brain and Development 5 (1983) 555-563.
2. Mori, A., Ichimura, T. and Matsumoto, H., Gas chromatography-mass spectrometry of guanidino compounds in brain, Anal. Biochem. 89 (1978) 393-399.
3. Marescau, B., Lowenthal, A., Terheggen, H.G., Esmans, E. and Alderweireldt, F., Guanidino compounds in hyperargininemia, in: "Guanidines," A. Mori, B.D. Cohen and A. Lowenthal Eds., Plenum Press, New York (1985) pp. 427-434.
4. Marescau, B., Lowenthal, A., Esmans, E., Luyten, Y., Alderweireldt, F. and Terheggen, H.G., Isolation and identification of some guanidino compounds in the urine of patients with hyperargininemia by liquid chromatography, thin-layer chromatography and gas chromatography-mass spectrometry, J. Chromat. 224 (1981) 185-195.
5. Marescau, B., Pintens, J., Lowenthal, A., Esmans, E., Luyten, Y., Lemiere, G., Dommisse, R., Alderweireldt, F. and Terheggen, H.G., Isolation and identification of 2-oxo-5-guanidinovaleric acid in urine of patients with hyperargininemia by chromatography and gas chromatography/mass spectrometry, J. Clin. Chem. Clin. Biochem. 19 (1981) 61-65.

6. Pisano, J.J., Abraham, D. and Udenfriend, S., Biosynthesis and disposition of γ-guanidinobutyric acid in mammalian tissues, Arch. Biochem. Biophys. 100 (1963) 323-329.
7. Perez, G., Rey, A. and Schiff, E., The biosynthesis of guanidinosuccinic acid by perfused rat liver, J. Clin. Invest. 57 (1976) 807-809.
8. Yokoi, I., Toma, J. and Mori, A., The effect of homoarginine on the EEG of rats, Neurochem. Pathol. 2 (1984) 295-300.
9. Jinnai, D., Sawai, A. and Mori, A., γ-Guanidinobutyric acid as a convulsive substance, Nature 217 (1966) 617.
10. Shindo, S., Tsuruta, K., Yokoi, I. and Mori, A., Synthesis of δ-guanidinovaleric acid and its effect on EEG of rats, Neurosciences 10 (1984) 177-182.
11. Matsumoto, M., Kobayashi, K., Kishikawa, H. and Mori A., Convulsive activity of methylguanidine in cats and rabbits, IRCS Med. Sci. 4 (1976) 65.

SERUM GUANIDINOSUCCINIC ACID LEVELS IN UREA CYCLE DISEASES

Bart Marescau, Ijaz A. Qureshi*, Peter P. De Deyn, Jacques
Letarte*, Nestor Chamoles**, Makoto Yoshino***, Werner P.
De Potter and Armand Lowenthal

Laboratory of Neurochemistry, Born-Bunge Foundation and
Laboratory of Neuropharmacology, U.I.A., 2610 Antwerp,
Belgium
*Pediatric Research Center Ste. Justine Hospital
Montreal, Canada H3T 1C5
**Laboratory of Neurochemistry, Uriarte 2383, 1425 Buenos
Aires, Argentina
***Department of Pediatrics, Kurume University School of
Medicine, Kurume-Shi, Fukuoka 830, Japan

INTRODUCTION

Guanidinosuccinic acid (GSA) was first isolated from urine of uremic
patients and identified by Natelson et al. in 1964[1]. Two pathways have been
proposed for the biosynthesis. In 1968 Cohen suggested that GSA would be
formed by transamidination of arginine to aspartic acid[2]. A second hypothesis, proposed by Natelson and Sherwin suggested that GSA could be synthesized
as an inert overflow product through the guanidine cycle. The guanidine cycle
is an alternate cycle proposed for the conversion of urea nitrogen to
creatine[3].

Up to now, reports in the literature have indicated the absence or only
trace amounts of urinary GSA, determined by colorimetric detection, in patients with urea cycle disorders[4-6].

In order to obtain a better understanding of the biosynthesis of GSA in
urea cycle diseases we determined the serum levels of GSA using a sensitive
method based on liquid chromatography with fluorescence detection. In this
study we included only serum samples from urea cycle patients on a free diet.
However urea cycle patients mostly impose a low protein diet on themselves.
Therefore we have also included in this study serum samples of patients on a
low protein diet or better a controlled protein diet.

PATIENTS

The urea cycle diseases considered in this study, the age and sex of the patients and the diet at the period of sampling are listed in Table 1. Most of the patients suffered from chronic and infantile forms of the diseases and presented with clinical symptoms during and after the first year of life.

The biochemical data obtained in this study were compared with data from a control group of the same age (n=30). Children with neurological, metabolic, renal and hepatic diseases were excluded from the control group.

METHODS

The concentration of the guanidino compounds was determined using a Biotronic LC 6001 amino acid analyser adapted for guanidino compound determination. The guanidino compounds were separated over a cation-exchange column using sodium citrate buffers and were detected with the fluorescence ninhydrin method as has been reported in detail earlier[7]. The procedure of sample preparation has also been described[8].

Serum urea nitrogen was measured by the technique of Ceriotti[9] and ammonia by a miniaturization of the ion exchange resin technique[10] as adapted for the "Hyland" blood ammonia kit (Travenol Laboratories, Cosa Mesa, CA, USA). Urinary orotate was measured by the method of Adachi et al.[11] with a non-brominated blank for each sample tube.

Table 1. Diagnosis, form, age, sex and diet of the studied urea cycle patients.

Diagnosis	Patient	Form	Age*	Sex	Diet**
OTC deficiency	C.	***	4 years	girl	free
	C.B.	heteroz.sympt.	5 years	girl	free
	M.D.	male variant	7 years	boy	C.P.D.
Citrullinemia	D.	benign	16 years	boy	free
	N.C.	chronic	6 years	girl	C.P.D.
Argininosuccinic aciduria	L.	chronic	3 years	girl	free
	M.L.	infantile	1-2 years	boy	C.P.D.
	C.M.	infantile	4 months	girl	C.P.D.
	I.L.	chronic	4 years	girl	C.P.D.
	E.L.	infantile	3 years	boy	C.P.D.
Hyperargininemia	C.		6 years	girl	C.P.D.
	L.C.		15 years	girl	free
	A.Y.		4 years	girl	C.P.D.
	F.F.L.		4 years	boy	free
HHH syndrome	M.P.	chronic	11 years	boy	C.P.D.
	C.L.	chronic	1-2 years	girl	C.P.C.

*; age at the sampling moment, **; diet at the dampling moment, ***; liver OTC activity has not been determined.
C.P.D. ; controlled protein diet, OTC; ornithine transcarbamylase and HHH; Hyperammonemia hyperornithinemia homocitrullinemia.

For the calculation of mean values and standard deviations for GSA, 12.5nM has been used for levels lower than detection limit (in our analytical system the detection limit is 25nM when injecting 200 μl deproteinized serum corresponding to 100 μl undeproteinized serum). The used value has been calculated as a class mean of the highest and lowest possible value below the detection limit.

RESULTS AND DISCUSSION

The serum levels of arginine and urea were determined along with GSA levels because, according to Cohen's and Natelson's hypotheses, these two compounds could be the precursors of GSA. The levels are given in Tables 2 and 3. Serum ammonia and urinary orotic acid[12] were determined and used as parameters for the urea cycle ammonia detoxification (table 2).

In 1969, we described the first patients with hyperargininemia[13]. The study of the urinary excretion levels of the guanidino compounds in these patients, with very high serum arginine levels, clearly showed a diminished excretion of GSA[14]. This observation certainly did not plead in favour of the hypothesis that arginine is the precursor for GSA. This is confirmed in this study: data in Table 2 clearly show that patients with hyperargininemia have the lowest serum GSA levels of all urea cycle patients. Only in one of the four studied hyperargininemic patients, serum GSA was found (70nM). The serum guanidino compounds of the hyperargininemic patiens L.C. and F.F.L. were also studied during their different dietary therapies and only exceptionally trace amounts of GSA could be detected. In all the other samples GSA

Table 2. Serum levels of guanidinosuccinic acid (GSA), urea, arginine, ammonia and urinary orotic acid of the studied urea cycle patients.

Diagnosis	Patient	GSA nmol/L	Urea mmol/L	Arginine μmol/L	NH$_3$ μmol/L	Orotic acid nmol/mg CTN
OTC deficiency	C.	29	1.8	220.1	209	N.D.
	C.B.	170	3.5	44.8	286	6886
Citrullinemia	D.	72	3.4	62.3	64.7	N.D.
	N.C.	< 25	4.4	103.0	54.9	N.D.
Argininosuccinic aciduria	L.	25	2.2	78.1	50.2	N.D.
	M.L.	250	2.6	118.4	43.6	46
	C.M.	120	3.2	93.4	51.4	24
	I.L.	130	2.8	91.7	38.6	30
	E.L.	< 25	3.5	35.7	40.7	51
Hyperarginimemia	C.	< 25	2.6	463.9	30.2	N.D.
	L.C.	< 25	0.8	619.5	53.6	4370
	A.Y.	< 25	1.4	460.6	N.D.	N.D.
	F.F.L.	70	5.1	836.7	42.3	4853
HHH syndrome	M.P.	130	7.0	45.8	111	756
	C.L.	150	2.9	158.7	55.3	2683
Controls (n=30)		251±72	4.3±1.5	93.8±19.4	36±9	34±12

N.D. =not determined, OTC; idem Table 1 and HHH; idem Table 1.

Table 3. Mean values and standard deviations of serum guanidinosuccinic
          acid (GSA), urea and arginine of hyperargininemic and other urea
          cycle patients.

Diagnosis	GSA nmol/L	Urea mmol/L	Arginine μmol/L
Hyperargininemia (n=4)	27 ± 29**	2.5 ± 1.9*	595.2 ± 177.3**
Other urea cycle patients (n=11)	100 ± 77**	3.4 ± 1.4	99.4 ± 53.3
Controls (n=30)	251 ± 72	4.3 ± 1.5	93.8 ± 19.4

* = P < 0.05 compared to controls, ** = P < 0.001 compared to controls.

levels were under detection limit. The other urea cycle patients have higher
serum GSA levels (Table 2). The heterogenicity in the levels is also worth
noting. However, the mean value is stil significantly (p < 0.001) lower than
the mean value for controls (Table 3). Serum urea were also found to be
significantly (p < 0.05) lower in hyperargininemia as compared to controls.
The serum urea from the other urea cycle patients is, like GSA, higher than
those of hyperargininemic patients. Although lower they are not significantly different from controls. The serum arginine levels of the studied urea
cycle patients, except for hyperargininemia, are not significantly different
from controls (Table 3). Since we found a linear correlation between serum
GSA and serum urea (r=0.820; p < 0.001) in non-dialyzed patients with renal
insufficiency (n=50), we did the same study for controls and urea cycle
patients: for controls we found much less of a correlation (r=0.440; p < 0.0.
and for urea cycle patients no significant correlation was found.

     In conclusion, our results strongly suggest that GSA is probably not
formed by transamidination of arginine to aspartic acid. Hyperargininemic
patients have the lowest sera urea levels. They also have the lowest serum
GSA levels. With a sensitive fluorescence detection method we were able to
show that the other urea cycle patients have subnormal serum GSA levels,
they also have subnormal serum urea levels. We also think that the determination of serum GSA levels could be useful together with other biochemical
parameters, as serum urea, ammonia and the urinary orotic acid levels as a
parameter for urea biosynthesis dysfunction.

ACKNOWLEDGEMENTS

     This work was supported by the "Fonds voor Geneeskundig Wetenschappeli
Onderzoek (grant n° 3.00019.86), the "Ministerie van Nationale Opvoeding en
Nederlandse Cultuur", the "Nationale Loterij" (grant n° 9.0017.83), the NAT
Research Grant N° 83/0913, the "Concerted Action U.I.A., nr. 84/89-68, the
"Ministerie van de Vlaamse Gemeenschap", the "Fondation Justine Lacoste-
Beaubien", the Québec-Belgium exchange program, the University of Antwerp
and the Born-Bunge Foundation.

REFERENCES

1. Natelson, S., Stein, I. and Bonas, J.A., Improvement in the method of separation of guanidino amino acids by column chromatography. Isolation and identification of guanidinosuccinic acid from human urine, Microchem. J. 8 (1964) 371-382.
2. Cohen, B.D., Stein, I.M. and Bona, J.E., Guanidinosuccinic aciduria in uremia. A possible alternate pathway for urea synthesis, Am. J. Med. 45 (1968) 63-68.
3. Natelson, S. and Sherwin, J.E., Proposed mechanism for urea nitrogen re-utilization: Relationship between urea and proposed guanidine cycles, Clin. Chem. 25 (1979) 1343-1344.
4. Stein, J.M., Cohen, B.D. and Kornhauser, R.S., Guanidinosuccinic acid in renal failure, experimental azotemia and inborn errors of the urea cycle, New Eng. J. Med. 280 (1969) 926-930.
5. Lowenthal, A. and Marescau, B., Urinary excretion of monosubstituted guanidines in patients affected with urea cycle diseases, in: "Neuro-genetics and Neuro-ophthalmology," A. Huber and D. Klein Eds., Elsevier/North-Holland Biomedical Press, Amsterdam (1981) pp. 347-350.
6. Böhles, H., Cohen, B.D. and Michalk, D., Guanidinosuccinic acid excre-tion in argininosuccinic aciduria, in: "Urea Cycle Diseases," A. Lowen-thal, A. Mori and B. Marescau Eds. Plenum Press, New York (1982) pp. 443-448.
7. Marescau, B., Qureshi, I.A., De Deyn, P., Letarte, J., Ryba, R. and Lowenthal, A., Guanidino compounds in plasma, urine and cerebrospinal fluid of hyperargininemic patients during therapy, Clin. Chim. Acta 146 (1985) 21-27.
8. Marescau, B., De Deyn, P., Wiechert, P., Van Gorp, L. and Lowenthal, A., Comparative study of guanidino compounds in serum and brain of mouse, rat, rabbit and man, J. Neurochem. 46 (1986) 717-720.
9. Ceriotti, G., Ultramicro determination of plasma urea by reaction with diacetylmonoxime antipyrine without deproteinization, Clin. Chem. 17 (1971) 400-402.
10. Hutchinson, J.H. and Labby, D.H., New method for the microdetermination of blood ammonia by use of the cation exchange resin, J. Lab. Clin. Med. 60 (1962) 170-178.
11. Adachi, I., Tanimura, A. and Asahina, M., A colorimetric determination of orotic acid, J. Vitaminol. 9 (1963) 217-226.
12. Visek, W.J. and Shoemaker, J.D., Orotic acid, arginine and hepatotoxic-ity, J. Am. Coll. Nutr. 5 (1986) 153-166.
13. Terheggen, H.G., Schwenk, A., Lowenthal, A., Van Sande, M. and Colombo, J.P., Argininemia with arginase deficiency, Lancet II (1969) 748-749.
14. Terheggen, H.G., Lavinha, F., Colombo, J.P., Van Sande, M. and Lowenthal A., Familial hyperargininemia, J. Génét. Hum. 20 (1972) 69-84.

α-KETO-δ-GUANIDINOVALERIC ACID, A COMPOUND ISOLATED FROM HYPERARGININEMIC PATIENTS, DISPLAYS EPILEPTOGENIC ACTIVITIES

Peter P. De Deyn, Robert L. Macdonald, Bart Marescau and
Armand Lowenthal

Laboratory of Neurochemistry, Born-Bunge Foundation and
Laboratory of Neuropharmacology, U.I.A., 2610 Antwerp, Belgium
*Department of Neurology, University of Michigan Medical
Center, Neuroscience Laboratory Building, Ann Arbor, MI 48104
U.S.A.

INTRODUCTION

α-Keto-δ-guanidinovaleric acid (α-k-δ-GVA), a monosubstituted guanidino compound, is found to be present in the urine, serum and cerebrospinal fluid of patients affected with hyperargininemia[1-3].

Hyperargininemia is an autosomal recessive disease, characterized by a deficiency of arginase, the last enzyme of the urea cycle, which converts arginine to urea and ornithine. The first clinical and biochemical descriptions were published in 1969[4] and 1970[5,6]. Since then, 18 other cases have been reported in the literature[7-21]. Patients affected with this disease, display the following clinical symptoms: vomiting, irritability, a progressive spasticity and epilepsy. In addition, some patients have an episodic hyperammonemia and are mentally retarded.

The causative factors of the neurological symptomatology are not known as yet. The hyperammonemia or the accumulation of arginine and its catabolites, the guanidino compounds, could be responsible. The lack of hyperammonemia or the presence of only slightly elevated plasma ammonium levels in several patients[9,10,13,21] indicates the importance of arginine and its catabolites as the possible toxins. α-K-δ-GVA, one of the guanidino compounds accumulated in biological fluids of hyperargininemic patients, was experimentally shown to display epileptogenic effects in rabbit. Topical application of the keto-analogue of arginine, on the sensory motor cortex evoked tonic convulsions and epileptiform discharges which were recorded on the electroencephalogram[22]. The experimentally demonstrated epileptogenicity of these guanidino compounds might contribute to the epileptic symptomatology observed in hyperargininemia.

Several convulsants have previously been shown to reduce responses to iontophoretically applied γ-aminobutyric acid (GABA) on mammalian spinal cord neurons grown in tissue culture[23]. In an attempt to determine the mechanisms through which α-k-δ-GVA might produce seizures, we studied its influence on responses evoked by iontophoretically applied GABA and glycine (GLY) on mouse spinal cord neurons in primary dissociated cell culture. Intracellular microelectrode recording techniques were used.

MATERIALS AND METHODS

Primary Dissociated Cell Culture

Cultures of spinal cord neurons were prepared from spinal cords and attached dorsal root ganglia from 12-14 day old fetal mice as described previously[24]. The dissociated cells were suspended in culture medium and then plated on sterile collagen-coated 35 mm dishes. Cultures were maintained for 4 to 9 weeks before electrophysiological experiments.

Experimental Procedures

Solutions: all recordings were made in a Dulbecco's phosphate buffered saline (DPBS) after removal of growth medium. The recording medium contained an elevated magnesium ion concentration in order to suppress spontaneous activity. Solutions of α-k-δ-GVA were always prepared on the day of the experiment in the following manner: dry, enzymatically synthetized, α-k-δ-GVA was dissolved in DPBS to form a 1 M stock solution, aliquots were removed and diluted in DPBS to give concentrations between 1 nM and 100 mM.

Experimental Apparatus

For experiments, the culture dish containing recording medium, was placed on a stage heated by a Pellitier device with temperature regulated at 34-35°C. The stage was mounted on a Leitz inverted microscope fitted with phase contrast optics to facilitate micropipette placement and to penetrate cells under direct visual control.

Electrophysiological Recordings

Intracellular recordings were made from the somata from the spinal cord neurons using glass micropipettes filled with 3 M KCl. An active bridge circuit (Dagan 8100 or W-P Instrument M707) was used. The preamplifier output signal was led to a 6-channel polygraph (Gould 2600S) for continuous recording.

GABA- and GLY-responses: GABA (0.5 M, pH 3.4) and GLY (0.5 M, pH 3.0) were applied iontophoretically using 500 msec duration current pulses at 5 sec intervals. The use of 3 M KCl-filled micropipettes shifted the chloride equilibrium potential from about -65 mV to about -20 mV. Under these experimental conditions, an incrase of chloride conductance results in an outward chloride current[25], giving depolarizing GABA- and GLY-responses. Responses of about 10-15 mV in amplitude were evoked following membrane hyperpolarization to ± -70 mV in order to avoid saturation at or near the reversal potential and to obtain a large transmembrane chloride ion gradient. Effects on GABA- and GLY-responses were accepted only if the responses returned to control amplitude within 5 min of removal of the α-k-δ-GVA containing micropipette.

Drug Application

For evaluation of α-k-δ-GVA effects on GABA-and GLY-responses, α-k-δ-GVA was applied by miniperfusion through a blunt tipped micropipette. The micropipette was positioned 15-30 μm from the soma of the cell under study. The open end of the miniperfusion pipette was connected to a pressure regulator, set between 2.8 and 5.5 kPa, by tight fitting polyethylene tubing. Pressure pulse duration was 10 sec. Under these conditions, no artifacts were produced and application of control solution (DPBS alone) was virtually free of effects. For assessment of the possible antagonist effect of CGS 9896 (a pure benzodiazepine receptor antagonist[26]) on α-k-δ-GVA-induced changes of GABA-responses, CGS 9896 (1μM) was applied by diffusion from large tipped micropipettes before application by miniperfusion of α-k-δ-GVA. As control, DPBS alone, applied by diffusion was without effect in this paradigm.

Drugs

CGS 9896 (2-(4-chlorophenyl)-2,5-dihydropyrazolo (4,3-C) quinoline-3(3-H)-one) was provided by Ciba-Geigy Corp. (Summit, New Jersey). GABA and GLY were purchased from Sigma Chemical Co. (St. Louis, Missouri). α-k-δ-GVA was synthesized enzymatically by a modification of the method of Meister[27]. The purity of the synthesized compound was controlled by liquid chromatography and nuclear magnetic resonance.

Algebraic Methods

At all applied concentrations, mean values and standard deviations were calculated for the effects on GABA-responses and GLY-responses. The effects were expressed as percentage decrease of GABA-responses or GLY-responses.

RESULTS

Miniperfusion of α-k-δ-GVA did not alter resting membrane potential or conductance. Miniperfusion of DPBS alone had virtually no effect on GABA-and GLY-responses. Reversible effects of α-k-δ-GVA on GABA- and GLY-responses are shown in Fig. 1.

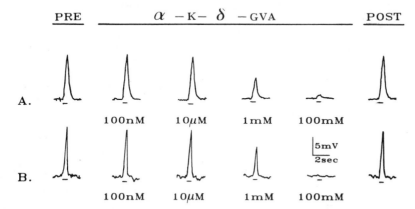

Figure 1. Reversible, dose dependent effects of α-keto-δ-guanidinovaleric acid (α-k-δ-GVA) on GABA-responses (A) and on glycine (GLY)-responses (B) on spinal cord neurons. PRE shows stable GABA- and GLY-responses before drug application. The four middle responses show the effect of the miniperfused guanidino compound. GABA- and GLY-responses returned to control values (POST) after removal of the α-k-δ-GVA containing micropipette. Iontophoretic application of GABA or GLY is indicated with a dash.

α-K-δ-GVA reversibly decreased GABA- (Fig. 1A) as well as GLY-responses (Fig. 1B) in a concentration dependent manner. Figure 2 further illustrates the concentration dependency of the effect of α-k-δ-GVA on GABA responses. A small reduction (3.6 ± 2.9%) was observed at 100 μM and a 94.4 ± 1.4% decrease was obtained at 100 mM. No clear effects were obtained at concentrations ranging from 10 nM to 10 μM.

The concentration dependency of the effects of α-k-δ-GVA on GLY-responses is illustrated in Fig. 3. A small decrease (2.5 ± 5.5%) of GLY-responses was observed at 100 μM. A 21.2 ± 8.6% reduction was obtained at 1 mM and 99.5 ± 1% decrease was seen at 100 mM.

Application by diffusion of CGS 9896, before application of α-k-δ-GVA by miniperfusion, did not influence the effect of the latter on GABA-responses. Indeed, α-k-δ-GVA 10 mM reduced GABA-responses to 18.0 ± 5.2% of their control value (n=5), while co-application of α-k-δ-GVA 10 mM and CGS 9896 1 μM gave a similar reduction of GABA-responses to 18.5 ± 10.0% (n=3) of control values.

DISCUSSION

Our experiments clearly demonstrate that α-k-δ-GVA reversibly decreased GABA-and GLY-responses on mouse spinal cord neurons, in a concentration dependent manner. Our findings furthermore indicate that α-k-δ-GVA does not exert its effects through interaction with the benzodiazepine receptor. In-

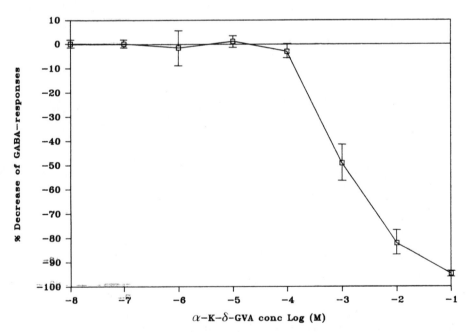

Figure 2. Concentration dependent decrease of GABA-responses in spinal cord neurons by α-keto-δ-guanidonovaleric acid (α-k-δ-GVA). Effects are expressed in percentage decrease of the original GABA-response. Data shown are means and standard deviations. Three to 6 cells have been studied for each tested concentration. α-K-δ-GVA concentration on the abscissa are logarithm molar.

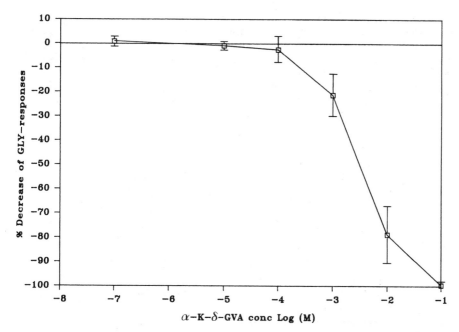

Figure 3. Concentration dependent decrease of the glycine (GLY)-responses in
spinal cord neurons by α-keto-δ-guanidinovaleric acid (α-k-δ-GVA).
Effects are expressed in percentage decrease of the original GLY-
response. Data shown are means and standard deviations. Three to 6
cells have been studied for each tested concentration. α-K-δ-GVA
concentrations on the abscissa are logarithm molar.

deed, CGS 9896, did not antagonize the inhibitory effect on GABA-responses
of α-k-δ-GVA. CGS 9896 was previously shown, in the same experimental paradigm, to be a pure antagonist at the benzodiazepine receptor, antagonizing
the effects of agonists as well as inverse agonists[26]. Earlier studies indicate that the inhibitory neurotransmitters GABA and GLY act through different receptors on the surface membranes of these cultured neurons[23,28,29].
Furthermore, activation of chloride conductance has been demonstrated to
underly the GABA-and GLY-receptor coupled events[30]. Given the knowledge that
GABA and GLY exert their effects by activation of chloride conductance
through interaction with different receptors, our results seem to indicate
that α-k-δ-GVA, shown here not to be a benzodiazepine receptor ligand, inhibits inhibitory amino acid-responses by blocking the chloride ionophore.

One could argue the pathophysiological importance of the observed inhibitory effects of α-k-δ-GVA on inhibitory neurotransmitter-responses. Indeed, the α-k-δ-GVA-induced decrease of GABA-and GLY-responses was obtained
only at levels higher than those earlier found in cerebrospinal fluid of
hyperargininemic patients. However, determinations of α-k-δ-GVA acid levels
have only been performed on cerebrospinal fluid of treated hyperargininemic
patients. Higher levels of α-k-δ-GVA acid are to be expected in untreated
patients. Furthermore, when evaluating the possible neurotoxicity of this
compound, one should take its possible accumulation in brain tissue into
account. Finally, a possible additive epileptogenic effect has been suggested for different guanidino compounds.

In conclusion, α-k-δ-GVA, a guanidino compound found to be increased in the biological fluids of hyperargininemic patients, inhibits responses to the inhibitory neurotransmitters GABA and GLY on mouse neurons in cell culture. The underlying mechanism is suggested to be blocking of chloride channels. This effect might explain the in vivo demonstrated epileptogenicity of this compound in rabbit. The possible pathophysiologic importance of this neurotoxin in hyperargininemic patients is still uncertain.

ACKNOWLEDGEMENTS

The authors wish to thank Ms. Nancy Fox for technical assistance and Ms. Barbara Pickut for editing the manuscript. Financial support was provided through the U.S. Public Health Service (Grant N° N.S.19613), the Rotary Foundation of Rotary International and the Born-Bunge Foundation.

REFERENCES

1. Marescau, B., Pintens, J., Lowenthal, A., Esmans, E., Luyten, Y., Lemière, G., Domisse, R., Alderweireldt, F. and Terheggen, H.G., Isolation and identification of 2-oxo-5-guanidinovaleric acid in urine of patients with hyperargininemia by chromatography and gas chromatography/mass spectrometry, J. Clin. Chem. Clin. Biochem. 19 (1981) 61-65.
2. Marescau, B. and Lowenthal, A., Isolation and identification of some guanidino compounds in the urine of patients with hyperargininemia by liquid chromatography, thin-layer chromatography and gas chromatography-mass spectrometry, J. Chromat. 224 (1981) 185-195.
3. Marescau, B., Qureshi, I.A., De Deyn, P., Letarte, J., Ryba, R. and Lowenthal, A., Guanidino compounds in plasma, urine and cerebrospinal fluid of hyperargininemic patients during therapy, Clin. Chim. Acta 146 (1985) 21-27.
4. Terheggen, H.G., Schwenk, A., Lowenthal, A., Van Sande, M. and Colombo, J.P., Argininemia with arginase deficiency, Lancet II (1969) 748-749.
5. Terheggen, H.G., Schwenk, A., Lowenthal, A., Van Sande, M. and Colombo, J.P., Hyperargininämia mit Arginase-defect. Eine neue familiäre Stoffwechselstörung. I. Klinische Befunde, Z. Kinderheilk. 107 (1970) 298-312.
6. Terheggen, H.G., Schwenk, A., Lowenthal, A., Van Sande, M. and Colombo, J.P., Hyperargininämia mit Arginase-defect. Eine neue familiäre Stoffwechseltstörung. II. Biochemische Untersuchungen, Z. Kinderheilk. 107 (1970) 313-323.
7. Terheggen, H.G., Lowenthal, A., Lavinha, F. and Colombo, J.R., Familial hyperargininemia, Arch. Dis. Child. 50 (1975) 57-62.
8. Cederbaum, S.D., Shaw. K.N.F. and Valente, M., Hyperargininemia, J. Pediatr. 90 (1977) 596-573.
9. Michels, V.V. and Beaudet, A.L., Arginase deficiency in multiple tissues in argininemia, Clin. Genet. 13 (1978) 61-67.
10. Snyderman, S.E., Sansaricq, C., Chen, W.J., Norton, P.M. and Phansalkar, S.V., Argininemia, J. Pediatr. 90 (1978) 563.
11. Snyderman, S.E., Sansaricq, C., Norton, P.M. and Goldstein, F., Argininemia treated from birth, J. Pediatr. 95 (1979) 61-63.
12. Qureshi, I.A., Letarte, J. and Quellet, R., Ammonia metabolism in a family affected by hyperargininemia, Diabete Metabol. 7 (1981) 5-11.
13. Qureshi, I.A., Letarte, J., Quellet, R. and Larochelle, J.K., A new French-Canadian family affected by hyperargininemia, J. Inher. Metab. Dis. 6 (1983) 179-182.
14. Yoshino, M., Kubota, K., Yoshida, I., Murakami, I. and Yamashita, F., Argininemia: Report of a new case and mechanisms of orotic aciduria and

hyperammonemia, in: "Urea cycle diseases," A. Lowenthal, A. Mori and B. Marescau Eds., Plenum Press, New York (1982) Vol. 153, pp. 121-125.

15. Sakiyama, T., Nakabayashi, H., Kondo, Y., Shimizu, H., Kodama, S. and Kitagawa, T., Argininemia: Clinical course and trial of enzyme replacement therapy, Biomed. Therapeut. (Tokyo) 8 (1982) 907-910. (in Japanese).

16. Mizutani, N., Maehara, M., Hayakawa, C., Kato, T., Watanabe, K. and Suzuki, S., Hyperargininemia: Clinical course and treament with sodium benzoate and phenylacetic acid, Brain Dev. 5 (1983) 555-563.

17. Kang, S.S., Wong, P.W.K. and Melyn, M.A., Hyperargininemia: Effect of ornithine and lysine supplementation, J. Pediatr. 103 (1983) 763-765.

18. Endres, W., Schaller, R. and Shin, Y.S., Diagnosis and treatment of argininaemia. Characteristics of arginase in human erythrocytes and tissues, J. Inher. Metab. Dis. 7 (1984) 8.

19. Bernar, J., Hanson, R.A., Kern, R., Phoenix, B., Shaw, K.N.F. and Cederbaum, S.D., Arginase deficiency in a 12-year-old boy with mild impairment of intelectual function, J. Pediatr. 108 (1986) 432-435.

20. Jorda, A., Rubio, V., Portoles, M., Vilas, J. and Garcia-Pino, J., A new case of arginase deficiency in a spanish male, J. Inher. Metab. Dis. 9 (1986) 393-397.

21. Cederbaum, S.D., Shaw, K.N.F., Spector, E.B., Verity, M.A., Snodgrass, P.J. and Sugarman, G.I., Hyperargininemia with arginase deficiency, Pediatr. Res. 13 (1979) 827.

22. Marescau, B., Hiramatsu, M. and Mori, A., α-Keto-δ-guanidinovaleric acid-induced epileptiform discharges in rabbits, Neurochem. Pathol. 1 (1983) 203-209.

23. Macdonald, R.L. and Barker, J.L., Specific antagonism of GABA-mediated postsynaptic inhibition in cultured mammalian spinal cord neurons: a common mode of convulsant action, Neurology 28 (1978) 325-330.

24. Ransom, B.R., Neale E., Henkart, M., Bullock, P.N. and Nelson, P.G., Mouse spinal cord in cell culture. I. Morphology and intrinsic neuronal electrophysiologic properties, J. Neurophysiol. 40 (1977) 1132-1150.

25. Nishi, S., Minota, S. and Karczmar, A.G., Primary afferent neurons: The ionic mechanism of GABA-mediated depolarization, Neuropharmacol. 13 (1974) 215-219.

26. De Deyn, P.P. and Macdonald, R.L., CGS 9896 and ZK 91296, but not CGS 8216 and Ro 15-1788, are pure benzodiazepine receptor antagonists on mouse neurons in culture, J. Pharmacol. Exp. Ther. 242 (1987) 48-55.

27. Meister, A., The α-keto analogues of arginase, ornithine and lysine, J. Biol. Chem. 206 (1954) 577-583.

28. Ransom, B.R., Bullock, P.N. and Nelson, P.G., Mouse spinal cord in cell culture. III. Neuronal chemosensitivity and its relationship to synaptic activity, J. Neurophysiol. 40 (1977) 1163-1177.

29. Nelson, P.G., Ransom, B.R., Henkart, M. and Bullock, P.N., Mouse spinal cord in cell culture. IV. Modulation of inhibitory synaptic function, J. Neurophysiol. 40 (1977) 1178-1187.

30. Barker, J.L. and Ransom, B.R., Amino acid pharmacology of mammalian central neurones grown in tissue culture, J. Physiol. 280 (1978) 331-354.

VI. INVOLVEMENT OF GIUANIDINO COMPOUNDS IN RENAL DYSFUNCTION

SERUM LEVELS OF GUANIDINO COMPOUNDS IN ACUTE RENAL FAILURE

Yasushi Suzuki, Fumitake Gejyo and Masaaki Arakawa

Department of Medicine (II), Niigata University School of
Medicine, Niigata 951, Japan

INTRODUCTION

There have been many reports dealing with the metabolism of guanidino
compounds in chronic renal failure (CRF). Howevever, the metabolism of these
compounds in acute renal failure (ARF) has not been adequately described.
These compounds are known to be metabolized in liver, kidney and pancreas.
The function of these organs in ARF may differ from that in CRF. The pathophysiology of ARF may be variable in each case, which will have some effect
on the metabolism of guanidino compounds. Severe liver damage has, for example, altered the metaboilism of guanidinoacetic acid (GAA) and guanidinosuccinic acid (GSA).

This paper reports serum levels of guanidino compounds in various types
of ARF including that with severe liver damage.

PATIENTS AND METHODS

According to clinical features, 33 undialyzed ARF patients were classified into 4 groups; 8 cases of pre-renal ARF (group A), 14 cases of renal ARF
(group B), 6 cases of post-renal ARF (group C), and 5 cases of ARF with
severe liver damage (group D). The clinical data of group D are shown in
Table 1.

Blood samples were obtained at the point of admission to our clinic.
Serial samples were followed in one case with severe liver damage. Separated
sera were stored at $-20°C$ until use.

Serum levels of GAA and GSA were measured by high performance liquid
chromatography (Auto-Guanidino Analyzer G-520, JASCO). The determination of
guanidino compounds were performed according to the method described by
Gejyo[1]. Serum creatinine (S-Cr) levels was also measured by an autoanalyzer
(JEOL, JCA-MS24).

Table 1.   Laboratory data in the patients with acute renal failure and severe liver damage.

Case	GOT (IU/1)	GPT (IU/1)	LDH (IU/1)	ALP (IU/1)	T–Bil (mg/dl)	D–Bil (mg/dl)	Ch–E (IU/1)	HPT (%)	AMMO (μg/dl)
1	8393	4061	15246	514	7.8	15.1	3200	7	
2	644	324		290	12.9			39	290
3	203	75	633	212	24.4	14.9	5100	36	257
4	361	126		1068	14.4				143
5	65	29	722	367	31.7		5100		

Table 2.   Serum levels of guanidinoacetic acid (GAA) and creatinine (Cr) in the patients with acute renal failure.

	group A	group B	group C	group D
GAA (nmol/ml)	2.16 ± 0.54	2.59 ± 1.54	1.95 ± 0.96	7.17 ± 8.95
Cr (mg/dl)	4.66 ± 2.76	6.10 ± 3.21	11.23 ± 4.77	5.24 ± 2.64
GAA/Cr	0.664 ± 0.418	0.517 ± 0.348	0.173 ± 0.046	1.815 ± 2.491

The serum levels of GAA in normal subjects was 3.00 ± 1.48, group A; prerenal acute renal failure (ARF), group B; renal ARF, group C; post-renal ARF and group D; ARF with severe liver damage

RESULTS

The mean level of creatinine in group C was significantly higher than that of the other groups. On the other hand, there was no significant difference in the serum GAA levels among the 4 groups, which was statistically same as that of normal subjects (3.00 ± 0.48) (Table 2).

GAA/Cr ratios were estimated in order to correct for the effect of renal dysfunction. The mean value of the GAA/Cr ratio of group C (0.173 ± 0.046) is significantly lower than those of group A (0.664 ± 0.418) or group B (0.517 ± 0.348). The time course of GAA levels in one case with severe liver damage is shown in Fig. 1. The ratio remained stable thoughout the course.

To analyze the decrease of GAA/Cr ratio in group C, the relationship between S–Cr and serum GAA is drawn in Fig. 2. There is no correlation between these two factors in group A and B. However, they show a positive correlation in group C (r=0.89, P < 0.02). The relationship between S–Cr and GAA/Cr ratio is also drawn in Fig. 3. A negative correlation is noticed in group A and B. However the GAA/Cr ratio in the group C shows nearly the same value regardless of S–Cr levels.

The serum GSA levels are increased in all 4 groups, as compared to those of normal subjects (0.511 ± 0.340) (Table 3). There is no significant difference among the 4 groups.

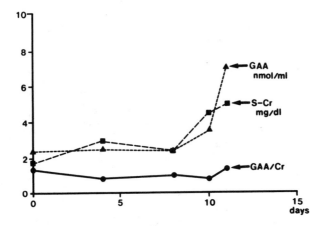

Figure 1.   Time course of serum guanidinoacetic acid (GAA) and creatinine
            (Cr) in a patient with severe liver damage.

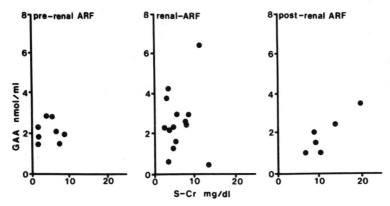

Figure 2.   Serum levels of creatinine (Cr) and guanidinoacetic acid (GAA) in
            acute renal failure (ARF).

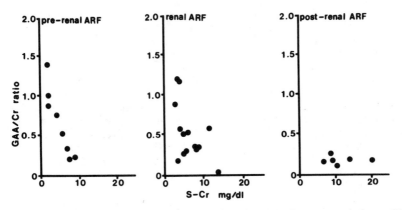

Figure 3.   Serum Cr level and GAA/Cr ratio in ARF. Cr; creatinine, GAA; guan-
            idinoacetic acid and ARF; acute renal failure.

Table 3. Serum levels of guanidinosuccinic acid (GSA) in the patients with
          acute renal failure.

	group A	group B	group C	group D
GSA (nmol/ml)	10.94 ± 6.99	24.39 ± 25.57	30.50 ± 31.16	19.90 ± 29.20

Serum levels of GSA in normal subjects was 0.511 ± 0.340.
The classification of group A,B,C,D is the same as that of table 2.

DISCUSSION

Cohen[2] reported that serum GAA levels in uremic patients increased and
urinary excretion of the guanidine decreased because of renal impairment. Van
Pilsum and Canfield[3] demonstrated that amidinotransferase relating to GAA
synthesis exists in the kidney cortex. Sasaki et al.[4] suggested that the
decrease of urinary GAA excretion might be due to decreased synthesis of GAA
in the kidney cortex. However, recent reports[5,6] demonstrate that serum GAA
levels show no difference between normal and uremic subjects.

When creatine was administrated to rats, the amidinotransferase activity
was supressed[7] due to negative feed-back. Creatine is known to be elevated
in uremic patients[2,8]. Kopple[9] thought that the decrease in GAA synthesis
could be due to both diminished renal parenchyma and inhibition of GAA synthesis by increased creatine. Furthermore, Sims and Seldin[10] report that
creatine inhibits the reabsoption of GAA by the tubules.

In the present study, the mean levels of GAA in all 4 groups were statistically the same as that of normal subjects. However, the GAA/Cr ratio of
4 groups was significantly different between group A or B and group C. The
GAA/Cr ratio of group C was lower than that of the others because of higher
S-Cr levels. (Table 2).

In group A and B (pre-renal and renal ARF), no correlation was found
between S-Cr and GAA. However, there was a negative correlation between S-Cr
and GAA/Cr in these two groups. These findings suggest that GAA production
may be maintained in the early stages and depressed gradually according to
the degree of renal dysfunction. On the other hand, in group C (post-renal
ARF), serum GAA levels were elevated while S-Cr increased, therefore, the
GAA/Cr ratio had been stable regardless of the S-Cr level. These findings
suggest that GAA production may be disturbed even in the early stages of
renal failure, although further study will be nessesary in patients showing
mild azotemia (S-Cr; 2-5 mg/dl).

Muramoto[11] demonstrated that when ARF was complicated with liver damage,
serum GAA levels would be higher than those without liver damage. Liver is
the site of metabolism of GAA to creatine. Cohen et al.[12] reported that serum
GSA levels were depressed in hepatorenal syndrome. However, our data showed
that there was no difference in serum GAA and GSA levels among the 4 groups.
This may be due to the wide standard deviation in group D.

REFERENCES

1. Gejyo, F., Baba, S., Watanabe, Y., Kishore, B.K., Suzuki, Y. and Arakawa, M., Possibility of a common metabolic pathway for the production of methylguanidine and dimethylamine in chronic renal failure, in: "Guanidines," A. Mori, B.D. Cohen and A. Lowenthal Eds., Plenum Press, New York (1985) pp. 295-308.
2. Cohen, B.D., Guanidinosuccinic acid in uremia, Arch, Intern. Med. 126 (1970) 846-850.
3. Van Pilsum, J.F. and Canfield, T.M., Transamidinase activities, in vitro, of kidneys from rats fed diets with nitrogen-containing compounds, J. Biol. Chem. 237 (1962) 2574-2577.
4. Sasaki, M., Takahara, K. and Natelson, S., Urinary guanidinoacetate/guanidinosuccinate ratio: An indicator of kidney dysfunction, Clin. Chem. 19 (1973) 315-321.
5. Shitomoto, M., Otsuji, S., Changes of guanidino compounds in chronic renal failure, Jpn. J. Nephr. 21 (1979) 33-49.
6. Tofuku, Y., Kuroda, M., Muramoto, H. and Takeda, R., A study of guanidin metabolism in uremia, with special reference to serum guanidinoacetic acid in anephric patients, Jpn. J. Nephr. 26 (1984) 33-43.
7. Walker, J.B., Metabolic control of creatine biosynthesis, J. Biol. Chem. 235 (1960) 2357-2361.
8. Bolliger, A. and Carrodus, A.L., Creatine retention in blood and cerebrospinal fluid, Med. J. Aust. 1 (1938) 69.
9. Kopple, J.D., "Clinical aspects of uremia and dialysis," C.C. Thomas, Illinois (1976).
10. Sims, E.A. and Seldin, D.W., Reabsorption of creatine and guanidinoacetic acid by renal tubules, Am. J. Physiol. 157 (1949) 14-20.
11. Muramoto, H., Tofuku, Y., Kuroda, M. and Takeda, R., The role of the liver as a significant modulator of the serum guanidinoacetic acid level in man, in: "Guanidines," A. Mori, B.D. Cohen and A. Lowenthal Eds. Plenum Press, New York (1985) pp. 105-112.

IS GUANIDINE A MARKER OF PEROXIDATION IN UREMICS?

Sohji Nagase, Kazumasa Aoyagi, Masako Sakamoto, Mitsuhiro
Miyazaki, Shoji Ohba, Mitsuharu Narita and Shizuo Tojo

Department of Internal Medicine, Institute of Clinical
Medicine, University of Tsukuba, Tsukuba 305, Japan

INTRODUCTION

There are many reports concerning the toxicity of oxygen in various
disease states[1-4]. Recently, attention has been called to the unfavorable
effects of active oxygen in renal diseases[5-9]. We have reported that the
hydroxyl radical plays an important role in vitro[10] and in isolated rat
hepatocytes[11] in the peroxidation of creatinine (CRN) to methylguanidine
(MG), a guanidino compound known to be a potent uremic toxin. In this study,
we investigate the correlation between some markers of the peroxidative state
and the concentration of various guanidino compounds in the sera of patients
undergoing regular hemodialysis.

MATERIALS AND METHOD

Fifty patients (37 males, 13 females), aged $45.6 \pm 15.3y$ (mean $\pm$ S.D.),
who had been undergoing maintenance hemodialysis for 403 ± 357 times were
studied. The orgins of chronic renal failure were chronic glomerulonephritis
in 33 patients, diabetic nephropathy in 4, polycystic kidney in 4, nephrosclerosis in 2, urological renal disease in 2, gouty kidney in 1 and unknown
in 4.

Blood samples were drawn from the arterial side of the arteriovenous
fistula before hemodialysis. Serum levels of total protein, albumin, total
cholesterol, urea nitrogen, CRN, uric acid, total calcium and inorganic
phosphorus were determined by an autoanalyzer (Hitachi 716, Japan). Blood
hemoglobin levels were determined by an autoanalyzer (Coulter S plus 5,
USA). Serum guanidinosuccinic acid (GSA), guanidinoacetic acid (GAA), arginine, guanidine (Gua), and MG were determined by high-pressure liquid chromatographic analysis (HPLC), using 9-10 phenanthrenequinone as described
previously[12]. Serum iron levels were measured essentially according to the
method of Ramsay[13]. Serum ferritin levels were measured with a commercial
radioimmunoassay kit (Baxter-Travenol, USA). Serum concentrations of malondialdehyde (MDA) were measured fluorometrically as the thiobarbituric acid
reactive material according to Yagi et al[14].

RESULTS

The correlation coefficients between various guanidino compounds and oth
er laboratory findings are summarized in Table 1. As reported previosly[15-17]
serum GSA and MG concentrations are correlated with the serum urea nitrogen
and CRN levels, respectively as shown in Fig. 1. We reported, however, that
MG was derived from CRN by the action of hydroxyl radicals, however serum
concentrations are not correlated with the concentrations of serum iron,
ferritin or MDA as shown in Fig. 2. On the other hand, the concentratins of
Gua are significantly correlated with serum levels of iron, ferritin and
MDA. The correlation between Gua and iron is shown in Fig. 3. In addition,
the correlation coefficient between serum Gua and ferritin is 0.685. Though
this value is already significant, it grows more significant as shown is
Fig. 4 if we disregard the 3 cases that have extremely high concentrations
of ferritin because of the over administration of iron (Fig. 4). The concentration of serum Gua is significantly correlated with the level of MDA as
shown in Fig. 5.

DISCUSSION

Recently, there have been many reports concerning unfavorable biologica
effects of active oxygen[1-4,8,9]. In uremia, elevated levels of thiobarbituri
acid reactive material in sera were reported[5] and it is reported that MDA is

Table 1. Correlation between various guanidino compounds and other labora-
          tory findings including markers of peroxidative state in the sera
          of 50 patients undergoing regular hemodialysis.

	Age	Duration	TP	Alb	T.Cho	BUN	CRN	UA	Ca	P
GSA	-0.297	0.066	0.136	-0.022	0.289	0.406	0.293	0.106	0.122	0.416
GAA	-0.025	0.150	-0.040	-0.065	0.006	0.192	0.404	0.029	0.193	0.280
Arg	0.061	-0.356	0.078	0.044	-0.023	0.202	0.313	0.023	0.182	-0.074
Gua	-0.119	0.535	-0.080	0.091	0.163	0.345	-0.111	0.056	-0.072	0.268
MG	-0.549	0.386	0.271	-0.006	0.027	0.086	0.672	-0.042	-0.080	0.042

	GSA	GAA	Arg	Gua	MG	Hb	Fe	Ferritin	MDA
GSA	-----	0.242	-0.113	0.389	0.334	-0.050	0.261	0.221	0.078
GAA	0.242	-----	0.469	0.085	0.181	-0.102	0.197	-0.034	-0.198
ARG	-0.113	0.469	-----	-0.307	-0.036	-0.325	-0.111	-0.229	-0.230
Gua	0.389	0.085	-0.307	-----	0.173	0.145	0.498	0.685	0.630
MG	0.334	0.181	-0.036	0.173	-----	-0.091	0.235	0.005	-0.032

Age: age of patients, Duration : times of underwent hemodialysis,
TP : total protein, Alb : albumin, T.Cho : total cholesterol,
BUN: blood urea nitrogen, CRN : creatinine, UA : uric acid,
Ca : calcium, P : total inorganic phosphorus,
GSA: guanidinosuccinic acid. GAA : guanidinoacetic acid,
Arg: arginine, Gua : guanidine, MG : methylguanidine,
Hb : hemoglobin, Fe : serum iron, Ferritin : serum ferritin,
MDA: malondialdehyde

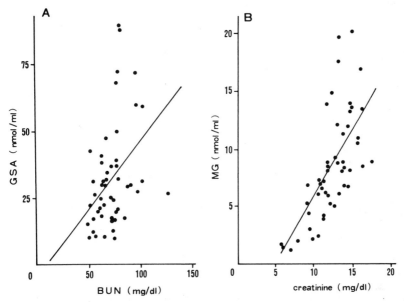

Figure 1.   A) Correlation between serum guanidinosuccinic acid (GSA) and
urea nitrogen (BUN) concentration in the patients undergoing
regular hemodialysis. (r=0.406, n=50, p < 0.001)
B) Correlation betwen serum methylguanidine (MG) and creatinine
concentration in the patients undergoing regular hemodialysis.
(r=0.672, n=50, p < 0.001)

elevated in the red blood cell membranes of hemodialysis patients[6]. In addition, it has been shown that serum levels of antioxidant activity are significantly lower in patients with chronic renal failure[7].

We have studied MG, a potent uremic toxin, and reported that the liver
is one of the organs which synthesizes MG and that CRN is the precursor[12].
In addition, we showed that Gua was synthesized from GAA in isolated rat
hepatocytes[10]. Furthermore, we described the mechanism of MG formation in
vitro, and reported that MG was the peroxidative product of CRN mainly
through the action of the hydroxyl radical[10]. Following these results, we
confirmed the role of active oxygen in the biosynthesis of MG from CRN using
isolated hepatocytes[11]. Our hypothesis that MG may be a useful indicator of
peroxidation in vivo, because MG is the peroxidative product of CRN, is
compatible with the aforementioned.

In this study, we examine the correlation between various guanidino
compounds and other laboratory findings in the sera of hemodialysis patients, and show that MG is not correlated with iron, ferritin or MDA, as had
been reported[7]. However, significant correlation was found between Gua and
iron, ferritin, and MDA. Iron is possibly a source of the hydroxyl radical
via Fenton's reaction with hydrogen peroxide[18,19], and also provides perferryl ion in a water solution[18,20]. Serum ferritin levels are better correlated with iron reserves than serum iron levels[21] while serum MDA levels
are generally accepted as an appropriate marker reflecting lipid peroxidation[22]. Serum MDA levels in this study are correlated with serum ferritin
levels, the correlative coefficient being 0.761. The fact that iron stimu-

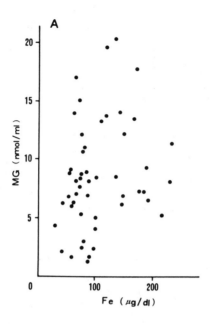

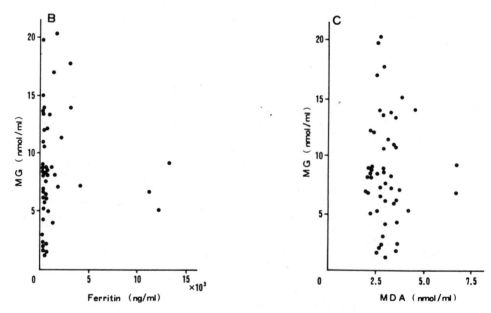

Figure 2.   A) Correlation between serum methylguanidine (MG) and iron (Fe)
            concentration in the patients undergoing regular hemodialysis.
            (r=0.235, n=50)
            B) Correlation between serum methylguanidine (MG) and ferritin
            concentration in the patients undergoing regular hemodialysis.
            (r=0.005, n=50)
            C) Correlation between serum methylguanidine (MG) and malondial-
            dehyde (MDA) concentration in the patients undergoing regular
            hemodialysis. (r=-0.032, n=50)

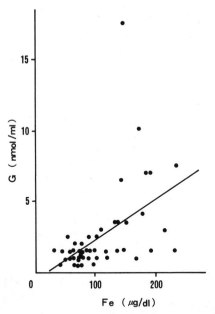

Figure 3.   Correlation between serum guanidine (G) and iron (Fe) concentration in the patients undergoing regular hemodialysis. (r=0.498,
n=50, p < 0.001)

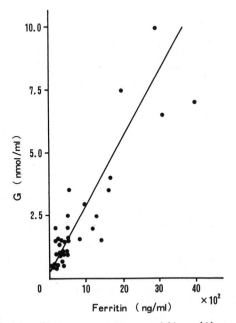

Figure 4.   Correlation between serum guanidine (G) and ferritin concentration
in the patients undergoing regular hemodialysis.(r=0.882, n=50,
p < 0.001)

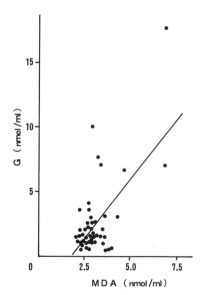

Figure 5.   Correlation between serum guanidine (G) and malondialdehyde
            concentration in the patients undergoing regular hemodialysis.
            (r=0.630, n=50, P < 0.001)

lates lipid peroxidation in vitro[23] and in vivo[24,25] suggests that Gua correlates with the source of active oxygen.

    Why does Gua and not MG correlate with these markers of peroxidation?
Further study is required to clarify this issue. One possible explanation is
that different species of active oxygen are concerned in Gua and MG synthesis. In other words, Gua synthesis may be mainly affected by the species of
active oxygen relating to iron, i.e. perferryl ion or hydroxyl radical,
while MG synthesis may be affected by iron-related species and others, i.e.
the superoxide radical or hydrogen peroxide. Also it is reported that Gua is
removed from red blood cells more easily than MG by hemodialysis, and the
opposite may be said about removal rates of Gua and MG from plasma[26]. In
other words, the plasma concentration of MG is modified strongly by hemodialysis, while that of Gua tends to reflect the tissue concentration more
precisely. Therefore, serum MG may not correlate with peroxidative markers.

    Elevated serum concentrations of Gua were reported in patient with
hypertension[27,28], renal diseases[29-31], epilepsy[32], and unidentified myopa-
thy[33]. Some of these reports are unreliable because the methods for measurement of guanidino compounds were somewhat primitive. It does appear, however, that the concentration of Gua in uremic serum is extremely high compared with that in normal subjects, measured by HPLC as in this study. As
for its biological effects, it has been reported that administration of Gua
causes the uncoupling of oxidative phosphorylation and inhibition of oxygen
uptake in rabbit kidney and rat liver mitochondria[34]. It is also known that
Gua is a potent inhibitor of many mammalian enzymes[35].

    There are many reports concerning the mechanism of formation of Gua.
Reductive cleavage of canavanine to Gua and homoserine by certain bacteria
is reported[36]. In addition, it is shown that Gua derives from hydroxyguanidine by "hydroxyguanidine reductase" and the activity was found in

Escherichia coli, mammalian kidney, pigeon liver and Streptomyces griceus[37]. In both reports, purification of biological activity was not carried out, so the existence of enzymes which catalyze these reactions is still unclear. Furthermore, there are reports that the reductive cleavage of canavanine to Gua and homoserine occur in the presence of liver extract, ferrous ion and reduced lipoate, and ferritin. It was also revealed that this activity in liver homogenate is nonenzymatic and essentially identical with the activity of ferrous ion either alone or with reduced lipoate[38]. Subsequently, there was a report that the guanidino carbon of either arginine or canavanine is a precursor of the carbon atom of Gua in the rat, though the biosynthetic pathway for Gua is unclear[39]. Recently, it was reported that Gua production is strongly increased by the intravenous administration of GAA in the uremic rat, whereas it is increased by the administration of GAA or arginine in the normal rat[40]. In our previous study, it was pointed out that Gua is synthesized in the liver and kidney of the rat by the intraperitoneal infusion of CRN, and it is also synthesized in isolated rat hepatocytes by the addition of GAA to the incubation medium[12].

The positive correlation between Gua and those factors which suggest oxidative risk in vivo leads to the possibility that Gua may be a useful indicator of peroxidation in uremics. Additional studies are necessary to determine the biological precursor of Gua and to clarify the species of active oxygen which affects Gua formation.

ACKNOWLEDGEMENTS

These research findings were presented, in part, at the 29th Annual Meeting of the Japanese Congress of Nephrology in November 1986, in Tokyo, and at the 9th Annual Meeting of the Japanese Guanidino Compounds Research Association in October 1986, in Tokyo. This study was supported in part by a research grant from University of Tsukuba Project Reseach and Intractable Diaease Division, Public Health Bureau, Ministry of Health and welfare, Japan. We are indebted to Mrs. Satomi Kawamura for her valuable technical assistance.

REFERENCES

1. Misra, H.P. and Fridovich, I., Superoxide dismutase and the oxygen enhancement of radiation lethality, Arch. Biochem. Biophys. 176 (1976) 577–581.
2. Totter, J.R, Spontaneous cancer and its possible relationship to oxygen metabolism, Proc. Natl. Acad. Sci. USA 77 (1980) 1763–1767.
3. Weitzman, S.A. and Stossel, T.P., Effects of oxygen raidcal scavengers and antioxidants on phagocyte-induced mutagenesis, J. Immun. 128 (1982) 2770–2772.
4. Freeman, B.A. and Crapo, J.D., Free radical and tissue injury, Lab. Invest. 47 (1982) 412–426.
5. Fillet, H., Elion, E., Sullivan, J., Sherman, R. and Zabriskie, J.B., Thiobarbituric acid reactive material in uremic blood, Nephron 29 (1981) 40–43.
6. Giardini, O., Taccone-Gallucci, M., Lubrano, R., Ricciardi-Tenore, G., Bandino, D., Silvi, I., Ruberto, U. and Casciani, C.U., Evidence of red blood cell membrane lipid peroxidation in haemodialysis patients, Nephron 36 (1984) 235–237.
7. Kuroda, M., Asaka, S., Tofuku, Y. and Takeda, R., Serum antioxidant activity in uremic patients, Nephron 41 (1985) 293–298.

8. Rehan, A., Johnson, K., Wiggins, R.C., Kunkel, R.G. and Ward, P.A., Evidence for the role of oxygen radical in acute nephrotoxic nephritis, Lab. Invest. 51 (1984) 396–403.

9. Paller, M.S., Hoidal, J.R. and Ferris, T.F., Oxygen free radical in ischemic acute renal failure in the rat, J. Clin. Invest. 74 (1984) 1156–1164.

10. Nagase, S., Aoyagi, K., Narita, M. and Tojo, S., Active oxygen in methylguanidine synthesis, Nephron 44 (1986) 299–303.

11. Aoyagi, K., Nagase, S., Narita, M. and Tojo, S., Role of active oxgen on methylguanidine synthesis in isolated rat hepatocytes, Kidney Int. 32 (1987) S229–S233.

12. Nagase, S., Aoyagi, K., Narita, M. and Tojo, S., Biosynthesis of methylguanidine in isolated rat hepatocytes and in vivo, Nephron 40 (1985) 470–475.

13. Ramsay, W.N.M., The determination of iron in blood plasma or serum, Biochem. J. 53 (1953) 227–231.

14. Yagi, K., A simple fluorometric assay for lipoperoxide in blood plasma, Biochem. Med. 15 (1976) 212–216.

15. Baker, L.R.I. and Marshall, R.D., A reinvestigation of methylguanidine concentration in sera from normal and uraemic subjects, Clin, Sci. 41 (1971) 563–568.

16. Giovannetti, S., Balestri, P.L. and Barsotti, G., Methylguanidine in uremia, Arch. Intern. Med. 131 (1973) 709–713.

17. Kopple, J.D., Gordon, S.I., Wang, M. and Swendseid, M.E., Factors affecting serum and urinary guanidinosuccinic acid levels in normal and uremic subjects, J. Lab. Clin. Med. 90 (1977) 303–311.

18. Halliwell, B. and Gutterridge, M.C., Oxygen toxicity, oxygen radicals, transition metals and disease, Biochem. J. 219 (1984) 1–14.

19. Balinski, T., Krawiec, Z., Liczmanski, A. and Litwinska, J., Is hydroxyl radical generated by Fenton reaction in vivo? Biochem. Biophys, Res. Commun. 130 (1985) 533–539.

20. Nakano, H., Sugioka, K., Nakano, M., Mizukami, M., Kimura, H., Tero-Kubota, S. and Ikegami, Y., Importance of Fe^{2+}-ADP and the relative unimportance of ·OH in the mechanism of mitomycin C-induced Lipid peroxidation, Biochem. Biophys. Acta 796 (1984) 285–293.

21. Lipschitz, D.A., Cook, J.D. and Finch, C.A., A clinical evaluation of serum ferritin as an index of iron stores, New Engl. J. Med. 290 (1974) 1213–1216.

22. Yagi, K., Assay for serum lipid peroxidase level and its clinical significance, in: "Lipid peroxides in biology and medicine," K. Yagi Ed., Academic Press, New York (1982) pp. 223–242.

23. Bucher, J.R., Tien, M. and Aust, S.D., The requirement for ferric in the initiation of lipid peroxidation by chelated ferrous iron, Biochem. Biophys. Res. Commun. 111 (1983) 777–784.

24. Bacon, B.R., Tavill, A.S., Brittenham, G.M., Park. C.H. and Recknagel, R.O., Hepatic lipid peroxidation in vivo in rats with chronic iron overload, J. Clin. Invest. 71 (1983) 429–439.

25. Willmore, L.J. and Rubin, J.J., Formation of malonaldehyde and focal brain edema induced by subpial injection of $FeCl_2$ into rat isocortex, Brain Res. 246 (1982) 113–119.

26. Hatakeyama, Y., Chiba, E., Arai, S. and Suzuki, T., Removal rate of guanidino compounds in various method of hemodialysis, Proc. Annual Meeting Jpn. Guanidino Compounds Res. Assoc. 5 (1982) 39–42 (in Japanese).

27. Major, R.H. and Weber, C.J., The possible increase of guanidine in the blood of certain persons with hypertension, Arch. Int. Med. 40 (1927) 891–899.

28. De Wesselow, O.L.V.S. and Griffiths, W.G., The blood guanidine in hypertension, Brit. J. Exper. Path. 13 (1932) 428–434.

29. Andes, J., Lineger, C.R. and Myers, V.C., Guanidine-like substances in the blood. II. Blood guanidine in nitrogen retention and hypertension, J. Lab. Clin. Med. 22 (1937) 1209-1216.
30. Carr, M.H. and Schloerb, P.R., Analysis for guanidine and methylguanidine in uremic plasma, Anal. Biochem. 1 (1960) 221-227.
31. Menichini, G.C., Giovannetti, S. and Lupetti, S., A new method for measuring guanidine in uremia, Experientia 29 (1972) 506-507.
32. Murray, M. and Hoffman, C.R., The occurrence of guanidine-like substances in the blood in essential epilepsy, J. Lab. Clin. Med. 25 (1940) 1072-1073.
33. Greenblatt, I.J., Guanidinuria, J. Biol. Chem. 136 (1941) 791.
34. Hollunger, G., Guanidines and oxidative phosphorylations, Acta Pharmacol. Toxicol. 11 (1955) S1-S84.
35. Rajagopalan, K.V., Fridovich, I. and Handler, P., Inhibition of enzyme activity by urea, Fed. Proc. 19 (1960) 49.
36. Kihara, H., Prescott, J.M. and Snell, E.E., The bacterial cleavage of canavanine to homoserine and guanidine, J. Biol. Chem. 217 (1955) 497-503.
37. Walker, J.B. and Walker, M.S., The enzymatic reduction of hydroxyguanidine, J. Biol. Chem. 234 (1959) 1481-1484.
38. Takahara, K., Nakanishi, S. and Natelson, S., Study on the reductive cleavage of canavanine and canavaninosuccinic aicd, Arch. Biochem. Biophys. 145 (1971) 85-95.
39. Reiter, A.J. and Horner, W.H., Studies on the metabolism of guanidine compounds in mammals. Formation of guanidine and hydroxyguanidine in the rat, Arch. Biochem. Biophys. 197 (1979) 126-131.
40. Mikami, H., Orita, Y., Ando, A., Fujii, M., Kikuchi, T., Yoshihara, K., Okada, A. and Abe, H., Metabolic pathway of guanidino compounds in chronic renal failure, in: "Urea cycle diseases," A. Lowenthal, A. Mori and B. Marescau Eds., Plenum Press, New York (1983) pp 449-458.

DISTURBANCE OF CREATINE METABOLISM IN RATS WITH CHRONIC RENAL FAILURE

Masato Inouchi, Tomoya Fujino, Takeo Sato, Takashi Yasuda, Hitoshi Tomita, Tsukasa Kanazawa, Chikako Shiba, Sadanobu Ozawa, Shigeru Ohwada and Masashi Ishida

1st Department of Internal Medicine, St. Marianna University School of Medicine, 2-16-1 Sugao, Miyamae-Ku, Kawasaki, Japan

INTRODUCTION

Abnormalities in creatine metabolism have been reported in chronic renal failure (CRF)[1,2]. In chronic uremic patients, levels of guanidinoacetic acid (GAA) and creatine in plasma are decreased. The mechanism of this phenomenon is not entirely understood. The present report describes a study of creatine metabolism in rats with CRF. Creatine and GAA levels in plasma, kidney and liver were measured. Glycine amidinotransferase (GAT) activity in the kidney and guanidinoacetic acid methyltransferase (GAA-MT) activity in the liver were determined. A radioisotopic evaluation of the metabolites of the creatine pathway using ^{14}C-glycine and ^{14}C-GAA was also carried out.

MATERIALS AND METHODS

Animals

Male and female Wistar rats, about 150g in body weight, were used for a model of chronic renal failure (CRF). Two weeks after resection of 2/3 of the left kidney, the contralateral kidney was removed. Three to four months later, the animals were sacrified.

Chemical Analysis

Plasma creatinine and urea nitrogen (BUN) were measured by the Jaffe reaction and urease-indophenol method. Creatine and GAA levels in the plasma and tissues were measured by HPLC (ninhydrin method) employing a LC-6A liquid chromatography set (Shimadzu, Kyoto). Tissue homogenates and plasma were deproteinized by trichloroacetic acid (TCA) solution (final concentration 10%). Tissue homogenate was prepared as follow: the liver and kidney were removed, perfused with cold saline, weighed and used immediately. One gram of tissue was homogenized with two volume of cold water using a YAMATO ultra dispercer, then deproteinized with TCA (Fig. 1). After centrifugation for 10 min at 1,500 x g, the supernatants were analyzed. Eluent buffers were as

277

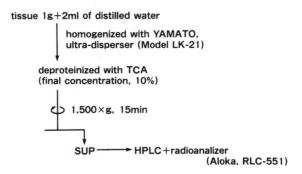

Figure   1.   Preparation of tissue homogenate. TCA; trichloroacetate

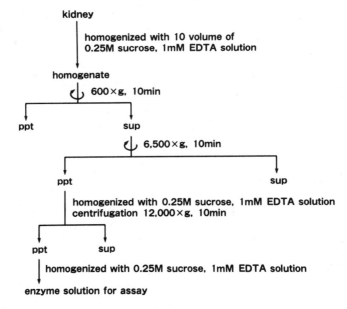

Figure   2.   Preparation of the mitochondrial fraction for kidney GAT assay.

follows: 1st buffer; 0.15 M Na-citrate, pH 3.4, 2nd buffer; 0.35 M Na-citrate, pH 5.0, 3rd buffer; 0.35 M Na-citrate, pH 6.0, 4th buffer; 0.35 M Na-citrate, 0.3 M NaCl, 0.2 M Na-borate, pH 11.4, 5th; 0.2 N NaOH and 6th; water. The gradient time program was as follows:

    6.5 min   8 min   2.5 min   18 min   3 min   3 min
    1st → 2nd   →   3rd   →   4th   →   5th   →   6th   →   1st

Determination of Enzyme Activities

Kidney glycine amidinotransferase (GAT). Kidney GAT is distributed mainly in the inner mitochondrial membrane[3], so the mitochondrial fraction was obtained according to Schneider's method[4] (Fig. 2). GAT activity in the kidney was measured according to the method of Van Pilsum[5] with the following modification. GAA which is the product of the GAT reaction was measured

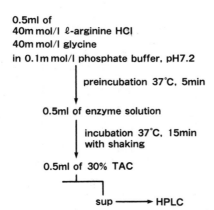

0.5ml of
40m mol/l ℓ-arginine HCl
40m mol/l glycine
in 0.1m mol/l phosphate buffer, pH7.2

preincubation 37°C, 5min

0.5ml of enzyme solution

incubation 37°C, 15min
with shaking

0.5ml of 30% TAC

sup ⟶ HPLC

Figure 3. The assay procedure for kidney glycine amidinotransferase activity.

instead of ornithine using HPLC. The specificity and sensitivity of the GAA quantitative analysis are excellent and also not affected by arginase activity, so the reaction time was shortened and Na-fluoride, as used in the original method, was omitted (Fig. 3).

Liver guanidinoacetate methyltransferase (GAA-MT). GAA-MT activity was measured according to the method of Im[6] and Ogawa[7]. Liver was homogenized with 5 vol. of cold Tris-HCl buffer (62.5 mM, pH 8.0), then after centrifugation for 30 min at 4°C, 100,000 x g, the supernatants were measured. Creatine which is the product of GAA-MT reaction was measured using HPLC. The assay procedure for liver GAA-MT activity is shown in Fig. 4.

Radioisotopic Study

Preparation of labeled guanidinoacetic acid. (2-^{14}C)-Glycine (50 Ci/mol, Amersham) were diluted with unlabeled glycine. Labeled GAA was synthe-

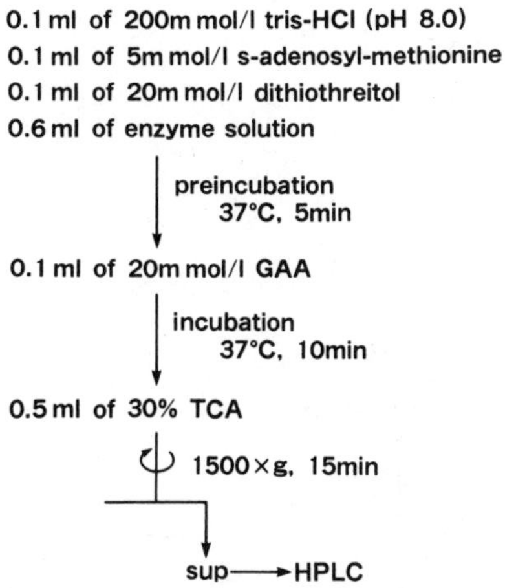

0.1 ml of 200m mol/l tris-HCl (pH 8.0)
0.1 ml of 5m mol/l s-adenosyl-methionine
0.1 ml of 20m mol/l dithiothreitol
0.6 ml of enzyme solution

preincubation
37°C, 5min

0.1 ml of 20m mol/l GAA

incubation
37°C, 10min

0.5 ml of 30% TCA

1500×g, 15min

sup⟶HPLC

Figure 4. The assay procedure for liver guanidinoacetate methyltransferase activity.

sized from glycine and S-methylisothiourea sulfate in alkaline solution. The
mixture was stirred at room temperature for 2-3 days. The pH was adjusted to
3, and the sample was placed on a 1 x 50 cm column of Dowex 50 W-X4 (Na⁺ form,
200-400 mesh) and then eluted with 0.1 M sodium cirtrate buffer, pH 3.43, at
room temperature. The synthesized GAA was purified according to the method
of Daly[8] using ion-exchange chromatography. Four milliliter fractions were
mixed with Biofluor (NEN), and their radioactivity was measured in a liquid
scintillation spectrometer (LSC-753). Fractions in the radioactive peak
corresponding to GAA were collected and desalted by passage through a 1 x 20
cm column of Dowex 50W-X4 (H⁺ form, 50-100 mesh), the column was washed with
10 ml water and eluted with 5N NH₄OH. After removal of ammonia by evapora-
tion, the sample was disolved in saline. The labeled GAA specific activity
was about 12.5 Ci/mol.

[14]C-glycine metabolism in vivo. Two control rats and two CRF rats were
used for this experiment. After 18hrs. fasting, 12.5µCi of (2-^{14}C)-glycine
(specific activity, 58 Ci/mol, Amershal) was injected into the femoral vein,
and blood was taken with heparinized syringes from the abdominal aorta and
renal vein at 5, 10, 20, 30, 45 and 60 min after injection. Following centrifugation for 10 min at 1,500 x g, the plasma radioactivity was counted.
The plasma and tissue homogenates were deproteinized using TCA (final concen
tration 10%), and centrifuged for 10 min at 1,500 x g, following analysis of
the supernatnat using HPLC, the radioactivity of the separated fractions wer
counted serially using a radioanalyzer (Aloka, RLC-551).

[14]C-GAA metabolism in vivo. This experiment was performed after dividin
the animals into two groups. First, ^{14}C-GAA was administered to controls and
CRF rats, both male and female. Five minutes after injection, the rats were
sacrificed and blood and tissue were collected. Similar samples were collected 10 min after injection of 12.5 µCi per animal. Samples were treated i
the same way as those obtained from the ^{14}C-glycine experiment.

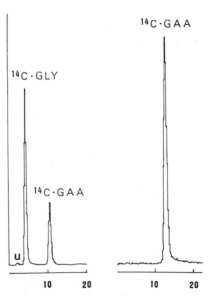

Figure 5. The radiochromatograms of ^{14}C-glycine and ^{14}C-guanidinoacetate
            (^{14}C-GAA).

RESULT

Fig. 5 shows the radiochromatogram of ^{14}C–glycine and ^{14}C–GAA using the HPLC–radioanalyzer system. The retention time for ^{14}C–GAA was 12 min. Mixed samples of ^{14}C–glycine and ^{14}C–GAA were clearly separated. The retention time was 4 min and 11 min respectively.

Table 1 shows concentrations of GAA in plasma, kidney and liver. In male rats, levels of plasma GAA were significantly lower in CRF, but not in females. In both male and female CRF rats, the levels of kidney GAA were significantly lower than in controls. The liver GAA levels did not differ between control and CRF rats, but tended to decrease slightly in CRF rats.

Table 2 shows the kidney GAT activity. In male CRF rats, the kidney GAT activity, expressed as unit per gram of mitochondrial protein, was slightly higher than in control rats but not significantly. In famale CRF rats, this enzyme activity was significantly higher than in control rats.

Fig. 6 shows the plasma radioactivity in control and CRF male rats following the intravenous injection of ^{14}C–glycine. In controls, the plasma radioactivity decreased rapidly then increased 30 min after injection. In CRF rats, this phenomenon was not observed.

Table 1. GAA levels in plasma and tissues in control and CRF rats

	N	Plasma	Kidney	Liver
Male				
Control	7	5.03 ± 1.51	0.36 ± 0.14	0.06 ± 0.05
CRF	9	2.66 ± 1.04*	0.19 ± 0.06**	0.04 ± 0.02
Female				
Control	9	2.59 ± 0.70	0.33 ± 0.17	0.06 ± 0.05
CRF	10	2.41 ± 0.94	0.20 ± 0.06***	0.01 ± 0.02
		(μM)	(μmol/g tissue)	

Values are the means ± SD.
N; number of rats studied, GAA; guanidinoacetic acid and CRF; chronic renal failure.
Significant difference in control and CRF: $(p < 0.005)$*, $(p < 0.01)$**. $(p < 0.05)$***

Table 2. Kidney GAT activity in control and CRF rats

	N			N	
Male			Female		
Control	7	5.38 ± 0.74	Control	9	4.25 ± 1.05
CRF	9	6.88 ± 2.68	CRF	10	6.20 ± 1.62*
			(U/g protein)		

Values are the means ± SD.
N; number of rats studied, GAT: glycine amidinotransferase and CRF; chronic renal failure.
Significant difference in control and CRF: $(P < 0.01)$*

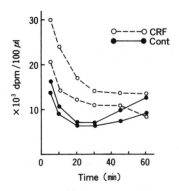

Figure 6.   Time-courses of plasma radioactivity in control and CRF male
             rats following intravenous injection of ^{14}C-glycine. Cont;
             control and CRF; chronic renal failure.

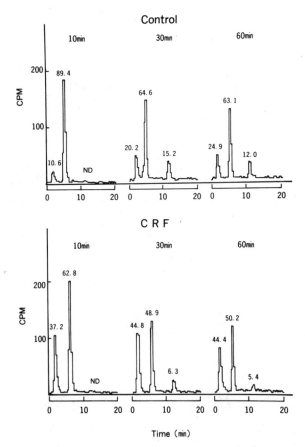

Figure 7.   Chromatograms of radioactive metabolites in 10, 30, and 60-min
             plasma samples collected after intravenous injection of ^{14}C-
             glycine. CRF; chronic renal failure.

Fig. 7 shows the chromatograms of radioactive metabolites in 10, 30 and 60 min. Plasma samples collected after the intravenous injection of ^{14}C-glycine. ^{14}C-GAA derived from ^{14}C-glycine was recognized 12–13 min after injection. The ^{14}C-GAA peak area rate to total radioactivity was larger in control than in CRF rats, but the unknown peak (retention time 1–2 min.) area rate was larger in CRF rats.

Table 3 shows the concentration of creatine in plasma, kidney and liver. In both male and female CRF rats, the levels of creatine in plasma were significantly lower than in controls. The levels of kidney creatine in male and female CRF rats were also significantly lower than in controls. In the liver, the levels of creatine in CRF male rats were significantly lower, but in female CRF rats were not significantly lower than in control female rats. However, the values tended to be slightly lower in female CRF rats.

Table 4 shows the liver GAA-MT activity. In male rats, no significant differences were found in control and CRF rats. In female rats, similar results were obtained but the activity in CRF male rats tended to be higher.

Fig. 8 shows plasma radioactivity in control and CRF rats following the intravenous injection of ^{14}C-GAA. The half-life of injected ^{14}C-GAA was shorter in CRF both in male and female rats.

Table 5 shows the rate of conversion of ^{14}C-GAA to creatine in plasma. In the 5 min after ^{14}C-GAA injection groups, the rate of ^{14}C-GAA and ^{14}C-

Table 3. Creatine levels in plasma and tissues in control and CRF rats

	N	Plasma	Kidney	Liver
Male				
Control	7	193.5 ± 61.9	2.22 ± 0.74	0.54 ± 0.22
CRF	9	56.2 ± 30.9*	0.69 ± 0.49**	0.25 ± 0.15
Female				
Control	9	252.7 ± 61.3	2.84 ± 1.37	0.54 ± 0.23
CRF	10	63.9 ± 34.8*	0.93 ± 0.54**	0.30 ± 0.29
		(µM)	(µmol/g tissue)	

Values are the means ± SD.
N; number of rats studied, and CRF; chronic renal failure. Significant difference in control and CRF: $(p < 0.001)$*, $(p < 0.01)$**

Table 4. Liver GAA-MT activity in control and CRF rats

	N			N	
Male			Female		
Control	7	130 ± 31	Control	9	125 ± 37
CRF	9	120 ± 12	CRF	10	170 ± 46
					(mU/g protein)

Values are the means ± SD.
N; number of rats studied, GAA-MT; guanidinoacetate methyltransferase and CRF; chronic renal failure.

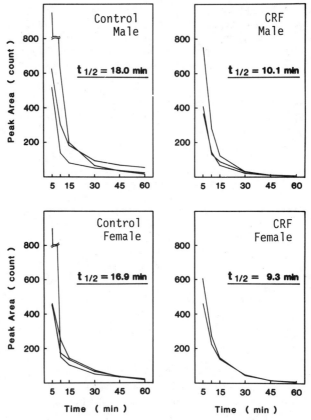

Figure 8. Time-courses of plasma radioactivity in control and CRF rats
following intravenous injection of ^{14}C-GAA. CRF; chronic renal
failure and ^{14}C-GAA: ^{14}C-guanidinoacetate.

creatine in the plasma collected from the abdominal aorta were similar in
control and CRF rats, both in males and females. In the 10 min groups,
similar results were obtained except in a female CRF-3 rat.

Table 6 shows the rate of accumulation of radioactive metabolites in
the kidney. The convension rate of ^{14}C-glycine from ^{14}C-GAA showed no significant difference between male and female rats in control and CRF. But in
both male and female CRF rats ^{14}C-glycine formation tended to be slightly
disturbed.

Table 7 shows the rate of accumulation of radioactive metabolites in
the liver. In the group at 5 min after injection of ^{14}C-GAA, the conversion
rate of ^{14}C-creatine from ^{14}C-GAA was lower in both male and female CRF rats
In the 10 min groups, in both male and female CRF rats, these rates were
slightly lower than in control rats.

DISCUSSION

Recent studies report disturbances in creatine metabolism in CRF patients and rats with experimentally induced CRF[1,2,9,10]. In order to eluci-

Table 5. Conversion rate of injected ^{14}C–GAA in plasma from control and CRF rats

Group of rats	5 min CTN	GAA	10 min CTN	GAA
Male				
Cont 1	38.3	58.5		
Cont 2	32.7	64.9		
CRF 1	39.7	54.7		
CRF 2	46.8	49.8		
Cont 1			50.6	49.4
Cont 2			50.0	50.0
Cont 3			65.2	34.8
CRF 1			68.8	31.3
CRF 2			78.6	21.4
CRF 3			74.9	25.1
Female				
Cont 1	51.6	48.4		
Cont 2	46.6	53.4		
Cont 3	55.2	44.8		
CRF 1	54.9	45.1		
CRF 2	60.9	39.1		
CRF 3	68.0	26.0		
Cont 1			75.7	24.3
Cont 2			77.2	22.8
Cont 3			91.4	8.6
CRF 1			88.0	12.0
CRF 2			98.0	2.0
CRF 3			37.6	62.4

Values are percentage to total radioactivity.
CTN; creatine, GAA; guanidinoacetate, Cont; control and CRF: chronic renal failure.

date the mechanisms of abnormal creatine metabolism in CRF, the present radioisotopic experiment on the metabolites of creatine was carried out.

Creatine is formed by the methylation of GAA in the liver. GAA is formed from arginine and glycine in a reaction catalyzed by GAT mainly in the kidney in the rat. In this study, plasma and kidney GAA levels were significantly lower in CRF male rats. In female rats, plasma GAA level was not significantly lower in CRF, but kidney GAA level was significantly lower than in controls. In female rats, both in control and CRF, plasma GAA levels were lower than in male control rats. This suggests that there is sex-dependent difference in creatine metabolism in the rat. There were no statistically significant differences found in liver GAA levels between control and CRF rats, but we reported previously that liver GAA content was significantly lower in male CRF rats[2]. This difference was thought to depend on the severity of the state of ranal failure, because the plasma creatinine level was lower in the present study (mean; 1.8mg/dl vs 2.7mg/dl).

Table 6. Conversion rate of injected ^{14}C–GAA in kidneys from control and GRF rats

| group of rats | | 5 min | | | 10 min | |
	Gly	CTN	GAA	Gly	GTN	GAA
Male						
Cont 1	15.8	10.6	71.1			
Cont 2	29.2	15.8	52.0			
CRF 1	21.8	19.0	59.2			
CRF 2	17.0	33.2	49.8			
Cont 1				18.6	27.8	53.6
Cont 2				11.1	32.0	57.0
Cont 3				8.5	28.2	63.3
CRF 1				0	34.1	65.9
CRF 2				0	0	100.0
CRF 3				20.1	20.1	59.8
Female						
Cont 1	11.8	15.9	72.9			
Cont 2	23.8	10.2	66.0			
Cont 3	15.6	15.2	69.2			
CRF 1	0	39.3	60.7			
CRF 2	11.1	10.6	78.3			
CRF 3	5.1	22.0	72.9			
Cont 1				16.5	27.5	51.8
Cont 2				9.5	32.1	56.0
Cont 3				17.7	24.5	57.7
CRF 1				5.5	23.3	71.7
CRF 2				15.7	25.2	59.1
CRF 3				0	21.5	78.5

Values are percentage to total radioactivity.
Gly; glycine, CTN; creatine, GAA; guanidinoacetate, Cont; control and CRF: chronic neral failure.

The kidney GAT activity expressed as units per gram mitochondrial protein tended to be slightly higher in male CRF rats and significantly higher in female CRF rats. However, the total renal mass in the CRF rats was less than half that of control rats, so we expected that the total activity of kidney GAT would be significantly lower in CRF rats than in control rats. From the results of radioisotope analysis of ^{14}C-glycine in vivo, we find that the conversion rate of ^{14}C-GAA from ^{14}C-glycine is decreased in CRF male decreased in CRF male rats. The above findings suggest that the capacity for GAA production is decreased in CRF rats, the cause of which is the decreased renal mass.

GAA produced in the kidney is converted to creatine in the liver by the catalytic action of GAA-MT. The plasma levels of creatine in both male and female rats were significantly decreased in the CRF groups, while the kidney and liver creatine levels were also decreased in the CRF groups compared to controls. Decreased creatine levels in plasma and tissues in male CRF rats

Table 7. Conversion rate of injected ^{14}C–GAA in livers from control and CRF
          rats

Group of rats	5 min CTN	5 min GAA	10 min CTN	10 min GAA
Male				
Cont 1	63.5	25.1		
Cont 2	75.7	13.7		
CRF 1	37.6	56.7		
CRF 2	48.1	48.0		
Cont 1			100.0	0
Cont 2			100.0	0
Cont 3			90.5	9.5
CRF 1			71.2	28.8
CRF 2			76.5	23.5
CRF 3			84.6	15.4
Female				
Cont 1	91.1	8.1		
Cont 2	85.0	15.0		
Cont 3	84.9	15.1		
CRF 1	81.5	18.5		
CRF 2	61.9	38.1		
CRF 3	45.0	55.0		
Cont 1			100.0	0
Cont 2			100.0	0
Cont 3			100.0	0
CRF 1			94.0	2.9 (MG) 3.1
CRF 2			95.0	3.0
CRF 3			80.5	19.5 2.0

Values are percentage to total radioactivity.
CTN; creatine, GAA; guanidinoacetate, MG, methylguanidine, Cont; control and
CRF; chronic renal failure.

were reported by some authors[2,11]. It is thought that this decreased creatine production is the result of the impaired capacity for GAA production in
CRF rats. In vitro the GAA–MT activity in the liver was similar in the CRF
and control rats. GAA–MT is an SH–enzyme. Cantoni[12] reported that the GAA–MT
activity is disturbed in the absence of SH metabolites and recovered with
the addition of these substances. Ogawa et al.[7] reported that the regulation
of in vivo GAA–MT activity is controlled by glutathione, a di-thiol compound.
Ohwada et al.[13] reported that liver GSH content was significantly decreased
in CRF rats. These findings suggest that a decrease in GSH content in the
liver leads to the disturbance of GAA–MT activity in vivo.

From the results of the radioisotopic analysis of ^{14}C–GAA in vivo, the
conversion rate of ^{14}C–creatine from ^{14}C–GAA was shown to be decreased in
both male and female CRF rats in the 5 min after injection groups. The above
findings suggest that creatine production is disturbed in CRF rats, probably

due to diminished production of GAA in the kidney and disturbed activity of liver GAA-MT.

REFERENCES

1. Owada, S., Ozawa, S., Inouchi, M., Kimura, Y. and Ishida, M., A study of creatine metabolism in chronic renal failure rats, in: "Guanidines," A. Mori, B.D. Cohen and A. Lowenthal Eds., Plenum Press, New York (1985) pp. 277-285.
2. Inouchi, M., A study of creatine metabolism in chronic renal failure, St. Marianna Med. J. 12 (1984) 502-513.
3. Magri, E., Baldoni, G. and Grazi, E., On the biosynthesis of creatine: Intramitochondrial localization of transamidinase from rat kindney, FEBS Lett. 55 (1975) 91-93.
4. Schneider, W.C., Intrcellular Distribution of Enzymes: The oxidation of octanoic acid by rat liver fractions, J. Biol. Chem. 176 (1984) 259-266.
5. Van Pilsum, J.F., Taylor, D., Zakis, B. and McCormick, P., Simplified assay for transamidinase activities of rat kidney homogenates, Anal. Biochem. 35 (1970) 277-286.
6. Im, Y.S., Cantoni, G.L. and Chiang, P.K., A radioactive assay for guanidinoacetate methyltransferase, Anal. Biochem. 95 (1979) 87-88.
7. Ogawa, H., Ishiguro. Y. and Fujioka, M., Guanidinoacetate methltransferase from rat liver: Purification, properties, and evidence for the involvement of sulfhydryl group for activity, Arch. Biochem. Biophys. 226 (1983) 265-275.
8. Daly, M.M., Guanidinoacetate methyltransferase in tissues and cells, Arch. Biochem. Biophys. 236 (1985) 576-584.
9. Tofuku, Y., Muramoto, H., Kuroda, M. and Takeda, R., Impired metabolism of guanidinoacetic acid in uremia, Nephron 41 (1985) 174-178.
10. Tsubakihara, Y., Iida, N., Yuasa, S., Kawashima, T., Nakanishi, I., Tomobuchi, M., Yokogawa, T., Ando, A., Orita, Y., Abe, H., Kikuchi, T. and Okamoto, H., Guanidinoacetic acid (GAA) deficiency and supplmentatio in rats with chronic renal failure (CRF), in: "Guanidines," A. Mori, B.D Cohen and A. Lowenthal. Eds., Plenum Press, New York (1985) pp. 373-379.
11. Kanazawa, T., Disturbance of creatine metabolism in experimental chronic renal failure rats, St. Marianna Med. J. 13 (1985) 618-631.
12. Cantoni, G.L. and Vignos, P.J., Jr., Enzymatic mechanism of creatine synthsis, J. Biol. Chem. 209 (1954) 647-659.
13. Owada, S., Takemura, H., Kosoto, H., Mochizuki, N., Hasegawa, R., Soga, T., Haranaka, R. and Kobayashi, M., Glutathion metabolism in chronic renal failure-Part II: Glutathion and its related enzymes in experimenta rats, Nihon Uviv. J. Med. 24 (1982) 427-438.

CHANGES IN URINARY METHYLGUANIDINE IN CASES WITH END-STAGE RENAL DISEASE

Makoto Ishizaki, Hiroshi Kitamura, Hisako Sugai, Kazuyuki
Suzuki, Kosei Kurosawa, Gen Futaki, Takao Sohn, Yoshio Taguma,
Hisashi Takahashi and Motoo Nakajima*

Kidney Center, Sendai Shakai Hoken Hospital, Sendai
*Bioscience Research Laboratory, Kikkoman Co., Noda, Japan

INTRODUCTION

Methylguanidine (MG), one of the most detrimental uremic toxins, is a
product of creatinine oxidation. Although the amount of MG output depends on
the concentration of creatinine, it is thought that the participation of
active oxygen in the process of MG production markedly boosts its productivity[1]. In patients with end-stage renal disease (ESRD), the level of MG output
may depend on the formation of active oxygen, since such patients have an
increased amount of ceratinine which is a precursor of MG[2-6]. In fact, some
patients who have more than 10 mg/dl of plasma creatinine (P-CR) respond well
to conservative therapy and do not require hemodialysis treatment.

In order to analyze such differences, we measured the urinary MG concentration for the purpose of evaluating MG productivity in patients with
ESRD, and we investigated the relationship between active oxygen and the
manifestations of uremia.

MATERIALS AND METHODS

The subjects, consisting of 145 patients whose P-CR was more than 2
mg/dl (Fig. 1), were either admitted to or treated as outpatients at Sendai
Shakai Hoken Hospital between September 1986 and March 1987. Among those
patients, there were 76 men and 69 women with a mean age of 53.6 ± 14.6,
ranging from 20 to 81 years of age, the mean body weight being 57.6 ± 9.8 kg,
ranging from 36 to 85 kg. Their original diseases were chronic glomerulonephritis in 103 patients, diabetes (DM) in 35, tuberculosis in 3, polycystic
kidneys in 2, and lupus nephritis in 2. All of the diabetic patients underwent conservative therapy, and none of them received renal replacement
therapy (RRT). Among 110 nondiabetic patients, 14 underwent RRT and all the
remaining patients received conservative therapy.

The urine samples represent spot collection and blood sampling was
carried out just prior to collection of the urine. Both P-CR and urine

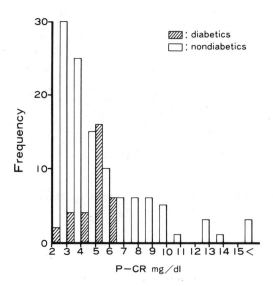

Figure 1. Histograms of plasma creatinine (P-CR) levels in patients with chronic renal disease.

creatinine were evaluated using a Technicon Autoanalyzer (SMAC II).

The urinary MG concentration was determined by the enzymic analysis shown in Fig. 2. A deproteinization reagent (0.2 ml) consisting of 7.5% trichloroacetic acid was put into each test tube containing a 1.0 ml urine sample; the contents were mixed and kept at room temperature for 5 min. Each mixture was then centrifuged (1,000 x g, 15 min) to yield a supernatant which was free from protein. The supernatant (0.5 ml) was put into each of duplicate test tubes, and 0.5 ml of a carbonate buffer solution (0.5 M. pH 10.2) was added to each of them, the contents being then mixed. Enzymic reagent (0.1 ml) containing methylguanidine amidinohydrolase and methylamine oxidase was added to one test tube to examine MG, while 0.1 ml of another enzymic reagent containing methylamine oxidase was added to the other test tube, a

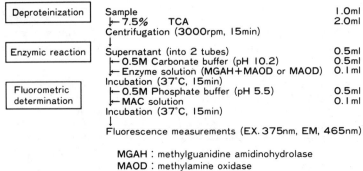

Figure 2. Enzymic analysis procedure for methylguanidine in uremia. TCA; trichloroacetic acid.

sample blank tube. The contents in each test tube were mixed and incubated at 37°C for 15 min.

Phosphate buffer solution (0.5 M, pH 5.5) in the amount of 0.5 ml and 0.1 ml of methyl 3-aminocrotonate reagent were successively added to both test tubes. The contents were mixed and incubated at 37°C for 15 min. Both test tubes were then cooled in running water for 1 min.

The fluorescence of both test tubes was measured at an excitation wavelength of 375 nm and at an emission wavelength of 465 nm, the values of MG and of the sample blank being compared. The MG concentration in urine was determined by the use of a standard curve.

RESULTS

The urine MG concentration (U-MG), urine creatinine concentration (U-CR), and the level of plasma creatinine were measured in nondiabetic patients and the relationship between U-MG/U-CR and 1/P-CR was examined as shown in Fig. 3. A significant negative correlation was observed (r= -0.682, P < 0.01). A drastic increase in the value of U-MG/U-CR was especially evident after the level of P-CR exceeded 8 mg/dl. In the figure, the open circles indicate patients started on RRT, while the closed circles are those treated by conservative therapy. The patients whose P-CR was below 8 mg/dl with a high level of U-MG/U-CR, were those who had lupus nephritis (1 and 2) and nephrotic syndrome complicated by pulmonary infection (3 and 4), as shown in the figure.

Since the value of the urinary concentration drastically changed as the level of P-CR exceeded 8 mg/dl (1/P-CR=1.05 x 10^{-1}), we divided the nondiabetic patients into two groups, those who had P-CR > 8 mg/dl and those who

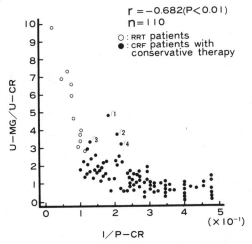

Figure 3.   Relationship between U-MG/U-CR and the reciprocal of P-CR in non-diabetic patients with chronic renal failure (CRF). Open circles show the patients who initiated RRT (n=14) and closed circles show the patients treated with conservative therapy (n=96). (1) and (2) were lupus nephritis, (3) and (4) show nephrotic patients complicated by pulmonary infections. RRT; renal replacement therapy, U-MG; urine methylguanidine, U-CR; urine creatinine, and P-CR; plasma creatinine.

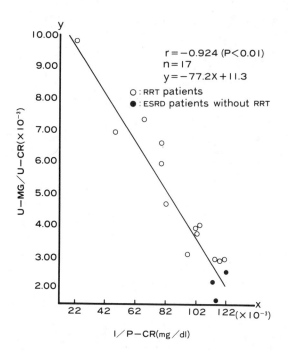

Figure 4.   Relationship between U–MG/U–CR and 1/P–CR in nondiabetic
            patients with end-stage renal disease (ESRD)(P–CR > 8 mg/dl.)
            RRT; renal replacement therapy, U–MG; urine methylguanidine,
            U–CR; urine creatinine and P–CR; plasma creatinine.

had 2 mg/dl < P–CR ≦ 8 mg/dl. The relationship of U–MG/U–CR and 1/P–CR and
that of U–MG and 1/P–CR were also examined.

Fig. 4 Shows the relationship between U–MG/U–CR and the reciprocal of
P–CR in 17 patients having P–CR > 8 mg/dl. A significant negative correlation
was noted (r = –0.924, P < 0.01). Assuming U–MG/U–CR to be y and the reciprocal of P–CR to be x, a linear regression equation, y = –77.2 x + 11.3, was
obtained. The 14 patients who underwent RRT (open circles) discharged greater
amounts of urinary MG than those treated by conservative therapy (closed
circles). This suggests that the MG output in uremic patients is a function
of the degree of uremia.

Fig. 5 shows the relationship between U–MG, without considering the
renal tubular concentrating capacity, and the reciprocal of P–CR in the group
of P–CR > 8 mg/dl. A good negative correlation was also evident (r = –0.854,
P < 0.01). In this group having P–CR > 8 mg/dl, no significant difference in
the coefficients of regression between U–MG/U–CR (r = –0.924) and U–MG (r =
–0.854) was recognized. However, U–MG/U–CR tends to have a better correlation
with the reciprocal of P–CR than does U–MG which does not take into consideration the renal tubular concentrating capacity.

Fig. 6 shows the relationship between U–MG/U–CR (y) and the reciprocal
of P–CR (x) in nondiabetic patients in the group with 2 mg/dl < P–CR ≦ 8
mg/dl. Although a significant negative correlation was evident (r = –0.609,
P < 0.01), it was less significant as compared to that of patients having
the higher level of P–CR. The regression line was y = –5.13 x + 2.70, whose

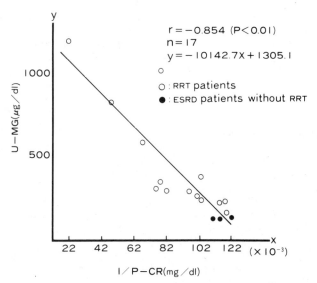

Figure 5.   Relationship between urine methylguanidine (U-MG) and the reciprocal of plasma creatinine (P-CR) in nondiabetic patients with
end-stage renal disease (ESRD) (P-CR > 8 mg/dl.) RRT; renal
replacement therapy.

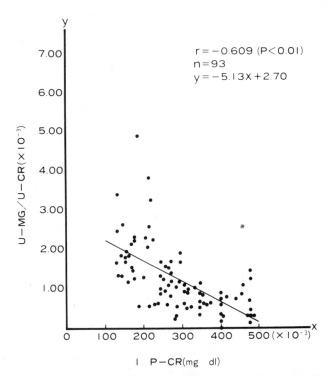

Figure 6.   Relationship between U-MG/U-CR and the reciprocal of P-CR in
nondiabetic patients with chronic renal failure (2 mg/dl < P-
CR ≤ 8 mg/dl) U-MG; urine methylguanidine, U-CR; urine creatinine
and P-CR; plasma creatinine.

gradient was lower than that of the patients with high P-CR values. In other words, in the group of patients having a low level of P-CR, the overall MG output correlated less with the degree of uremia; however, in patients among them with autoimmune diseases such as lupus nephritis or those with acute infections, MG output was augmented.

Fig. 7 shows the relationship between U-MG (y) and 1/P-CR (x) in nondiabetic patients in the group of 2 mg/dl < P-CR ≦ 8 mg/dl. A significant negative correlation was also noted (r = -0.471, P < 0.01), the regression line being y = -386 x + 204.

All 35 diabetic patients, with values of 2 mg/dl < P-CR ≦ 7 mg/dl, were treated by conservative therapy. As none of these patients received RRT, it is difficult to analyze the ESRD accompanying DM. The coefficients of correlation between U-MG/U-CR and 1/P-CR (Fig. 8) and between U-MG and 1/P-CR (Fig. 9) were -0.438 and -0.538, respectively. In both cases, a significant negative correlation was noted (P < 0.01); however, there is little difference in these values. In diabetic patients, U-MG tends to have a better negative correlation with the reciprocal of P-CR than it does in nondiabetics. The regression line between U-MG (y) and the reciprocal of P-CR (x) was y = -543 x + 209, showing a greater gradient than that on nondiabetic patients at the same level of P-CR.

DISCUSSION

Uremic symptoms include a variety of systemic disorders. Studies on the pathogenesis of uremia have been carried out focusing on single toxic substance. Giovannetti et al.[8] were able to create a syndrome similar to uremia in laboratory animals by chronically administering MG alone. However, the amount of MG administered far exceeded the estimated amount that uremic patients produced.

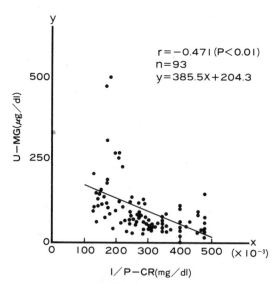

Figure 7.   Relationship between urinemethylguanidine (U-MG) and the reciprocal of plasma creatinine (P-CR) in nondiabetic patients with chronic renal failure (2 mg/dl < P-CR ≦ 8 mg/dl).

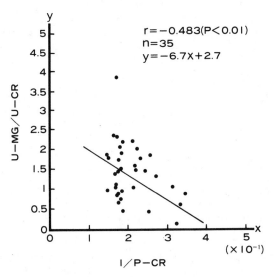

Figure 8.   Relationship between U-MG/U-CR and the reciprocal of P-CR in
            diabetic patients. U-MG; urine methylguanidine, U-CR; urine
            creatinine and P-CR; plasma creatinine.

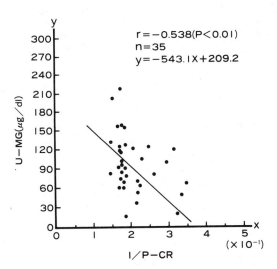

Figure 9.   Relationship between urine methylguanidine (U-MG) and the recipro-
            cal of plasma creatinine (P-CR) in diabetic patients.

        Recently, there have been reports on peroxidative cell damage[9,10] in
patients with uremia. Some researchers suggest that active oxygen participates
in the manifestations of uremic symptoms seen in patients with ESRD[1,10,11].
In addition, active oxygen is thought to have a close association with chronic inflammation, immunodeficiency, carcinogenesis, arteriosclerosis, and
amyloidosis[12-16]. However, the measurement of active oxygen is clinically
difficult because of its unstable nature.

Nagase et al. reported that MG synthesis induced by creatinine oxidation is augmented by active oxygen[1]. Based on Nagase's study, we measured the amount of urinary MG to determine whether or not MG output increased in ESRD and, by means of the increased MG production, we extrapolated to the participation of active oxygen in uremia.

We have previously reported[17] that patients with ESRD who received RRT due to their severe uremic symptoms showed a twofold increase in MG output as compared to those who responded well to conservative therapy, in spite of the fact that both patients had the same level of P-CR. For the purpose of this previous analysis, we used 24-hour urines; the value of MG was measured by using high performance liquid chromatography (HPLC). However, those who needed RRT came to the hospital in an emergent condition and there was rarely time for collection of 24-hour urine before RRT was initiated. In the present study, we used spot urines which avoided this problem and we applied enzymic analysis which was easier to employ than HPLC and was useful for mass study.

In nondiabetic patients with ESRD, urinary MG increased markedly as the level of P-CR exceeded 8 mg/dl. When MG output increased, uremic symptoms became even more severe. Conservative therapy can restrict the increase in MG output to some extent; however, there seems to be a limit to its ability as it is mainly based on a low protein diet.

Since active oxygen has an effect on MG output in ESRD, further investigation of the role of active oxygen could lead to an effective treatment of uremia. The combined use of scavengers of active oxygen and conservative therapy using a low protein diet might be effective in preventing uremic symptoms.

Since none of the diabetic patients received RRT, it is impossible to examine MG productivity. In general, diabetic patients with ESRD are started on RRT when the level of P-CR is over 8 mg/dl. In the present study, we recognized that the MG output was accelerated when the level of P-CR was close to 7 mg/dl. It is known that diabetic patients presenting with nephrotic syndrome often tend to develop pulmonary infections because of a degraded immune function. We noticed that a slight increase in MG output in such patients decreased when the pulmonary infection was cured by the administration of antibiotics. This implies that active oxygen mediated by pulmonary infections exerts an influence on MG output.

SUMMARY

The urinary methylguanidine (MG) concentration in patients with chronic renal failure with more than 2 mg/dl of plasma creatinine was measured by enzymic analysis. In patients with end-stage renal disease who did not have diabetes mellitus, MG output increased with the appearance of uremic symptoms. Since creatinine oxidation may accelerate MG output, it is conceivable that active oxygen participates in the manifestations of uremia. A new therapy which takes into consideration the influence of active oxygen should be implemented for the treatment of uremia.

REFERENCES

1. Nagase, S., Aoyagi, K., Narita, M. and Tojo, S., Active oxygen in methylguanidine synthesis, Nephron 44 (1986) 299-303.
2. Cohen, B.D., Uremic toxins, Bull. N.Y. Acad. Med. 51 (1975) 1228-1234.
3. Orita, Y. Tsubakihara, Y., Ando, A., Nakata, K., Takamitsu, Y., Fukuhara, Y. and Abe, H., Effect of arginine or creatinine administration on the urinary excretion of methylguanidine, Nephron 22 (1978) 328-336.
4. Mikami, H., Orita, Y., Ando, A., Fujii, M., Kikuchi, T., Yoshihara, K., Okada, A. and Abe, H., Metabolic pathway of guanidino compounds in chronic renal failure, in: "Urea cycle diseases," A. Lowenthal, A. Mori and B. Marescau Eds., Plenum Press, New York (1983) pp. 449-458.
5. Nagase, S., Aoyagi, K., Narita, M. and Tojo, S., Biosynthesis of methylguanidine in isolated rat hepatocytes and in vivo, Nephron 40 (1985) 470-475.
6. Gejyo, F., Baba, S., Watanabe, Y., Kishore, B.K., Suzuki, Y. and Arakawa, M., Possibility of a common metabolic pathway for the production of methylguanidine and dimethylamine in chronic renal failure, in: "Guanidine," A. Mori, B.D. Cohen and A. Lowenthal Eds., Plenum Press, New York (1985) pp. 295-308.
7. Nakajima, M., Nakamura, K. and Shirokane, Y., Enzymic determination of methylguanidine in urine, in: "Guanidine," A. Mori, B.D. Cohen and A. Lowenthal Eds., Plenum Press, New York (1985) pp. 39-46.
8. Giovannetti, S., Biagini, M., Balestri, P.L., Navalesi, R., Giagnoni, P., de Matteis, A., Ferro-Milone, P. and Perfetti, C., Uremia-like syndrome in dogs chronically intoxicated with methylganidine and creatinine, Clin. Sci. 36 (1969) 445-452.
9. Giardini, O., Taccone-Gallucci, M., Lubrano, R., Ricciardi-Tenore, G., Bandino, D., Silvi, I., Ruberto, U. and Casciani, C.U., Evidence of red blood cell membrane lipid peroxidation in haemodialysis patients, Nephron 36 (1984) 235-237.
10. Kuroda, M., Asaka, S., Tofuku, Y. and Takeda, R., Serum antioxidant activity in uremic patients, Nephron 41 (1985) 293-298.
11. Kobayashi, M., Nagase, S., Aoyagi, K., Koyama, A., Narita, M. and Tojo, S., Improvement of renal function in AA type amyloidosis by prolonged treatment with dimethylsulfoxide, Jpn. J. Nephrol. 29 (1987) 97-101.
12. Halliwell, B., Gutteridge, J.M.C. and Blake, D., Metal ion and oxygen radical reactions in human inflammatory joint disease, Phil. Trans. R. Soc. Lond. B 311 (1985) 659-671.
13. Harman, D., Free radical theory of aging : Beneficial effect of antioxidants on the life span of male NZB mice; Role of free radical reactions in the deterioration of the immune system with age and in the pathogenesis of systemic lupus erythematosus, Age 3 (1980) 64-73.
14. Endo, H. and Takahashi, K., Methylguanidine, a naturally occurring compound showing mutagenicity after nitrosation in gastric juice, Nature 245 (1973) 325-326.
15. Harman, D., Heidrick, M.L. and Eddy, D.E., Free-radical-reaction inhibitors on the immune response, J. Am. Geriatrics Soc. 25 (1977) 400-407.
16. Harman, D., Eddy, D. and Noffsinger, J., Free radical theory of aging inhibition of amyloidosis in mice by antioxidants; Possible mechanism, J. Am. Geriatrics Soc. 24 (1976) 203-210.
17. Kitamura, H. Ishizaki, M., Kitamoto, Y., Futaki, G., Ueda, H., Taguma, T., Momma, H. and Takahashi, H., Urinary excretion of methylguanidine in patients with end stage renal diseases, Proc. Annual Meeting Jpn. Guanidino Compounds Res. Assoc. 7 (1986) 40-41 (in Japanese).

THE ROLE OF NON-PROTEIN NITROGEN IN PROGRESSIVE RENAL FAILURE

Burton D. Cohen and Harini Patel

Bronx-Lebanon Hospital, Bronx, NY 10456

INTRODUCTION

Of the 10-15 guanidino compounds currently isolated from blood and urine of uremic subjects the most prominently and consistently increased is guanidinosuccinic acid (GSA)[1]. The toxicity resulting from this increase has been elusive and, to date, the principal and most reproducible effect reported has been a reduction in platelet microtubular contractibility which may contribute to the bleeding diathesis which accompanies uremia[2]. The origin of this material has been shown to be the mixed function oxidation of urea occurring in the liver and driven by the substrate, urea[3]. Fig. 1 diagrams a part of this alternate pathway showing those substances which, injected intraperitoneally in rats, increase urinary output of GSA. Injecting a little more than a millimole of methionine results in a profound reduction in GSA output (64%) with homocysteine, a methionine precursor, producing a lesser decline (18%). That this is not an effect of the sulfhydril group is shown by the lack of effect by either cysteine or acetyl cysteine (chosen for its greater cellular permeability).

While there is some dispute whether serum levels of guanidinoacetic acid (GAA) increase in uremia, as would be anticipated from the increased conversion of urea to canavanine, our studies show a consistent rise both in animals (0.054 ± .021 to 0.101 ± .033) and humans (0.058 ± .008 to 0.098 ± .031) following methionine dietary supplementation. Methionine serves multiple metabolic functions one of which is to provide sulfur for cysteine synthesis in organisms incapable of fixing inorganic sulfur. It keeps this sulfur sequestered behind a methyl group and must be demethylated to yield its precious cargo. In fact, it is proposed that higher organisms invest in creatine, when lesser phosphagens would do, primarily as a depository for methyl groups so that the sulfhydril necessary to cysteine synthesis becomes available[4]. This appears to be the case with methionine stimulating the synthesis of GAA from arginine and canavanine and, thereby, reducing production of GSA. The following study was undertaken to determine the effect of methionine dietary supplementation in uremia.

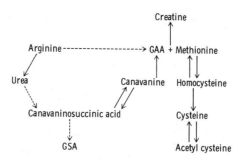

Figure 1. The guanidine cycle an alternate pathway for the production of crea
tine which begins with the oxidation of urea. Substances which significantly increase urinary GSA ourput are arginine (3%), urea
(100%), and canavanine (90%). GAA: guanidinoacetic acid and GSA:
guanidinosuccinic acid.

MATERIALS AND METHODS

Three patients with concurrent uremia and nephrosis were selected and
hospitalized for six days on a metabolic unit. The etiology of the renal
failure was unknown in all three but, based upon history, felt to be heroin
associated nephropathy (HAN). Creatinine clearance was 2.5, 8.0 and 30.0
ml/min respectively in the three subjects and maintained constant throughout the period of observation. Patients were placed on a 40 g protein diet
supplemented with 20 g of gelatine to assure sufficient substrates of arginine and glycine to prevent methionine toxicity[5]. Following three control
days, the diets were further supplemented with 1 g three times daily of
L-methionine in 500 mg tablets of the free base. Twenty-four urines were
analyzed for protein by the standard Esbach reaction and for GSA by HPLC
using a reverse phase column and detecting fluorescence with phenanthoquinone
as previously described[6]. Analyses were performed in triplicate with the
reported result being the mean of the best two.

Table 1. The relationship of proteinuria to urinary guanidinosuccinic acid
(GSA) after methionine

Patient	period	Urine GSA (mg/d)	Proteinuria (g/d)
IC	control	77.3 ± 6.8	4.5 ± .8
	methionine	48.4 ± 1.7	2.0 ± .3
RB	control	39.9 ± 5.9	9.6 ± .3
	methionine	34.6 ± 2.8	7.5 ± 1.1
SN	control	30.3 ± 4.9	2.8 ± .2
	methionine	24.2 ± 5.4	1.8 ± .3

(mean ± S.D.)

RESULTS

Despite the short duration of observation, the mean urinary GSA excretion fell in all three instances as shown in Table 1. An unanticipated result was the statistically significant (P=.002) decline in urinary protein.

DISCUSSION

The precise factors governing proteinuria or the ultrafiltration coefficient of the glomerular basement membrane are not yet clear but among the candidates is the contractility of the mesangial cell. Contracted, the mesangial cell opens the glomerular capillary loop increasing its volume and reducing the hydrostatic or ultrafiltration pressure. Failing to contract produces the opposite effect and, hence, proteinuria. The effect of GSA upon platelets is to reduce microtubular contraction and this is a linear function which can be overcome by increasing the contractile stimulus (adenosine diphosphate, epinephrine or synthetic aggregating agents). While the fall in filtered GSA induced by methionine is not great, if this same linearity applies and if the mesangial cell behaves in a manner similar to the platelet exposed to GSA, then the decline in proteinuria is not so startling and the clinical implications of these somewhat preliminary findings are of profound significance.

REFERENCES

1. DeDeyn, P., Marescau, B., Lornoy, W., Becaus, I., Van Leuven, I., Van Gorp, L. and Lowenthal, A., Serum guanidino compound levels and the influence of a single hemodialysis in uremic patients undergoing maintenance hemodialysis, Nephron 45 (1987) 291-295.
2. Horowitz, H.I., Stein, I.M., Cohen, B.D. and White, J.G., Further studies on the platelet-inhibitory effect of guanidinosuccinic acid and its role in uremia bleeding, Am. J. Med. 49 (1970) 336-345.
3. Natelson S. and Sherwin, J.E., Proposed mechanism for urea nitrogen re-utilization: relationship between urea and proposed guanidine cycles, Clin. Chem. 25 (1979) 1343-1344.
4. Huxtable, R.J., "Biochemistry of Sulfur," Plenum Press, New York (1986).
5. Harper, A.E., Benevenga, N.J. and Wohlhueter, R.M., Effects of ingestion of disproportionate amounts of amino acids, Physiol. Rev. 50 (1970) 428-464.
6. Baker M.D., Mohammed, H.Y. and Veaning, H. Reversed-phase ionpairing liquid chromatographic separation and fluorometric detection of guanidino compounds, Anal. Chem. 53 (1981) 1658-1662.

URINARY GUANIDINOACETIC ACID EXCRETION AS AN INDICATOR OF GENTAMICIN

NEPHROTOXICITY IN RATS

Shuei Nakayama, Masahiro Junen, Ikuo Kiyatake and Hikaru Koide

Division of Nephrology, Department of Medicine, Juntendo
University School of Medicine, Tokyo 113, Japan

INTRODUCTION

Guanidinoacetic acid (GAA) is a guanidino compound which is formed from
arginine in the presence of glycine catalyzed by glycine amidinotransferase
(GAT) in the kidney and the pancreas[1]. Previous studies have demonstrated a
marked decrease in the amount of urinary GAA excretion in cases of uremia[2-4].

Recently, nephrotoxicity has emerged as a problem resulting from the
frequent use of antibiotics and other drugs. Measurements of urinary β_2-
microglbulin (β_2-MG) and/or N-acetyl-β-D-glucosaminidase (NAG) have been
widely used as indicators for detecting tubular damage. These conventional
indicators, however, are not sensitive enough to detect early stages of
renal damage.

The kidney is the main site for the production of GAA[1]. This study is
designed to determine whether measurement of urinary GAA is useful for the
early detection of gentamicin (GM) nephrotoxicity, and to test whether superoxide dismutase (SOD) is protective against the nephrotoxicity.

MATERIALS AND METHODS

Experimental GM-induced Nephropathy

Male Fischer 344 rats weighing 190-210 g were used in all studies. To
induce an early stage of renal damage the animals were given a single inter-
venous injection of 5, 10 or 30 mg of GM per kilogram of body weight in 0.4
ml of 0.9 % saline. After injection of GM, the animals were housed singly in
metabolic cages and fasted but allowed free access to water. Twenty-four-hour
urine specimens were collected. Twenty-four-hours after injection, all animals were killed and blood was collected for the determination of serum urea
nitrogen (BUN), creatinine and guanidino compounds. The kidneys and the liver
were immediately removed and frozen in an acetone-dry ice mixture for assays.

Pretreatment of Biological Samples

        Serum was centrifuged at 1,000 x g for 30 min in a Centriflo membrane
CF-25 (Amicon Corp). The ultrafiltrate was adjusted to pH 2.2 and used for
the determination of guanidino compounds. The tissues were deproteinized
using ice cold trichloroacetic acid (TCA) at a final concentration of 10%.
After centrifugation at 2,000 x g for 10 min, TCA was removed from the supernatant using water-saturated ethyl ether. The urine specimens were passed
through 45 μm membrane filters and urinary creatinine excretion in the
twenty-four-hour sample was measured by the method of Jaffe[5]. A part of the
urine obtained by filtration was deproteinized by Centriflo membrance CF-25
at 1,000 x g for 30 min. The supernatant was diluted twenty fold with 0.01 N
HCl for urinary GAA determination. The amount of GAA excreted was expressed
as μg/mg creatinine. The remaining urine specimens were dialyzed against
Tris-HCl buffer, pH 7.4 for 2h at 4°C, and analyzed for enzyme activities.

Analysis of Guanidino Compounds

        The concentrations of guanidinosuccinic acid (GSA) and GAA in serum,
urine and tissues were determined by the automated guanidine analyzer G-520
(JASCO Ltd., Japan). The concentrations of arginine, ornithine, citrulline
and argininosuccinic acid in tissues were determined simultaneously by a high
speed amino acid analyzer 835 (Hitachi Ltd., Japan).

Assays of Enzyme Activities

        GAT activities of the kidney and liver were assayed using a 10% homo-
genate with glycine and canavanine as substrates according to Van Pilsum et
al[6,7]. SOD activity of the kidney was assayed by the nitrous acid method[8].

        The urinary activities of NAG, leucine aminopeptidase (LAP) and γ-
glutamyltranspeptidase (γ-GTP) were determined using m-cresolsulfonphthal-
einyl-N-acetyl-ß-glucosaminide, L-leucine-p-nitroanilide and γ-glutamyl-p-
nitroanilide as substrates and expressed in international units (IU) or mIU
creatinine per twenty-four-hour urine.

Assay of ß$_2$-MG

        The urinary level of ß$_2$-MG was determined by radioimmunoassay.

Treatment of Rats by SOD

        SOD (4,000 U/g) was purchased from Sigma Chemical Co., (U.S.A.). Rats
were injected intravenously with a total of 40,000 units of SOD 30 min before
and 1 h after the injection of gentamicin.

RESULTS

Serum Concentrations of Guanidino Compounds in Rats injected with GM

        Serum levels of BUN and creatinine were not changed in any of the GM-
injectd group, suggesting that these rats did not have renal failure. Serum
concentrations of GAA and GSA also were not changed as compared with controls
(Table 1).

Table 1.   Effect of gentamicin (GM) on BUN, creatinine, guanidinoacetic acid
           (GAA) and guanidinosuccinic acid (GSA) in rat serum.

	BUN (mg/dl)	Creatinine (mg/dl)	GAA (μg/dl)	GSA (μg/dl)
Control	19.3 ± 1.3	0.71 ± 0.07	85.0 ± 1.7	1.99 ± 0.24
GM 10mg	19.6 ± 1.5	0.69 ± 0.08	83.3 ± 4.5	1.99 ± 0.05
GM 30mg	19.6 ± 1.9	0.80 ± 0.18	80.9 ± 3.9	2.07 ± 0.21

Urinary GAA Excretion in GM Nephrotoxicity

     Fig. 1 shows the urinary activity of NAG in rats injected with various
doses of GM. As compared with controls, the enzyme activity in the group
receiving either 10 or 30 mg/kg of GM was significantly increased, whereas
it was not increased in the group receiving 5 mg/kg GM. Urinary activities
of γ-GTP and LAP were also significantly increased (Fig. 2). However, as
shown in Fig. 3, urinary GAA excretion had a different pattern. The excretion was significantly decreased even after injection of 5 mg/kg of GM.

     These results suggest that urinary GAA might reflect GM nephrotoxicity
more sensitively than the other urinary enzyme activities tested.

Effects of GM on Renal Content of GAA in Rats

     To determine a possible mechanism of decreased excretion of urinary GAA,
the tissue content of GAA were measured. The results are shown in Fig. 4.
The renal content of GAA was significantly decreased in the 5 to 30 mg/kg
dosage groups and the decrease was dose-dependent. The hepatic content of
GAA showed a tendency to decrease also but the decrease was not significant.

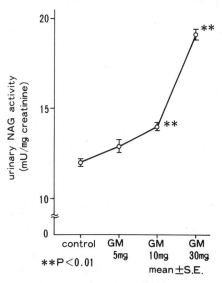

Figure 1.   Changes in urinary N-acetyl-ß-D-glucosaminidase (NAG) activity in
            rats injected with gentamicin (GM).

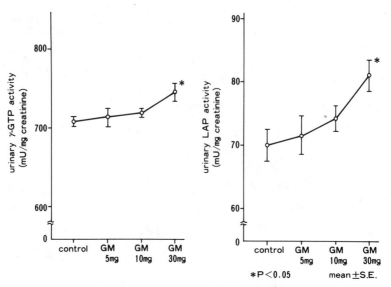

Figure 2.   Changes in urinary γ-glutamyltranspeptidase (γ-GTP) and leucine
            aminopeptidase (LAP) activities in rats injected with gentamicin
            (GM).

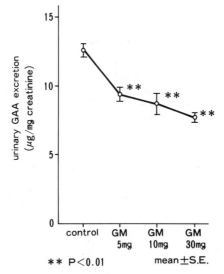

Figure 3.   Changes in urinary guanidinoacetic acid (GAA) excretion in rats
            injected with gentamicin (GM).

The content of GSA and amino acids including arginine in the kidney were not
changed.

Effects of GM on Renal GAT Activity in Rats

        To elucidate the decreased renal content of GAA, renal GAT activity was
assayed. The activity was significantly decreased in 5 to 30 mg/kg dosage
groups (Fig. 5).

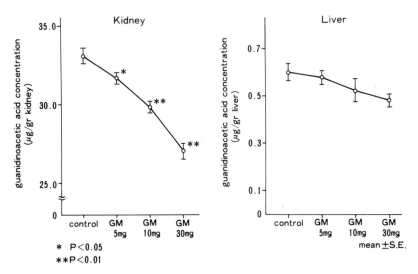

Figure 4. Changes in the guanidinoacetic acid content in rat tissues after
           the injection of gentamicin (GM).

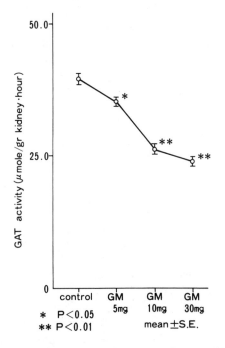

Figure 5. Changes in glycine amidinotransferase (GAT) activity in rat
           kidneys after the injection of gentamicin (GM).

The Effect of SOD on GM Nephrotoxicity

        Before the experiment, we assayed SOD activity in the kidneys from the
rats injected with GM. The activity was significantly increased in the 5 to
30 mg/kg dosage groups, and the increase was dose-dependent (Fig. 6).

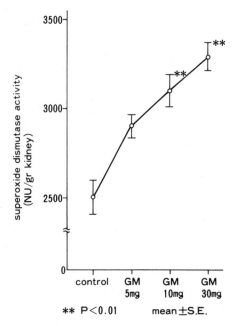

Figure 6.   Changes in the superoxide dismutase activity of rat kidney after
            the injection of gentamicin (GM).

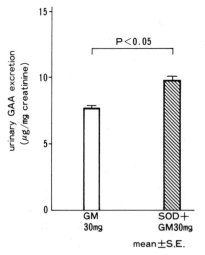

Figure 7.   The effect of superoxide dismutase (SOD) on urinary guanidino-
            acetic acid (GAA) excretion.

These results suggest that oxygen free radicals are in part responsible
for GM nephrotoxicity.

We have previously repored that treatment with coenzyme $Q_{10}(CoQ_{10})$ can
protect rats against GM nephrotoxicity[9]. We tested whether SOD would be protective against the nephrotoxicity as well as CoQ_{10}.

Fig. 7 shows the effect of SOD on urinary GAA excretion in rats. Treatment of rats by SOD increased significantly the urinary excretion of GAA as compared with controls.

Electron microscopy showed degeneration of the proximal tubular cells after 30 mg/kg of GM. Treatment with SOD reduced these abnormalities of proximal tublar structure.

The Profiles of Urinary GAA Excretion in Clinical Cases

Fig. 8,9 and 10 show the serial alteration of urinary GAA in patients who received antibiotic treatment. As shown in Fig. 8, patient 1 had a decrease in urinary GAA excretion from 88 to 58 μg/mg creatinine when cephazolin (CEZ) and cephalexin (CEX) were prophylactically administrated after renal biopsy. In patient 2 with acute pyelonephritis, there was a preferential decrease in urinary excretion of GAA without any alteration of the urinary NAG activity or β_2-MG when cefotiam (CTM) was administrated. The urinary NAG activity and β_2-MG began to increase when another antibiotic, tobracin (TOB) was added (Fig. 9). In patient 3 with interstitial nephritis induced by propylthiouracil (PTU), prednisolone treatment was instituted, and then CoQ_{10} (3 mg/kg/day) was administered intravenously for 3 days. Treatment of this patient with CoQ_{10} resulted in a dramatic increase of urinary GAA excretion. Creatinine clearance also returned to normal (Fig. 10).

DISCUSSION

Fischer 344 rats used in this study are reported to be greatly susceptible to aminoglycoside-induced nephrotoxicity[10]. Rats injected with 5 mg/kg of GM showed a preferential decrease in urinary excretion of GAA without any alteration of the urinary NAG, γ-GTP and LAP activities. This result indicates that urinary GAA is a more sensitive indicator than other urinary

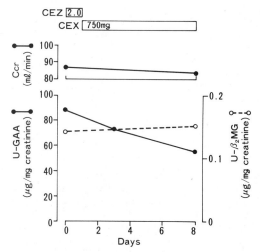

Figure 8. The serial alteration of urinary guanidinoacetic acid (GAA) excretion in patient 1 who received antibiotic treatment. (Patient J.T., F 40y, chronic glomerulonephritis) Ccr; creatinine clearance, CEZ; cephazolin, CEX; cephalexin and β_2-MG; β_2-microglobulin.

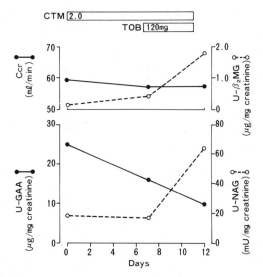

Figure 9. The serial alteration of urinary guanidinoacetic acid (GAA)
           excretion in patient 2 who received antibiotic treatment.
           (Patient E.M., F 50y, hyperthyroidism, pyelonephritis). Ccr;
           creatinine clearance, CTM; cefotiam, TOB; tobracin, ß2-mg; ß2-
           microglobulin and NAG; N-acetyl-ß-D-glucosaminidase.

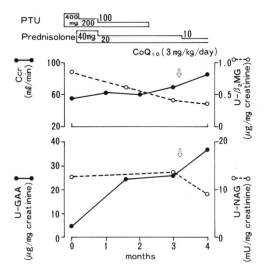

Figure 10. The serial alteration of urinary guanidinoacetic acid (GAA)
           excretion in patient 3 with interstitial nephritis induced by
           propylthiouracil. (Patient Y.F., F 25y, hyperthyroidism). Ccr;
           creatinine clearance, PTU; propylthiouracil, CoQ$_{10}$; coenzyme Q$_{10}$,
           ß2-MG; ß2-microglobulin and NAG; N-acetyl-ß-D-glucosaminidase.

enzyme activities for detecting this type of nephrotoxicity. The reduction
in urinary GAA excretion is closely associated with decreased renal content
of GAA and decreased renal activity of GAT. It has previously been shown tha
GAA is formed from arginine by GAT in the kidney[1]. Therefore, the decreased

excretion of urinary GAA is the result of the decreased renal activity of GAT, probably induced by relatively small-doses of GM.

Furthermore, in this study we examined the effect of SOD on GM-induced nephrotoxicity. Treatment of rats with SOD did restore the urinary excretion of GAA, whereas the renal SOD activity did not increase. Recently, it has been shown that oxygen free radicals mediated the renal damage in acute renal failure in the rat, because several different free radical scavengers including SOD provided functional and histological protection against ischemic damage[11]. Therefore, the restored excretion of urinary GAA may have come from the ability of SOD to scavenge oxgen free radicals.

CONCLUSION

Measurement of urinary GAA excretion is useful for the early diagnosis of drug-induced nephrotoxicity. Superoxide dismutase has a protective effect on GM nephrotoxicity in the rat.

REFERENCES

1. Walker, J.B., Formanidine group transfer in extracts of human pancreas, liver and kidney, Biochem. Biophys. Acta 73 (1963) 241–247.
2. Bonas, J.E., Cohen, B.D. and Natelson, S., Separation and estimation of certain guanidino compounds. Application to human urine, Microchem. J. 7 (1963) 63–77.
3. Cohen, B.D., Stein, I.M. and Bonas, J.E., Guanidinosuccinic aciduria in uremia. A possible alternate pathway for urea synthesis, Am. J. Med. 45 (1968) 63–68.
4. Koide, H. and Azushima, C., Metabolic profiles of guanidino compounds in various tissues of uremic rats, in: "Guanidines," A. Mori, B.D. Cohen and A. Lowenthal Eds, Plenum Press, New York (1985) pp. 365–372.
5. Jaffe, M., Uber den Niedershlag, welchen Pikrinsäure in normalem Harn erzeugt und über eine neue Reaktion des kreatinins, Hoppe-Seyler Z. Physiol. Chem. 10 (1986) 391.
6. Van Pilsum, J.F., Berman, D.A. and Wol, E.A., Assay and some properties of kidney transamidinase, Proc. Soc. Exp. Biol. Med., 95 (1957) 96–100.
7. Van Pilsum, J.F., Olsen, B., Taylor, D., Rozycki, T. and Pierce, J.C., Transamidinase activities, in vitro, of tissues from various mammals and from rat fed protein-free, creatine-supplemented and normal diets, Arch. Biochem. Biophys. 100 (1963) 520–524.
8. Oyanagi, Y., Reevaluation of assay methods and establishment of kit for superoxide dismutase activity, Anal. Biochem. 142 (1984) 290.
9. Itabashi, H. Rinno, H. and Koide, H., Urinary guanidinoacetic acid excretion in gentamicin nephrotoxicity, in: "Guanidines," A. Mori, B.D. Cohen and A. Lowenthal Eds, Plenum Press, New York (1985) pp. 327–344.
10. Kosek, J.C., Mazze, R.I. and Cousins, M.J., Nephrotoxicity of gentamicin, Lab. Invest. 30 (1974) 48–57.
11. Paller, M.S., Hoidal, J.R. and Ferris, T.F., Oxygen free radicals in isehemic acute renal failure in the rat, J. Clin, Invest. 74 (1984) 1156–1164.

EFFECT OF ANTIBIOTIC ADMINISTRATION ON URINARY GUANIDINOACETIC ACID EXCRE-
TION IN RENAL DISEASE

Shuei Nakayama, Ikuo Kiyatake, Yoshio Shirokane* and
Hikaru Koide

Division of Nephrology, Department of Medicine, Juntendo
University School of Medicine, Tokyo 113; *Bioscience Research
Laboratory, Kikkoman Corporation, Chiba 278, Japan

INTRODUCTION

It is known that there is a marked decrease in urinary guanidinoacetic
acid (GAA) excretion in patients with uremia[1-3]. We have previously reported
that urinary GAA (U-GAA) excretion is significantly decreased in experimental
animals injected with 20 to 50 mg/kg of gentamicin[4]. Our further study also
showed that such a reduction in GAA excretion is seen even in animals injected with 5 mg/kg, a relatively small dose of gentamicin[5]. These findings have
led some to suggest that GAA would be useful for the early diagnosis of
antibiotic-induced nephrotoxicity.

We studied: a) the relationship between the concentration of U-GAA and
levels of N-acetyl-ß-D-glucosaminidase (NAG) and ß$_2$-microglobulin (ß$_2$-MG) in
urine, b) the excretion profile of U-GAA in patients receiving prophylactic
antibiotic treatment.

MATERIALS AND METHODS

Patients

Twenty-six patients with various renal diseases were studied. The patients' ages ranged from 14 to 50 y (average, 27.2 ± 9.7). They met the following criteria: they received prophylactic intravenous injections of antibiotics [1 to 2 g/day of cefmenoxime (CMX) in 25 patients or ceftizoxime
(CZX) in one patient] for 4 to 5 days after renal biopsy. They had creatinine
clearances (Ccr) ≥ 80 ml/min (average, 112 ± 22).

Renal histology included 3 cases of minimal change, 16 cases of IgA
nephropathy, 3 cases of membranous nephropathy, 1 case of focal glomerulosclerosis, 1 case of chronic glomerulonephritis categorized into
unclassified, 1 case of benign recurrent hematuria and 1 case of Wegener's
granulomatosis.

Determination of GAA in Urine

The concentration of GAA in urine was determined using a new enzymic method, which was established by Shirokane et al[6]. To 1.3 ml urine samples in test tubes, 0.2 ml urease reagent was added in order to eliminate endogeneous urinary urea. The mixture was incubated at 37°C for 15 min. The reaction mixture was centrifuged at 1,000 x g for 20 min using a Centriflo membrane CF-50 to yield an ultrafiltrate free from urease. Into each of duplicate test tubes, 0.5 ml of ultrafiltrate was placed, then 0.1 ml of GAA amidinohydrolase reagent was added to one tube and distilled water to the other for a sample blank. The mixture was incubated at 37°C for 15 min. Into both tubes were added successively 2.0 ml of N-(1-naphthyl)-N'-diethylethylenediamine reagent and 2.0 ml of o-phthalaldhyde reagent, mixed, and incubated at 37°C for 15 min. The tubes were cooled in running water for 2 min and the absorbance measurted at 465 nm vs the sample blank. The GAA concentration was determined by the use of a standard curve.

Other Laboratory Tests

The urinary activity of NAG was determined using a NAG-Kit (Shionogi, Japan), and the urinary level of β_2-MG was determined by radioimmunoassay.

RESULTS

Relationship between GAA Concentrations in Urine determined by the Enzymic and Chromatographic Method

The enzymic method was compared with a high performance liquid chromato graphic (HPLC) method with the use of 100 urine samples obtained from 48 healthy volunteers (30 men and 18 women, ages 23 to 43y). Samples of twenty-hour urine were also collected from patients with suspected or proven renal failure. These samples were stored at -20°C until analysis. Linear regression analysis revealed a good correlation (r=0.986, p < 0.001) between the two methods (Fig. 1). The mean (±SD) concentration of U-GAA in healthy volunteers was 88.1 ± 50.3 mg/day when GAA was determined by the enzymatic method. The concentration of U-GAA was stable for one year when urine samples were stored at -20°C, and was not influenced by the degree of urinary protein or sediment. We determined the concentration of U-GAA using the enzymic method in the present study.

Relationship between the Urinary Excretion of GAA and the Level of Ccr

There was a good correlation (r=0.705, p < 0.01) between the concentration of U-GAA and the level of Ccr. (Fig. 2).

Relationship between the Urinary Excretion of GAA and Levels of NAG, β_2-MG

There was an inverse correlation (r=0.329, p < 0.05) between the concentration of U-GAA and the level of urinary NAG activity (Fig. 3). However no correlation was observed between the concentration of U-GAA and the leve of urinary β_2-MG (Fig. 4).

The Effect of Preservatives on the Enzymic Assay System

Before measurement of urine obtained from patients with various renal diseases receiving antibiotics after renal biopsy, we tested the effect of several preservatives on the enzymatic assay system because a preservative

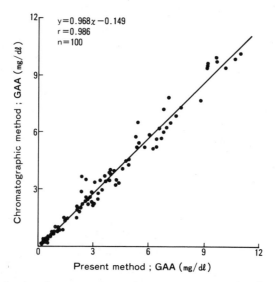

Figure 1.   Correlation between guanidinoacetic acid (GAA) concentration in
            urines determined by the enzymic and chromatographic methods.

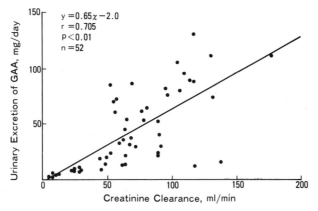

Figure 2.   Correlation between urinary excretion of guanidinoacetic acid
            (GAA) and creatinine clearance.

is widely used for twenty-four-hour urine collection. As shown in Table 1,
all preservatives tested did not affect the assay system. From these results,
we chose xylene as a preservative which was added to urine, making the final
concentration 0.5%.

The Effect of Antibiotics on the Enzymic Assay System

    We also tested the effect of several antibiotics on the snzymic assay
system. As shown in Table 2, all antibiotics including CMX did not affect the
assay system.

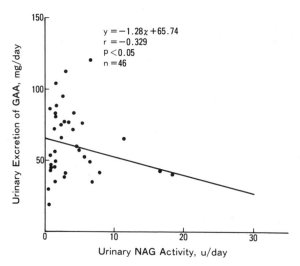

Figure 3.   Correlation between urinary excretion of GAA; guanidinoacetic
            acid and NAG; N-acetyl-ß-D-glucosaminidase.

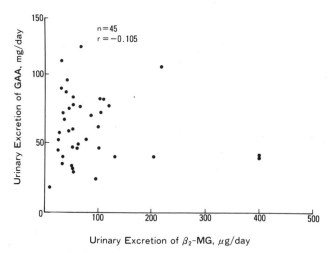

Figure 4.   Relationship between urinary excretion of GAA and ß₂-MG. GAA;
            guanidinoacetic acid and ß₂-MG; ß₂-microglobulin.

The Effect of Antibiotics on Urinary GAA Excretion in Renal Diseases

We examined the effect of prophylactic doses of antibiotics on urinary
excretion of GAA in patients with early-stage renal diseases. The GAA concentration obtained from urine on a day before administration of antibiotics
was used as a baseline value. As compared with baseline values, the concentrations of urinary GAA were significantly decreased on days 5 and 7 after
the administration of antibiotics. Decreased concentrations of U-GAA were
restored to normal on day 14 (Fig. 5). However, as seen in Fig. 6, levels of
NAG and ß₂-MG in urine were not changed by day 14.

Table 1. The effect of several different preservatives on the enzymic assay system

Preservatives	Recovery of GAA added (10 mg/dl)	Normal subject A		Normal subject B	
		GAA content	Recovery	GAA content	Recovery
None	100 %	10.55 mg/dl	100 %	14.44 mg/dl	100 %
Xylene	98	10.52	100	14.26	99
Toluene	97	10.48	99	14.65	101
Thymol	98	10.57	100	14.23	99

To guanidinoacetic acid (GAA) solutions or urine samples from normal subjects, we added each preservative to adjust the final concentration to 0.5 %, and then GAA was determined.

Table 2. The effect of several diferent antibiotics on the enzymic assay system

Antibiotics	Concentration	Normal subject A		Normal subject B	
		GAA content	Recovery	GAA content	Recovery
Gentamicin	9.6 mg/dl	9.19 mg/dl	97 %	12.86 mg/dl	99%
Tobracin	2.4	9.27	98	12.76	99
Cefmenoxime	0.1	9.16	97	12.86	99
Ceftizoxime	1.0	9.02	95	12.86	99
Piperacillin	10.0	9.12	96	12.71	98
Ampicillin	1.0	9.02	95	13.19	102
None	-	9.46	100	12.94	100

To urine samples from normal subjects, added each of drug, and then guanidinoacetic acid (GAA) was determined.

These results indicate that U-GAA might reflect antibiotic nephrotoxicity more sensitively than the other urinary parameters tested.

The Profiles of U-GAA Excretion in Clinical Cases

Basically, excretion profiles of U-GAA in 26 patients could be classified into three patterns. As shown Fig. 6, pattern I is that in which GAA is decreased as compared with the baseline value, but remains within the normal range, whereas urinary NAG, β_2-MG, serum creatinine and BUN were not changed. Eleven patients had pattern I. Pattern II is that in which GAA is preferentially decreased below the normal level and remains decreased by day 7 without any alteration of other parameters in the urine and serum (Fig.7). Nine patients had pattern II. Pattern III is that in which all parameters including GAA were not changed (Fig. 8). Six patients had pattern III. To characterized patterns I, II and III, patient's age, Ccr, cumulative dose of antibiotics and renal histology were evaluated. The cumulative dose of antibiotics in pattern I and II were higher than that in pattern III (Table 3). However, no difference in the other parameters could be found among the three patterns.

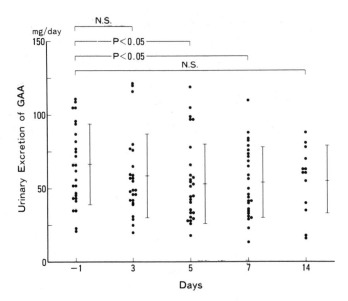

Figure 5. The effect of antibiotics on urinary guanidinoacetic acid (GAA)
           excretion in renal diseases. N.S.; not significant.

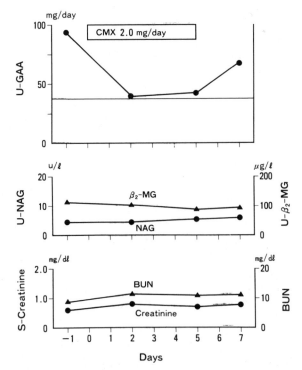

Figure 6. The profile of urinary GAA excretion in a clinical case, pattern
           I. (Patient Y.I., F. 22y, membranous nephropathy).
           U-GAA; urinary guanidinoacetic acid, U=NAG; urinary N-acetyl-ß-D-
           glucosaminidase, BUN; serum urea nitrogen, U-ß2-MG, urinary ß2-
           microglobulin and CMX; cefmenoxime.

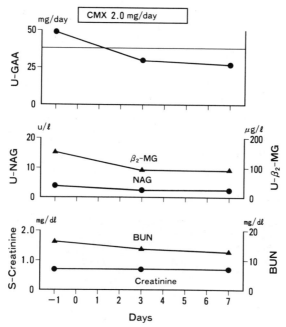

Figure 7. The profile of urinary GAA excretion in a clinical case, pattern
II. (Patient M.N., F. 19y, IgA nephropathy).
Abbreviations are the same as shown in Figure 6.

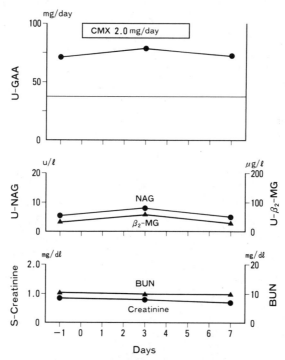

Figure 8. The profile of urinary GAA excretion in a clinical case, pattern
III. (Patient M.N., M. 29y, IgA nephropathy).
Abbreviations are the same as shown in Figure 6.

Table 3.   Characterization of each pattern of urinary guanidinoacetic acid

Pattern	N	Age (years)	Creatinine clearance (ml/min)	Cumulative dose (g)	Histological diagnosis
I	11	27.4 ± 9.3	108.8 ± 21.2	8.0 ± 2.1	IgA-N(6)*, MN(2), UNC(1), WG(1), BRH(1)
II	9	28.5 ± 12.2	118.4 ± 23.0	10.4 ± 8.4	IgA-N(5), MC(2), MN(1), FGS(1)
III	6	24.6 ± 6.4	105.3 ± 23.1	7.5 ± 5.8	IgA-N(5), MN(1)

IgA-N; IgA nephropathy, MC; minimal change, MN; membranous nephropathy, UNC; chronic glomerulonephritis categorized into unclassified, WG; Wegener's granulomatosis, BRH; benign recurrent hematuria, FGS; Focal glomerulosclerosis. Numbers in parentheses indicate the numbers of patients with each renal histology.

DISCUSSION

Early diagnosis and treatment of antibiotic-induced nephrotoxicity may prevent irreversible renal injury that often develops after nephrotoxic antibiotics. In order to detect nephrotoxicity, tests of renal function are routinely performed after administration of the aminoglycoside antibiotics. Of these, serum cratinine has been the most useful test, but lags behind histologic changes of nephrotoxicity[7]. Recently, it has been reported that urinary enzymes are particularly helpful in detecting subtle tubular disfunction before renal damage has proceeded to the point of producing a decrease in glomerular filtration rate[8]. The increased excretion of urinary enzyme has been noted after high-doses of salicylates[9] and aminoglycosides[10,11], although its sensitivity for detecting nephrotoxicity has not been demonstrated to be superior to other indices.

GAA is formed mainly in the kidney and is excreted into the urine[12]. Urinary enzyme activities including NAG, U-GTP and LAP reflect the escape of cytosol enzymes from the cells, whereas U-GAA reflects impaired metabolism in tubular cells by renal tubular injury. The mechanisms involved could be quite different with regard to excretion into the urine.

In the present study, the concentration of U-GAA was determined using a new enzymic method. The present enzymic method for GAA in urine is rapid and simple, and also suitable for use in analysis of clinical samples[6]. As shown in Fig. 1, the precision of this method is similar to that of HPLC, and there is good correlation between GAA as determined by the enzymic and HPLC methods. The concentration of U-GAA was well correlated with Ccr, inversely correlated with urinary NAG. However, no correlation was recognized between U-GAA and β_2-MG.

Our interest in GAA monitering was to determine if quantitating U-GAA would make an early diagnosis of a nephrotoxic episode prior to changes in urinary enzyme activities. Before examining the profiles of U-GAA excretion in patients receiving antibiotics, the effects of preservatives and antibiotics on the enzymic assay system were tested. Virtually, no effect was demonstrated. These observations show that the enzymic method is suitable for evaluating U-GAA excretion.

In patients receiving prophylactic doses of antibiotics, serial determinations of U-GAA revealed a decrease in excretion of U-GAA in 12 of 26 patients (76.9%) without any alteration of urinary NAG, ß$_2$-MG or serum creatinine. The reduction in U-GAA excretion was restored to notmal 9 to 10 days after the antibiotic administration was discontinued. These results indicate that decreased U-GAA is associated with administration of antibiotics, and that U-GAA is a more sensitive indicator than urinary NAG or ß$_2$-MG for detecting the early-stages of antibiotic-induced nephrotoxicity. Gonick et al. have proposed that if it is to be useful in the diagnosis of renal diseases, a urinary indicator should have several properties: 1) it should be present in high concentration in the kidney; 2) it should be stable for several days at freezer temperatures; 3) contributions to it from bacteria and urinary sediment should be minimal; 4) inhibitors or activators of it should be absent from the urine or, if present, should be consistently and reproducibly removed by dialysis; 5) the determination should be accurate and rapid[13]. U-GAA examined in the present study would seem to fulfill these criteria most adequately.

CONCLUSION

We have found U-GAA to be a useful procedure for monitoring after antibiotic administration. It can be a very sensitive indicator and helpful for the early diagnosis of antibiotic-induced nephrotoxicity.

REFERENCES

1. Bonas, J.E., Cohen, B.D. and Natelson, S., Separation and estimation of certain guanidino compounds: Aplication to human urine, Microchem. J. 7 (1963) 63-77.
2. Cohen, B.D., Stein, I.M. and Bonas, J.E., Guanidinosuccinic aciduria in uremia. A possible alternate pathway for urea synthsis, Am. J. Med. 45 (1968) 63-68.
3. Koide, H. and Azushima, C., Metabolic profiles of guanidino compounds in various tissues of uremic rats, in "Guanidines," A. Mori, B.D. Cohen and A. Lowenthal Eds., Plenum Press, New York (1985) pp. 365-372.
4. Itabashi H. Rinno, H., and Koide, H., Urinary guanidinoacetic acid excretion in gentamicin nephrotoxicity, in "Guanidines," A. Mori, B.D. Cohen and A. Lowenthal Eds, Plenum Press, New York (1985) pp. 327-334.
5. Nakayama, S., Junen, M., Kiyatake, I. and Koide, H., Urinary guanidinoacetic acid excretion as an indicator of gentamicin nephrotoxicity in the rats, (in this book) pp 303-311.
6. Shirokane, Y., Utsushikawa, M. and Nakajima, M., A new enzymic determination of guanidinoacetic acid in urine, Clin. Chem. 33 (1987) 394-397.
7. Schweizer, R.T., Moore, R., Bartus, S.A., Bow, L. and Hayden, J., Beta$_2$-microglobulin monitoring after renal transplantation, Transplant. Proc. 13 (1981) 1620-1623.
8. Mondorf, A.W., Zegelman, M., et al., Comparative studies on the action of aminoglycosides and cephalosporins on the proximal tubule of the human kidney, Proceedings of the 10th International Congress of Chemotherapy, Zurich, 1977.

9. Proctor, R.A. and Kunin, C.M., Salicylate-induced enzymuria, Am. J. Med., 65 (1978) 987–993.

10. Beck, P.R., Thompson, R.B. and Chadhuri, A.K.R., Aminoglycoside antibiotics and renal function: Changes in urinary glutamyltransferase excretion, J. Clin, Pathol., 30 (1977) 432–437.

11. Wellwood, J.M., Lovell, D., Thompson. A.E., et al., Renal damage caused by gentamicin: A study of the effects on renal morphology and urinary enzyme excretion, J. Pathol., 118 (1976) 171–182.

12. Walker, J.B., Formanidine group transfer in extracts of human pancreas, liver and kidney, Biochem. Biophys. Acta 73 (1963)241–247.

13. Gonick, H.C., Kramer, H.J. and Schapiro, A.E., Urinary ß-glucuronidase activity in renal disease, Arch. Intern. Med. 132 (1973) 63–69.

METABOLISM OF GUANIDINOACETIC ACID IN RENAL TUBULAR TISSUE CULTURE: A MODEL SYSTEM FOR CISPLATIN NEPHROPATHY

Kazuyuki Tasaki, Kazukiyo Yoshida, Fumitake Gejyo and Masaaki Arakawa

Department of Medicine (II), Niigata University School of Medicine, Niigata 951, Japan

INTRODUCTION

Guanidinoacetic acid (GAA) is produced in the kidney by the transamidination of glycine from arginine. The urinary excretion of GAA is reported to be diminished in renal failure or drug induced nephropathy[1,2]. This study was attempted to evaluate the effect of cisplatin (CDDP), a recently developed anti-cancer agent known to have nephrotoxicity, on GAA production in rat renal tubular tissue culture. In addition, the urinary excretion of GAA in patients with lung cancer treated with CDDP was also studied.

MATERIALS AND METHODS

Renal cortical non-glomerular tissue fractions of Wistar rats were obtained by a sieving method[3]. The entire process is summarized diagrammatically in Fig. 1. The tubular tissues obtained were cultured in RPMI 1640 containing 10-20% fetal bovine serum (FBS) for 2-14 days. The GAA level in the culture medium was measured by high-performance liquid chromatography (Jasco G-520 guanidino autoanalyzer, Japan Spectroscopic Co.).

The tubular tissue fragments obtained were distributed in GIBCO/culture wells and 2 ml of RPMI 1640 containing 10% FBS was added to each well as a culture medium. CDDP was added to the wells at a concentration of 5, 25 and 100 μg/ml. Six wells were used at each concentration of CDDP. The GAA level in the culture meidum was measured after 7 days' culture.

The urinary excretion of GAA in 5 patients with lung cancer (histologically adenocarcinoma) treated with CDDP was also estimated (Table 1). Their ages ranged from 36 to 69 y (average age 56y). Vindesine, etoposide, and occasionally mitomycine C were also administerd. CDDP was administered on the first day of treatment by intravenous drip infusion of 80 mg/m^2 for 3h. Sufficient hydration to maintain a urine output of at least 3 l/day was given to all patients with CDDP. Serum creatinine (Cr) remained under 1.3 mg/dl and serum urea nitrogen (UN) under 30 mg/dl during treatment in all patients. Serum Cr, UN, Mg, as well as urinary excretions of N-acetlyl-ß-glucosaminidase (NAG) (enzyme immunoassay), and ß$_2$-microglobulin (ß$_2$M) (latex agglutination method) were also measured.

RESULTS

Although GAA was not detected in the medium from the renal glomerular tissue fraction, a large amount of GAA was detected in that from the renal tubular tissue fraction. The GAA levels in the culture medium containing 25 and 100 µg/ml of CDDP (23.0 ± 3.6 nmol/ml and 13.9 ± 2.7 nmol/ml, respectively) were significantly lower than those in the CDDP free medium, although the level in the culture medium containing 5mg/ml of CDDP (28.7 ± 3.9 nmol/ml) was not changed significantly (Fig. 2).

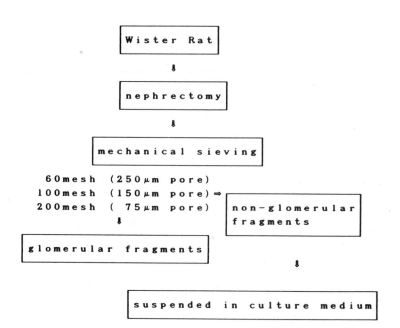

Figure 1.   Diagram of the sieving method for obtaining renal tubular fragments.

Table 1.   Clinical features in the 5 patients of lung cancer.

Pt.No.	Age	Sex	Diagnosis	Therapy
1	64	F	lung cancer (adenocarcinoma)	CDDP VDS VP16
2	50	F	lung cancer (adenocarcinoma)	CDDP VDS
3	63	F	lung cancer (adenocarcinoma)	CDDP VDS
4	36	M	lung cancer (adenocarcinoma)	CDDP VDS MMC
5	69	M	lung cancer (adenocarcinoma)	CDDP VDS

Abbreviations:
CDDP; cis-platinum, VDS; vindesine, VP16; etoposide, MMC; mitomycine C.

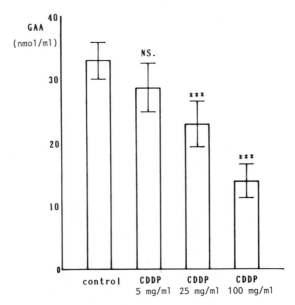

Figure 2. Guanidinoacetic acid (GAA) levels in culture medium.
(Each column represents mean ± SD)
(NS;not significant, *** P < 0.001)

The urinary excretion of GAA varied greatly among the patients, ranging from 140 μmol/day to 550 μmol/day before the CDDP exposure. After administration of CDDP, it decreased markedly in each case (Fig. 3) persisting from day 0 to day 14 (Fig.4). About 3 weeks after CDDP exposure, levels returned to the value before the exposure.

The urinary excretion of ß2M was measured in two patients. The peak levels were 0.15 mg/day on days 0 and 3. This is not significantly higher than before CDDP exposure (Fig. 5).

The urinary excretion of NAG was also measured in two patients. The highest excretion was 11.2 IU/day on day 7 (Fig. 6). Serum UN and Cr levels in each patient are shown in Fig. 7 and 8. All cases except case 5 showed no significant change in serum Cr level during the CDDP exposure as shown in Fig. 8. Serum Mg levels were lower than that seen in normal subjects (0.8 - 1.4 mg/dl).

DISCUSSION

L-Arginine-glycine amidinotrnsferase is found in renal proximal tubular cells, hepatocytes and α cells of the pancreatic islet in rats[4]. Large amounts of GAA are also detected in renal tubular tissue in this study.

In 1969, platinum compounds were reported by Rosenberg et al. to exert potent anti tumor activity in mice[5]. CDDP is now widely used as one of the anti-cancer agents, although it has rather severe nephrotoxicity. To prevent nephrotoxicity, it is important to keep the patients well hydrated during CDDP administration. Renal injury, when it occurs, is seen in the S_3 segment of the proximal tubule. Pathologically, epithelial cell degeneration, or

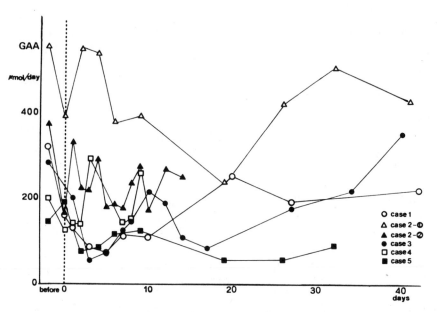

Figure 3. Urinary excretion of guanidinoacetic acid (GAA) after cisplatin
           exposure. (Case2 was treated twice)

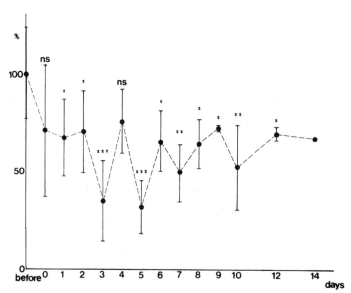

Figure 4. Urinary excretion of guanidinoacetic acid. The values before
           cisplatin exposure were estimated at 100%. (Each closed
           circle represented mean ± SD) (NS;not significant, * P < 0.05,
           ** p < 0.01, *** p < 0.001)

necrosis, and interstitial edema are observed. Significant glomerular changes are not seen[6,7]. In pharmacokinetic studies, 17–33% of CDDP is excreted
into the urine, and the urinary concentration is 3 to 4 times more than that

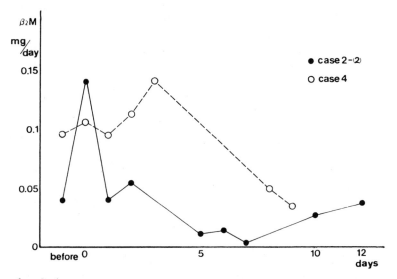

Figure 5. Urinary excretion of ß$_2$ microglubulin (ß$_2$M) after cisplatin
         exposure.

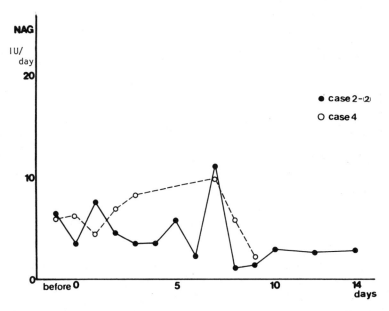

Figure 6. Urinary excretion of N-acetyl-ß-glucosaminidase (NAG) after
         cisplatin exposure.

of the plasma (Cmax in the urine was 15-50 µg/ml at 80 mg/m^{2}administration)[8].
The present study was attempted to measure GAA levels in tubular tissue
culture media containing various concentrations of CDDP as an in vitro model
of the nephrotoxicity of CDDP. GAA levels were significantly lower in the
culture media containing 25 µg/ml of CDDP. It seems to be a very sensitive
indicator of tubular cell damage in the culture system. Clinically, many
studies have shown a positive correlation between urinary excretion of NAG or
ß$_2$M and CDDP nephrotoxicity. In addition, the occurence of hypomagnesemia is
well known[9~11].

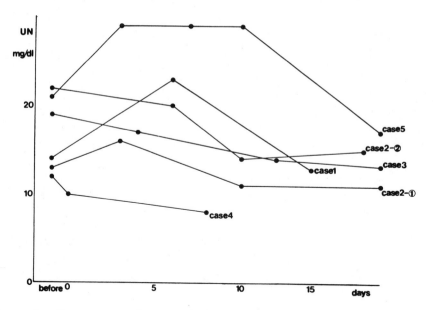

Figure 7.   Serum urea nitrogen (UN) levels after cisplatin exposure.

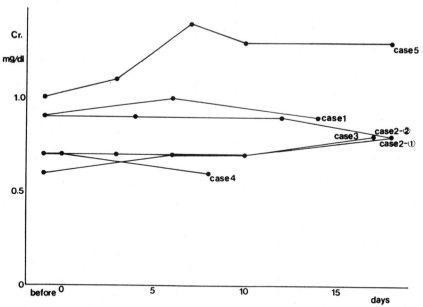

Figure 8.   Serum creatinine (Cr) levels after cisplatin exposure.

    The present study shows that the urinary excretion of GAA is signifi-
cantiy decreased from day 2 following CDDP administration. However, the
urinary excretion of NAG, and β_2M were not greatly changed. Although severe
renal failure did not occur in our 5 cases, the serum Mg level was decreased
in all cases. It is suggested that the measurement of the urinary GAA level
may be useful for detection of the early stage of CDDP nephropathy supple-
menting the serum Mg level.

SUMMARY

GAA is produced mainly in the renal proximal tubules by transamidination of glycine from arginine. This study was attempted to evaluate the effect of CDDP on GAA production in rat renal tubular tissue culture. The renal tubular tissue fraction obtained from Wistar rats by sieving was cultured in RPMI 1640 medium for 2-14 days. The GAA level in the culture medium was measured by HPLC. The level in the culture medium containing 25 and 100 µg/ml of CDDP was significantly lower than that found in the CDDP free medium.

The urinary excretion of GAA in 5 patients with lung cancer treated with CDDP was also studied. It decreased significantly after CDDP treatment and remained low for 7-14 days.

In conclusion, we show that GAA is produced mainly in renal tubular tissue and its production is depressd by CDDP exposure either in vitro or in vivo. Urinary GAA levels might therefore be useful for the early detection of CDDP nephropathy.

REFERENCES

1. Bonas, J.E., Cohen, B.D. and Natelson, S., Separation and estimation of certain guanidino compound, Microchem. J. 7 (1963) 63-77.
2. Itabashi, H., Rinno, H. and Koide, H., Urinary guanidinoacetic, acid excretion in gentamycin nephrotoxicity, in "Guanidines," A. Mori, B.D. Cohen and A. Lowenthal Eds., Plenum Press, New York (1985) pp 327-333.
3. Burlington, H. and Cronkite, E.P., Characteristics of cell culture derived from renal glomeruli, Proc. Soc. Exp. Biol. Med. 142 (1973) 143-149.
4. McGuire, D.M., Gross, M.D., Elde, R.P. and Von Pilsum, J.F., Localization of L-arginine-glycine amidinotransferase protein in rat tissues by immunofluorescence microscopy, J. Histochem. Cytochem. 34 (1986) 429-435.
5. Rosenberg, B., Trosko, J.E. and Mansour, V.H., Platinum compounds: A new class of potent antitumor agents, Nature 222 (1969) 385-386.
6. Ries, F. and Klastersky, J., Nephrotoxicity induced by chemotherapy with special emphasis on cisplatin toxicity, Am. J. Kid. Dis. 8 (1986) 368-379.
7. Safirstein, R., Winston, J., Goldstein, M., Model, D., Dikman, S. and Guttenplan, J., Cisplatin nephrotoxicity, Am. J. Kid. Dis. 8 (1986) 356-367.
8. Reece, P.A., Stafford, I., Russel, J., Khan, M. and Gill, P.G., Creatinine clearance as a predictor of ultrafilterable platinum disposition in cancer patients treated with cisplatin, J. Clin. Oncol. 5 (1987) 304-309.
9. Mavichack, V., Wong N.L.M., Quamme, G.A., Magil, A.B., Sutton, R.A. and Dirks, J.H., Studies on the pathogenesis of cisplatin-induced hypomagnecemia in rats, Kidney Int. 28 (1985) 914-921.
10. Gislain, C., Dumas, M., Lautissier, J.L. and Guerrin, J., Urinary β_2-microglobulin, Cancer Chemother. Pharmacol. 18 (1986) 276-279.
11. Kobayashi, H., Cisplatin for ovarian carcinoma-early detection of cisplatin induced nephrotoxicity, Acta Obst. Gynec. Jpn. 37 (1985) 888-896.

SHORT-TERM PROTEIN LOAD IN ASSESSMENT OF GUANIDINOACETIC ACID SYNTHESIS IN PATIENTS WITH CHRONIC RENAL FAILURE

Yoshiharu Tsubakihara, Eiji Yamato, Kenji Yokoyama, Eisaku Kitamura, Noriyuki Okada, Isao Nakanishi and Nobutoshi Iida

Kidney Disease Center, Osaka Prefectural Hospital, 3-1-56 Bandaihigashi, Sumiyoshi-ku, Osaka, Japan

INTRODUCTION

Guanidinoacetic acid (GAA) is the precursor of creatine, an essential element in the energy metabolism of muscle and nerve tissue. GAA has been demonstrated to be formed by the interaction of arginine and glycine mainly in the kidney.

We have already demonstrated[1] a deficiency of GAA and creatine in major organs and in the serum of rats with experimental chronic renal failure (CRF). Furthermore, GAA administration was shown to increase the creatine contents of these organs, and to improve muscular power and physical strength of these rats. In CRF patients, we have also demonstrated[2] that the serum concentration and the urinary excretion of GAA were lower than those in normal subjects. However, this led to a controversy about the deficiency of GAA in CRF patients.

Borsook et al. had shown[3] that a short-term protein load could increase the urinary excretion of GAA in normal subjects. This suggested that normal persons have a reserve of GAA production. Recently, a protein load was reported also to increase the glomerular filtration rate (GFR) in healthy subjects but not in CRF patients[4,5].

In this study, to evaluate the reserve or capacity of GAA synthesis in CRF patients, an oral load of bean protein was administered and the serum concentration and urinary excretion of GAA were subsequently measured.

MATERIALS AND METHODS

The subjects were six healthy volunteers and twelve CRF patients undergoing conservative therapy for chronic glomerulonephritis as shown in Table 1. Studies were performed both in the fasting state and under hydration.

Experimental protocol is shown in Fig 1. Bean protein powder (Meiji Pharm. Co.,Tokyo Japan) containing about 4.2% glycine and 7.8% arginine was

Table 1. Subjects.

	normal N=6 (male 3)		CRF N=12 (male 6)	
Cr (mg/dl)	1.0 ± 0.1	(0.9 - 1.2)	6.7 ± 2.4	(2.0 - 13.7)
BUN (mg/dl)	13 ± 2	(10 - 19)	57 ± 8	(17 - 101)
Ccr (ml/min)	89 ± 7	(74 - 117)	12 ± 2	(3.4 - 37.4)
U-Cr (mg/day)	1493 ± 174	(960 - 2232)	774 ± 153	(456 - 1248)
S-GAA (μg/dl)	31.4 ± 2.6	(20.8 - 38.6)	19.4 ± 2.1	(6.7 - 29.2)
U-GAA (μg/min)	31.9 ± 2.6	(25.3 - 41.1)	1.9 ± 0.5	(0.3 - 5.0)

CRF; chronic renal failure, Cr; creatinine, Ccr; creatinine clearance, U-Cr; urinary creatinine excretion, S-GAA; serum GAA concentration and U-GAA; urinary GAA excretion

administered orally with 5 ml/kg of water. Blood was sampled just before and 1, 2.5 and 4 h after protein loading. Urine was collected for 1 h before protein loading so as to obtain baseline values, and then at 1, 2.5, 4 h after loading. The urine was analysed for hours 0 - 1 (test period I), hours 1 - 2.5 (test period II) and hours 2.5 - 4 (test period III) (Fig. 1). GAA was measured by JASCO G-520 Guanidine Autoanalyser (Japan Spectroscopic Co., Tokyo Japan). The buffers and detection system used have been described[2-6].

Values were shown as mean ± SEM. Significance (P < 0.05) between means was calculated using the Student's t-test.

RESULTS

Fig. 2 shows the pattern of creatinine clearance (Ccr) in response to protein loading in normal subjects and CRF patients. In normal persons, the Ccr of test periods I and II showed a significant increase from the baseline. There were no significant differences in any period in the CRF patients.

The changes in the serum concentration and urinary excretion of GAA after the protein loading are shown in Fig. 3. The baseline serum concentration and urinary excretion of GAA in CRF patients were significantly lower

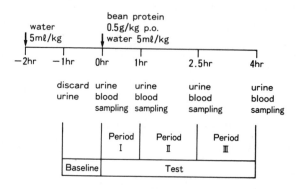

Figure 1. Experimental protocol.

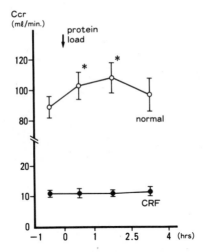

Figure 2. Response of creatinine clearance (Ccr) to the protein load in
normal and CRF patients. (*; significant difference to the
baseline)

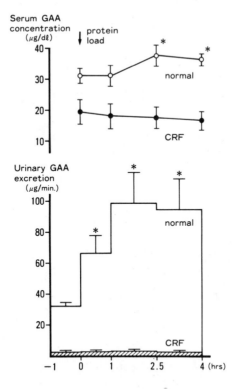

Figure 3. Effect of protein loading on the serum concentration and urinary
excretion of GAA in normal and CRF subjects.
(*; significant difference to the baseline)

than those in normal subjects. The serum concentrations at 2.5 and 4 h after
protein loading and the urinary excretions in all test periods were significantly elevated from the baseline values in normal subjects. However, in CRF
patients, there were no significant changes in these values. The urinary
excretion of GAA corrected by Ccr (U_{GAA}.V/Ccr) was also increased by protein
loading in normal subjects (Fig. 4), indicating that the increment of urinary
GAA excretion by the protein load was not due to the rise of Ccr.

Table 2 summarizes some parameters of the baseline and of test period
II which showed the maximum effects from the protein load. The mean Ccr of
normal subjects increased by 21%, indicating a reserve in the GFR. However,
there was no gain of Ccr in CRF patients. BUN and urinary excretion of urea
nitrogen (U-UN) were significantly increased in both normal and CRF subjects
by the catabolism of the protein load.

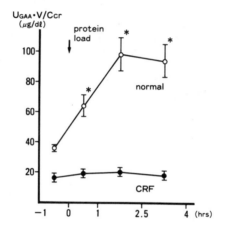

Figure 4. Response of the urinary GAA excretion (U_{GAA}.V) corrected by Ccr to
protein loading in normal and CRF subjects.
(*; significant difference to the baseline)

Table 2. Summary of some parameters in the baseline and in test period II
which showed the maximum effects from the protein load.

		normal			CRF		
		baseline		test period II	baseline		test period II
Cr	(mg/dl)	1.0 ± 0.1	–ns–	1.1 ± 0.1	6.7 ± 2.4	–ns–	6.6 ± 1.2
Ccr	(ml/min)	89 ± 7	–S–	108 ± 10	12 ± 2	–ns–	12 ± 2
BUN	(mg/dl)	13 ± 2	–S–	17 ± 1	57 ± 8	–S–	62 ± 8
U-UN	(mg/h)	275 ± 24	–S–	610 ± 71	123 ± 13	–S–	150 ± 16
S-GAA	(μg/dl)	31.4 ± 2.6	–S–	37.3 ± 3.4	19.4 ± 2.1	–ns–	17.6 ± 1.8
U-GAA	(μg/min)	31.9 ± 2.6	–S–	98.8 ± 20.6	1.9 ± 0.5	–ns–	2.3 ± 0.5

U-UN; urinary urea nitrogen excretion, S; significant, ns; not significant,
other abbreviations are shown in table 1.

The mean serum concentration of GAA (S-GAA) rose significantly (P < 0.05), from 31.4 ± 2.6 to 37.3 ± 3.4 µg/dl, and the mean urinary excretion of GAA (U-GAA) increased to about three times the baseline value (from 31.9 ± 2.6 to 98.8 ± 20.6 µg/min) after protein loading in normal subjects, which suggested that the renal synthesis of GAA could be stimulated by protein. However, in CRF patients, no increase at all was found.

DISCUSSION

Creatine (or creatine-phosphate) is an essential in the energy metabolism of muscle and nerve tissues. GAA is the precursor of creatine and is produced by the kidney. Goldman et al.[7] demonstrated that nephrectomized rats converted only 6-20% as much ^{14}C-labelled glycine to creatine as unoperated controls.

We have shown that the serum concentration and urinary excretion of GAA in CRF patients were significantly lower than those of normal subjects[2]. We have also demonstrated[1] that the GAA and the total creatine concentration in the serum, urine and major organs of experimental CRF rats were significantly lower than those in normal rats. Furthermore, we found[1] that long term GAA administration significantly improved muscular power, as measured by the inclined screen test, and physical strength, as evaluated by the forced swimming method (in which the length of time that the animals could survive in water was measured), and that the total creatine content of the major organs increased. All of these results indicate the deficiency of GAA in CRF.

Furthermore, Tofuku et al.[8] demonstrated that glycine amidinotransferase (GAA synthetase) activity in the remaining kidney of the rabbits with CRF induced by ligating the major branches of a renal artery followed by contralateral nephrectomy was significantly decreased. However, because of the detection of serum GAA concentration in anephric patients, they speculated that GAA deficiency in advanced renal failure was prevented by extrarenal production[8].

In this study, in order to evaluate the reserve or capacity of GAA production in CRF patients, bean protein powder (not including creatine or GAA) was administered orally and the GAA in the serum and urine were measured. We found that the serum concentration and urinary excretion of GAA and Ccr were significantly increased by the protein load in healthy subjects. These normal responses to protein load were completely absent in CRF patients (Table 2). Van Pilsum[9] demonstrated that protein restriction suppressed the activity of GAA synthetase in the kidney in normal rats, which might be the mechanism for the decrease of GAA production in CRF patients on conservative therapy including a low protein diet. However, from the results of this study, we speculated that GAA production had already been maximally activated before protein loading, and that there was no reserve of GAA synthesis in CRF patients.

Tofuku et al.[8] demonstrated that the serum GAA concentration of CRF patients undergoing conservative therapy was significantly lower, but increased to the nearly normal range with regular hemodialysis treatment. We have previously shown that the serum GAA concentration of CRF patients who were not on conservative therapy with a low protein diet was significantly higher than when conservative therapy was being performed (unpublished data). The flesh of mammals contains a large quantity of creatine. This led us to the further speculation that normal protein intake could partially improve the deficiency of creatine, and that the serum GAA concentration might increase because of the decreased GAA uptake into the liver for the formation of creatine.

CONCLUSION

In the study, we evaluated the reserve or capacity of GAA production by
an oral load of bean protein in normal and CRF subjects. Serum concentration
and urinary excretion of GAA and Ccr were significantly increased by the
protein load in healthy subjects, indicating the reserve or capacity of GAA
synthesis and GFR. In CRF patients, these baseline values were significantly
lower than those in normal subjects, and these normal responses to protein
loading were completely abolished.

In conclusion, the capacity of GAA synthesis as well as the reserve of
GFR is remarkably reduced and GAA is deficient in CRF patients.

REFERENCES

1. Tsubakihara, Y., Iida, N., Yuasa, S., Kawashima, T., Nakanishi, I.,
   Tomobuchi, M., Yokogawa, T., Ando, A., Orita, Y., Abe, H., Kikuchi, T.
   and Okamoto, H., Guanidinoacetic acid (GAA) deficiency and supplementation in rats with chronic renal failure (CRF), in:"Guanidines," A. Mori,
   B.D. Cohen and A. Lowenthal. Eds., Plenum Press, New York (1985) pp.
   373-379.
2. Tsubakihara, Y., Iida, N., Yuasa, S., Kawashima, T., Nakanishi, I.,
   Tomobuchi, M. and Yokogawa, T., Guanidinoacetic acid (GAA) in patients
   with chronic renal failure (CRF) and diabetes mellitus (DM), in: "Guanidines," A. Mori, B.D. Cohen and A. Lowenthal Eds., Plenum Press, New
   York (1985) pp. 309-316.
3. Borsook, H., Dubnoff, J.W., Lilly, J.C. and Marriott, W., The formation
   of glycocyamine in man and it's urinary excretion, J. Biol. Chem. 138
   (1941) 405-410.
4. Bosch, J.P., Lauer, A. and Glabman, S., Short-term protein loading in
   assessment of patients with renal disease, Am. J. Med. 77 (1984)
   873-879.
5. Iturbe, B.R., Herrera, J. and Garcia, R., Response to acute protein load
   in kidney donors and in apparently normal postacute glomerulonephritis
   patients; evidence for glomerular hyperfiltration, Lancet II (1985)
   461-464.
6. Yamamoto, Y., Manji, T., Saito, A., Maeda, K. and Ohta, K., Ion-exchange
   chromatographic separation and fluorometric determination of guanidino
   compounds in physiological fluids, J. Chromato. 162 (1979) 327-333.
7. Goldman, R. and Moss, J.X., Synthesis of creatine in nephrectomized
   rats, Am. J. Physiol. 197 (1959) 865-868.
8. Tofuku, Y., Muramoto, H., Kuroda, M. and Takeda, R., Impaired metabolism
   of guanidinoacetic acid in uremia, Nephron 41 (1985) 174-178.
9. Van Pilsum, J.F., Creatine and creatine phosphate in normal and protein
   depleted rats, J. Biol. Chem. 228 (1957) 145-148.

THE SIGNIFICANCE OF SERUM AND URINARY GUANIDINOACETIC ACID LEVEL FOR THE

RESTORATION OF RENAL METABOLIC FUNCTION IN PATIENTS WITH KIDNEY TRANS-

PLANTATION

Kazuo Isoda, Tetsuya Mitarai, Noritsugu Imamura, Hidehiko
Honda, Ryoji Nagasawa, Satoru Hirose, Kenichi Sugimoto,
Ryo Tanaka, Takeo Yokoyama* and Hidehiko Kashiwabara*

Department of Internal Medicine, Saitama Medical Center
Saitama Medical School, 1981 Kamoda, Kawagoe, Saitama 350,
Japan; *Department of Surgery, Sakura National Hospital
Sakura, Chiba 286, Japan

INTRODUCTION

It is well known that guanidinoacetic acid (GAA), precursor of creatine,
is mainly produced by the kidney and the concentration of GAA in serum and
urine and decreased in patients with chronic renal failure (CRF). However,
the mechanism for the excretion of this substance by the kidney still remains
to be clarified. Therefore, this study is aimed to assess the restoration of
the metabolic function of the transplanted kidney and the excretory manoeuvre
by means of measurements of serum and urinary guanidino compounds, especially
the GAA levels using fluorometric HPLC.

MATERIALS AND METHODS

Patients

Six patients with CRF due to chronic glomerulonephritis were studied.
Four were males and two were females. Ages ranged from 22 to 41y (average age
28.5y). Three patients received renal transplants from living donors and the
remainder were from cadaveric donors (Table 1).

Analysis

Serum and urine were obtained from patients on the dialysis program and
from transplanted patients at 6 A.M.. The deproteinizing of serum was performed by adding 0.5 ml of 30% trichloroacetic acid solution (TCA) to 1.0 ml
of serum, followed by centrifugation for 60 min at 10,000 rpm. Urine was
also deproteinized by adding 1.0 ml of 10% TCA to 4.0 ml of urine and supernatant was collected after 10,000 rpm, 60 min ppt. Guanidino compounds (GAA,
guanidinosuccinic acid (GSA) and methylguanidine (MG)) were measured by fluorometric HPLC (JASCO G-520 Guadinino autoanalizer, Japan Spectroscopic Co,
Tokyo Japan) at excitation 365 nm and emmision 495 nm. Analysis of guanidino

Table 1.   Patient profiles.

Patient	age	sex	disease	donor	result
S.J.	23y	F	CGN	LD (mother)	Live
H.T.	22y	M	CGN	LD (mother)	TxNx
M.A.	30y	M	CGN	LD (mother)	live
O.T.	31y	M	CGN	CD	live
W.J.	24y	F	CGN	CD	live
I.K.	41y	M	CGN	CD	live

TxNx; Graftectomy, CGN; Chronic glomerulonephritis, LD; Living donor and CD; Cadaveric donor

compounds were made by step gradient system using cation exchanger at reaction temperature of 60°C. The column eluent was mixed with stream of 2 N NaO and 9,10-phenanthrenequinone reagent. The time consumed was within 50 min fo each analysis.

According to the report of the Japanese Society of Guanidino Compounds, the normal value of serum and urinary GAA in healthy persons is 28.4 ± 9.9 μg/dl and 4676 ± 3647 μg/dl (58.43 ± 41.71 mg/day) N=120, respectively. The normal value for GSA is 2.94 ± 2.45 μg/dl in serum and 439.2 ± 284.4 μg/dl (5.56 ± 3.68 mg/day) in urine, N=127, MG concentration is 0.01 μg/dl in seru and it is not detectable in urine, N=37. We employed these values as normal levels of three kinds of guanidino compounds.

RESULTS

Changes of MG and creatinine (CRN) concentration in serum on the pre and post renal transplantation period are shown in Fig. 1 and 2. CRN and MG concentrations in serum were rapidly reduced and GAA levels in serum were conversely increased after renal transplantation at least for ten months without any rejection.

In the case of rejection after transplantation, GAA declined to low lev els in the serum a couple days before increasing serum CRN concentration and noticeable clinical signs developed (Fig. 3).

The urinary excretion of GSA decreased from 35 mg/day to negligible val ues in proportion to the duration of the post transplantation period. In con trast to GSA, increasing quantities of GAA in the urine was terminated by th graft rejection and this increasing tendency resumed after proper treatment of the rejetion episode (Fig. 4).

The relationship between the serum concentration and the quantity of GSA, MG and GAA in urine is shown in Fig. 5. It is worth noting that quantities of dialy excretion of those substances paralleled the serum concentrations. However, a decrease in urine GAA predominated on the developement of graft rejection (patient I.K. Fig. 5).

Comparing the ratio of GSA and GAA clearances to CRN clearance (Ccr) after renal transplantation in 4 cases yielded the following. C_{GSA}/Ccr ratio were all below 1.0, however C_{GAA}/Ccr ratios were above 1.0 in 3 cases and

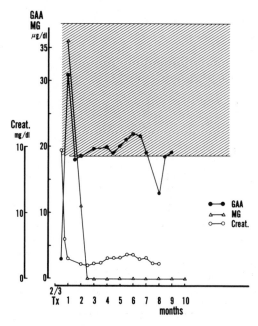

Figure 1. Guanidinoacetic acid (GAA), methylguanidine (MG), serum creatinine (Creat.) levels after renal transplantation. (Patient; O.M., M. 31y, cadaveric renal transplantation).

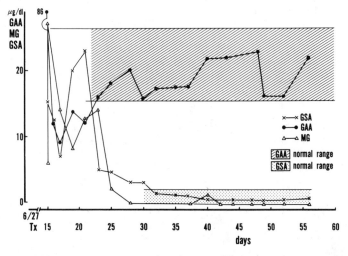

Figure 2. Guanidinosuccinic acid (GSA), guanidinoacetic acid (GAA), methyl-guanidine (MG) levels in serum after renal transplantation. (Patient W.J., F. 24y, cadaveric renal transplantation).

this latter increased from day to day after renal transplantation in all cases (Table 2). The ratio of C_{MG}/Ccr was not available due to very low concentrations of MG in serum and urine.

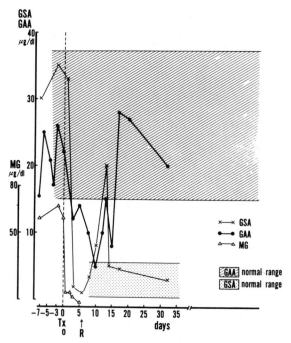

Figure 3. Guanidinosuccinic acid (GSA), guanidinoacetic acid (GAA), methylguanidine (MG) levels in serum before and after renal transplantation. (Patient M.A., M. 30y, live renal transplantation).
R; rejection.

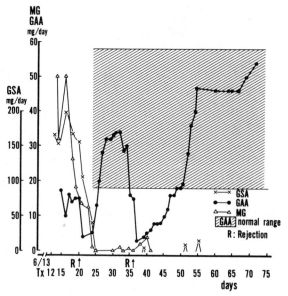

Figure 4. Urinary excretion of guanidinosuccinic acid (GSA), guanidinoacetic
acid (GAA), methylguanidine (MG) after renal transplantation.
(Patient W.J., F. 24y, cadaveric renal transplantation).

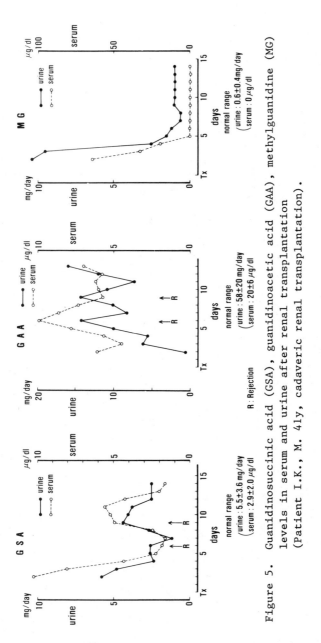

Figure 5. Guanidinosuccinic acid (GSA), guanidinoacetic acid (GAA), methylguanidine (MG) levels in serum and urine after renal transplantation (Patient I.K., M. 41y, cadaveric renal transplantation).

Table 2. Ratio of GSA, GAA, MG clearances to creatinine clearance after
          renal transplantation.

Patient	day aft.Tx.	C_{GSA}/C_{cr}	C_{GAA}/C_{cr}	C_{MG}/C_{cr}
W.J.	37	0.89	0.45	N.D.
24y	39	0.48	0.40	2.3
F	41	0.41	0.28	N.D.
CD	44	0.50	0.42	N.D.
	46	0.43	0.44	1.72
M.A.	42	0.10	4.31	N.D.
30y	43	0.10	2.57	N.D.
M	44	1.00	2.16	N.D.
LD				
I.K.	174	0.88	1.88	3.9
41y	175	0.89	4.49	5.0
M	176	0.97	1.67	2.3
CD	187	0.63	2.28	N.D.
	179	0.96	2.30	N.D.
	244	0.24	1.60	N.D.
	245	1.00	0.51	N.D.
	246	0.73	2.44	N.D.
O.T.	245	0.13	1.40	N.D.
31y	246	1.0	1.99	N.D.
M	247	1.0	6.99	N.D.
CD				

CD; Cadaveric donor, LD; Living donor, N.D.; not detectable, M; Male, GSA;
guanidinosuccinic acid, GAA; guanidinoacetic acid and MG; methylguanidine

DISCUSSION

It has been proposed that a series of guanidino compounds, which derive
from the ornithine cycle or other possible pathways, produce various effect
on patients with CRF[1,2].

An estimation of GSA and MG in serum and urine indicates a very high
concentration in patients with CRF under conservative therapy and relatively
low serum levels in patients treated with dialysis.

On the other hand, it is well known that GAA is synthesized from glycin
and arginine by transamidination in renal tissue and serves as a precursor o
creatine in humans.

In our study, serum and urinary levels of GAA were very low and also
decreased in patients on dialysis.

As shown in Fig. 1, serum and urinary GAA concentration rose from month
to month after renal transplantation without an episode of graft rejection

(Patient O.T. 31y female). From this fact, it seems to be correct that serum and urinary GAA levels may reflect renal function, that is, renal blood flow or GAA synthesizing power, glomerular filtration rate (GFR) or renal tubular function.

In patient W.J. a 24y female, who received a cadaveric renal transplant, serum GAA rose after three weeks of transplantation. Nevertheless, serum GSA and MG levels rapidly declined to normal range in a short time after transplantation (Fig. 2).

In patient M.A., a 30y male, who received a renal graft from a living donor, the serum GAA concentration abruptly decreased in association with an episode of graft rejection 6 days after transplantation. Conversely, the GSA level in serum increased due probably to a decrease of GFR (Fig. 3).

As indicated in Fig. 4, the same relationship between quantities of urinary GAA and GSA was detected. However, the decrement of GAA in urine was much more remarkable than that of GSA.

The relationship between the serum concentration and the daily out put of GAA, GSA and MG during the post-transplant period was examined. Both serum concentration and urinary excretion were parallel. Considering the fact that GAA levels declined in serum and urine with an episode of graft rejection and that the decrement of urinary GAA is the more remarkable, we can use the measurement of urinary GAA for the early detection of graft rejection.

Calculation of the ratios of GSA, GAA and MG clearances to Ccr revealed that GAA excretion is based on renal tubular secretion as seen in 4 cases of this study (Table 2). The ratio of GAA clearance to Ccr was above 1.0 in 3 cases and less than 1.0 in one case where the urinary GAA was below the normal range at the time of calculation. This case (patient W.J. 24y female) fully recovered renal function 50 days after transplantation. It is possible that the ratio of GAA clearance to Ccr might have achieved 1.0, if it were done over 50 days after the operation.

SUMMARY

Using fluorometric high pressure liquid chromatography, serum and urinary guanidinoacetic acid (GAA), guanidinosuccinic acid and methylguanidine were measured in order to estimate the restoration of renal metabolic function in patients received kidney transplantation.

GAA concentrations in serum and urine obtained from chronic dialysis patients were low and tended toward normal after renal transplantation. However, this normalization of the GAA levels in serum and urine was disrupted by the episode of graft rejection.

It is noteworthy that the estimation of GAA in serum and urine is valuable not only to detect rejection but also to anticipate the restoration of renal function, especially renal tubular metabolic function. When the ratio of GAA clearance to creatinine clearance rises above 1.0, the excretion of GAA is due to renal tubular secretion.

In conclusion, the normalization in serum and urine concentrations of GAA reflect restoration of renal tubular cell function after renal transplantation.

REFERENCES

1. Stein, M., Cohen, B.D. and Kornhauser, R.S., Guanidinosuccinic acid in renal failure, experimental azotemia and inborn errors of the urea cycle, New Engl. J. Med. 280 (1969) 926-930.
2. Cohen, B.D., Guanidino compounds: Implications in uremia, in: "Guanidines," A. Mori, B.D. Cohen and A. Lowenthal Eds., Plenum Press, New York and London (1985) pp. 265-276.
3. Ishizaki, M., Kitamura, H., Takahashi, H., Asano, H., Miura, K. and Okazaki, H., Evaluation of the efficacy of anti-rejection therapy using the quantitative analysis of guanidinoacetic acid (GAA) urinary excretion as a guide, in: "Guanidines," A. Mori, B.D. Cohen and A. Lowenthal Eds., Plenum Press, New York and London (1985) pp. 353-364.

CONTRIBUTORS

Aoyagi, K.	Univ. of Tsukuba, Tsukuba, Japan
Arakawa, M.	Niigata Univ. Sch. of Med., Niigata, Japan
Bonhaus, D. W.	Univ. of Arizona, Tucson, U.S.A.
Chamoles, N.	Lab. of Neurochem., Buenos Aires, Argentina
Cohen, B. D.	Bronx-Lebanon Hospital, Bronx, U.S.A.
De Deyn, P.P.	Univ. of Antwerp, Antwerp, Belgium
De Potter, W. P.	Univ. of Antwerp, Antwerp, Belgium
Deshmukh, D. R.	Univ. of Michigan, Ann Arbor, U.S.A.
Edaki, A.	Okayama Univ. Med. Sch., Okayama, Japan
Edamatsu, R.	Okayama Univ. Med. Sch., Okayama, Japan
Fujino, T.	St. Marianna Univ. Sch. of Med., Kawasaki, Japan
Futaki, G.	Sendai-Shakai-Hoken Hospital, Sendai, Japan
Gejyo, F.	Niigata Univ. Sch. of Med., Niigata, Japan
Gray, E. D.	Univ. of Minnesota, Minneapolis, U.S.A.
Gross, M. D.	Univ. of Michigan, Ann Arbor, U.S.A.
Guillou, Y.	Institut Pasteur, Paris, France
Hamana, K.	Gunma Univ., Maebashi, Japan
Haruki, K.	Kanazawa Univ., Kanazawa, Japan
Hashida, A.	Osaka Univ., Suita, Japan
Hatakeyama, S.	Kanazawa Univ., Kanazawa, Japan
Hayakawa, C.	Nagoya Univ. Med. Sch., Nagoya, Japan
Hiramatsu, M.	Okayama Univ. Med. Sch., Okayama, Japan
Hirasawa, Y.	Shinrakuen Hospital, Niigata, Japan
Hirata, M.	Kanazawa Univ., Kanazawa, Japan
Hirose, S.	Saitama Med. Sch., Kawagoe, Japan
Honda, H.	Saitama Med. Sch., Kawagoe, Japan
Huxtable, R. J.	Univ. of Arizona, Tucson, U.S.A.
Iida, N.	Osaka Prefectural Hospital, Osaka, Japan
Imamura, N.	Saitama Med. Sch., Kawagoe, Japan
Inouchi, M.	St. Marianna Univ. Sch. of Med., Kawasaki, Japan
Ishida, M.	St. Marianna Univ. Sch. of Med., Kawasaki, Japan
Ishizaki, M.	Sendai-Shakai-Hoken Hospital, Sendai, Japan
Isobe, K.	Amano Pharmaceutical Co. Ltd., Aichi, Japan
Isoda, K.	Saitama Med. Sch., Kawagoe, Japan

Jenny, R. J.   Univ. of Minnesota, Minneapolis, U.S.A.
Junen, M.   Juntendo Univ. Sch. of Med., Tokyo, Japan

Kabutan, K.   Okayama Univ. Med. Sch., Okayama Japan
Kanazawa, T.   St. Marianna Univ. Sch. of Med., Kawasaki, Japan
Kashiwabara, H.   Sakura National Hospital, Sakura, Japan
Kitamura, E.   Osaka Prefectural Hospital, Osaka, Japan
Kitamura, H.   Sendai-Shakai-Hoken Hospital, Sendai, Japan
Kitaura, M.   Okayama Univ. Med. Sch., Okayama, Japan
Kiyatake, I.   Juntendo Univ. Sch. of Med., Tokyo, Japan
Kohno, M.   JEOL Ltd., Akishima, Japan
Koide, H.   Juntendo Univ. Sch. of Med., Tokyo, Japan
Kosaka, F.   Okayama Univ. Med. Sch., Okayama, Japan
Kosogabe, Y.   Okayama Univ. Med. Sch., Okayama, Japan
Kurosawa, K.   Sendai-Shakai-Hoken Hospital, Sendai, Japan

Lehmann, A.   Univ. of Goteborg, Goteborg, Sweden
Letarte, J.   Ste-Justine Hospital, Montreal, Canada
Levy, M.   Univ. of Antwerp, Antwerp, Belgium
Lowenthal, A.   Univ. of Antwerp, Antwerp, Belgium

Macdonald, L. R.   Univ. of Michigan Med. Center, Ann Arbor, U.S.A.
Marescau, B.   Univ. Intelling Antwerpen, Antwerp, Belgium
Matsuzaki, S.   Gunma Univ., Maebashi, Japan
McGuire, D. M.   Univ. of Minnesota, Minneapolis, U.S.A.
Mitarai, T.   Saitama Med. Sch., Kawagoe, Japan
Miyazaki, M.   Univ. of Tsukuba, Tsukuba, Japan
Mizutani, N.   Nagoya Univ. Med. Sch., Nagoya, Japan
Mori, A.   Okayama Univ. Med. Sch., Okayama, Japan
Muramoto, H.   Kanazawa Univ., Kanazawa, Japan

Nagai, K.   Osaka Univ., Suita, Japan
Nagasawa, R.   Saitana Med. Sch., Kawagoe, Japan
Nagase, S.   Univ. of Tsukuba, Tsukuba, Japan
Nakagawa, H.   Osaka Univ., Suita, Japan
Nakajima, M.   Kikkoman Corporation, Noda, Japan
Nakamura, K.   Kikkoman Corporation, Noda, Japan
Nakanishi, I.   Osaka Prefectural Hospital, Osaka, Japan
Nakayama, S.   Juntendo Univ. Sch. of Med., Tokyo, Japan
Narita, M.   Univ. of Tsukuba, Tsukuba, Japan

Obata, T.   Univ. of Arizona, Tucson, U.S.A.
Ochiai, Y.   Okayama Univ. Med. Sch., Okayama, Japan
Ohba, S.   Univ. of Tsukuba, Tsukuba, Japan
Okada, N.   Osaka Prefectural Hospital, Osaka, Japan
Owada, S.   St. Marianna Univ. Sch. of Med., Kawasaki, Japan
Oya, Y.   Nagoya Univ. Med. Sch., Nagoya, Japan
Ozawa, S.   St. Marianna Univ. Sch. of Med., Kawasaki, Japan

Patel, H.   Bronx-Lebanon Hospital, Bronx, U.S.A.

Qureshi, I. A.   Ste-Justine Hospital, Montreal, Canada

Robin, Y.   Fac. Pharmacie, Chatenay-Malrbry, France

Sakamoto, M.   Univ. of Tsukuba, Tsukuba, Japan
Sandberg, M.   Univ. Goteborg, Goteborg, Sweden
Sato, T.   St. Marianna Univ. Sch. of Med., Kawasaki, Japan

Shiba, C.	St. Marianna Univ. Sch. of Med., Kawasaki, Japan
Shimizu, Y.	Okayama Univ. Med. Sch., Okayama, Japan
Shindo, S.	Univ. of Arizona, Tucson, U.S.A.
Shiraga, H.	Okayama Univ. Med. Sch., Okayama, Japan
Shirokane, Y.	Kikkoman Corporation, Noda, Japan
Simon, A. M.	Univ. of Minnesota, Minneapolis, U.S.A.
Sohn, T.	Sendai-Shakai-Hoken Hospital Sendai, Japan
Sugai, H.	Sendai-Shakai-Hoken Hospital Sendai, Japan
Sugi, H.	Okayama Univ. Med. Sch., Okayama, Japan
Sugimoto, K.	Saitama Med. Sch., Kawagoe, Japan
Sugimoto, S.	Okayama Univ. Med. Sch., Okayama, Japan
Suzuki, K.	Sendai-Shakai-Hoken Hospital Sendai, Japan
Suzuki, Y.	Niigata Univ. Sch. of Med., Niigata, Japan
Taguma, Y.	Sendai-Shakai-Hoken Hospital Sendai, Japan
Takahashi, H.	Sendai-Shakai-Hoken Hospital Sendai, Japan
Takeda, R.	Kanazawa Univ., Kanazawa, Japan
Takano, Y.	Niigata Univ. Sch. of Med., Niigata, Japan
Tanaka, R.	Saitama Med. Sch., Kawagoe, Japan
Tasaki, K.	Niigata Univ. Sch. of Med., Niigata, Japan
Toda, H.	Okayama Univ. Med. Sch., Okayama. Japan
Tofuku, Y.	Kanazawa Univ., Kanazawa, Japan
Tojo, S.	Univ. of Tsukuba, Tsukuba, Japan
Tomita, H.	St. Marianna Univ. Sch. of Med., Kawasaki, Japan
Tsubakihara, Y.	Osaka Prefectural Hospital, Osaka, Japan
Tsuji, C.	Okayama Univ. Med. Sch., Okayama, Japan
Van Gorp, L.	Univ. of Antwerp, Antwerp, Belgium
Van Pilsum, J. F.	Univ. of Minnesota, Minneapolis, U.S.A.
Watanabe, K.	Nagoya Univ. Med. Sch., Nagoya, Japan
Watanabe, Y.	Okayama Univ. Med. Sch., Okayama, Japan
Wiechert, P.	W. Pieck Univ., Rostok, G.D.R.
Yamato, E.	Osaka Prefectural Hospital, Osaka, Japan
Yamamura, H. I.	Univ. of Arizona, Tucson, U.S.A.
Yasuda, T.	St. Marianna Univ. Sch. of Med., Kawasaki, Japan
Yokoi, I.	Okayama Univ. Med. Sch., Okayama, Japan
Yokoyama, K.	Osaka Prefectural Hospital, Osaka, Japan
Yokoyama, T.	Sakura National Hospital, Sakura, Japan
Yoshida, K.	Niigata Univ. Sch. of Med., Niigata, Japan
Yoshino, M.	Kurume Univ. Sch. of Med., Kurume, Japan

Abbreviations used: ARF; acute renal failure, Arg; arginine, Ccr, creatinine
clearance, CRF; chronic renal failure, CRN; creatinine, CSF; cerebrospinal
fluid, CTN; creatine, GAA; guanidinoacetic acid (glycocyamine), GBA;
γ-Guanidinobutyric acid, GES; 2-guanidinoethane sulfonate (taurocyamine),
GEt; 2-guanidinoethanol, GPA; ß-guanidinopropionic acid, GSA; guanidinosuccinic scid, Gua; guanidine, HArg; homoarginine, KGVA; α-keto-δ-guanidinovaleric acid, MGua; methylguanidine, NAA; α-N-acetylarginine

Abbreviations used: ARF; acute renal failure, Arg; arginine, CRF; chronic renal failure, CRN; creatinine, CSF; cerebrospinal fluid, CTN; creatine, GAA; guanidinoacetic acid (glycocyamine), GBA; γ-guanidinobutyric acid, GES; 2-guanidinoethane sulfonate (taurocyamine), GEt; 2-guanidinoethanol, GPA; ß-guanidinopropionic acid, GSA; guanidinosuccinic scid, Gua; guanidine, HArg; homoarginine, KGVA; α-keto-δ-guanidinovaleric acid, MGua; methylguanidine, NAA; α-N-acetylarginine